TROPICAL MARINE MOLLUSKS

TROPICAL MARINE MOLLUSKS

AN ILLUSTRATED BIOGEOGRAPHICAL GUIDE

Edward J. Petuch
and
David P. Berschauer

CRC Press
Taylor & Francis Group
Boca Raton London New York

CRC Press is an imprint of the
Taylor & Francis Group, an **informa** business

First edition published 2021
by CRC Press
6000 Broken Sound Parkway NW, Suite 300, Boca Raton, FL 33487-2742

and by CRC Press
4 Park Square, Milton Park, Abingdon, Oxon OX14 4RN

CRC Press is an imprint of Taylor & Francis Group, an Informa business

Library of Congress Cataloging-in-Publication Data

Names: Petuch, Edward J., author. | Berschauer, David P., author.
Title: Tropical marine mollusks : an illustrated biogeographical guide / Edward J. Petuch, David P. Berschauer.
Description: Boca Raton : CRC Press, 2020. | Includes bibliographical references and index.
Identifiers: LCCN 2020030276 (print) | LCCN 2020030277 (ebook) | ISBN 9780367636388 (hardback) | ISBN 9781003120070 (ebook)
Subjects: LCSH: Mollusks--Geographical distribution.
Classification: LCC QL408 .P48 2020 (print) | LCC QL408 (ebook) | DDC 594--dc23
LC record available at https://lccn.loc.gov/2020030276
LC ebook record available at https://lccn.loc.gov/2020030277

ISBN: 978-0-367-63638-8 (hbk)
ISBN: 978-1-00-312007-0 (ebk)

Dedication

This book is dedicated to the following:

Linda J. Petuch, Eric and Rasa Petuch,

Brian Petuch and Kendra Berentsen, and Jennifer Petuch

and

Felicia Weisbrot Berschauer, Morgan Berschauer,

Jonathon Berschauer, Tawni Berschauer, and Nora Berschauer

Close up view of an oyster clump at low tide on Rabbit Key, Ten Thousand Islands, Florida, showing the resident ecosystem composed of barnacles, bright yellow *Halichondria* sponges, the black cerithiid *Batillaria minima*, and a rose-colored Tampa Top Shell, *Calliostoma tampaense*.

Tropical Marine Mollusks:
An Illustrated Biogeographical Guide

TABLE OF CONTENTS

"The zoologist, gathering shells or mollusks from the coast of eastern North America or Japan … sees zoological geography rising into one of the grandest of sciences."

James Dwight Dana, "Thoughts on Species", *American Journal of Science and Art*, 1857

INTRODUCTION

The science of marine biogeography, the study of the spatial distribution of organisms in the world's oceans, is one of the most fascinating branches of oceanography. The distributions of molluscan faunas, in particular, were first studied by biologists over 160 years ago, leading to the first insights into the higher order biotic patterns of community ecology and allopatric speciation. This book is meant to be a continuation of that pioneer research. As noticed by the early malacologists, three ecological factors limit the spatial distribution of marine mollusks: **water temperature**, **salinity**, and **substrate type**. Since most marine mollusks are physiologically constrained by both limited osmoregulatory abilities and temperature-sensitive metabolic enzyme systems, their distributions are arranged by latitude along the Neritic and Bathyal Zones of the continental margins. The spatial arrangements of malacofaunas, as shown in this book, can be considered to be proxies for worldwide oceanic conditions and can be used as tools for determining patterns of global climate change. Our documentation of evolutionary "hot spots" and geographically-restricted endemic faunas can also be used as a baseline for future studies on patterns of environmental deterioration and extinction in the marine biosphere.

As important as temperature, salinity, and substrate are in determining distributional patterns, sea level change (eustatic fluctuations) is a larger, overriding factor that supersedes these ecological limiting factors. Occurring in regular cycles throughout geologic time, these sea level changes were due primarily to the building-up and melting of continental glaciers. The eustatic fluctuations caused wide areas of the worldwide continental shelves, offshore banks, and island chains to either become completely dry and emergent (during eustatic lows) or completely submerged under deeper water (during eustatic highs). During extended eustatic lows, some with sea levels that dropped to 100-200 m below present level, spatial configurations were greatly diminished and some isolated basins, like the Mediterranean Sea, Black Sea, and Persian Gulf, actually dried out completely, becoming deserts. Extreme lowered sea levels like this were particularly damaging to the faunas of shallow banks and wide carbonate platforms, when the habitable areas shrank to only thin peripheral strips bordering the emergent continental shelves. In the time of a eustatic low stand, many species became extinct and others were separated into small, ecologically stressed populations. These sequestered faunas subsequently underwent rapid speciation due to genetic isolation (allopatric speciation). This process was reversed during times of eustatic highs, when the continental shelves, banks, and platforms were again submerged. The small numbers of surviving species repopulated the newly available habitats and underwent wave after wave of rapid speciation (referred to as the "Founder Effect"). The cycles of eustatic fluctuations, then, can be seen to act as forcing vectors for the evolution of new taxa, many of which can arise in very short periods of time (a few thousands of years in some groups such as the cone shells and cowries).

Besides the effects of sea level fluctuations, the most important component of molluscan speciation is vagility: the ability of an animal to disperse from the place where it was born. After hatching from its egg capsule, most mollusks undergo a free-swimming larval stage called a "veliger." Most often, these veliger larvae swim within the plankton, feeding on photosynthetic phytoplankton or, in some cases, other small planktonic animals (zooplankton). These plankton-dwellers are referred to as "planktotrophic larvae" and some, such as the tonnoidean gastropods, can actually stay suspended in the plankton for over a year and travel thousands of miles from where they hatched. These high-vagility, wide-ranging species all have long-lived planktotrophic veligers and are used here for defining super-regions, regions, and subregions. Many gastropods, on the other hand, have direct development; where the veliger stays sealed within the egg capsule and hatches out as a free-crawling miniature adult. These types of low-vagility mollusks do not disperse far from their birth places and often are confined to very small geographic localities such as isolated bays or oceanic islands. These sequestered organisms become the endemic species that we use here for defining provinces, subprovinces, and infraprovinces (highly localized evolutionary hot spots). These two larval strategies, coupled with sea level fluctuations and water temperature changes over a million years of earth history, have resulted in the evolution of the amazingly-rich worldwide gastropod fauna that we describe and illustrate here in this book.

Both of us have always found biogeography to be one of the most beautiful and intellectually-satisfying branches of marine biology. This is primarily due to the synthetic nature of the science and its inherent complexity. Any in-depth marine biogeographic study involves aspects of physical and chemical oceanography, systematic and evolutionary biology, historical geology, tectonics, paleontology, climatology, geomorphology, and physical geography. Each one of these disciplines, by itself, is a fascinating field of study, but the fusion of all of them into a single synthetic body of knowledge is aesthetically beautiful, rivaling elegant mathematical proofs. In the various sections of the following chapters, we incorporate aspects of all of these disciplines into a new classification system for the nomenclature of biogeographical spatial units found in tropical, subtropical, and warm temperate seas. To underscore the intense beauty present in the worldwide molluscan provinces and subprovinces, we also illustrate 1778 species of gastropods on 161 color plates, many of which are extremely rare and poorly-known endemic species that are illustrated for the first time outside of their original descriptions. These color illustrations can be used as synoptic references for identification of species in the field and also as documentation of the biodiversity of isolated islands and enclosed seas.

As can be seen from the color plates, we place emphasis on only a few marine gastropod families, primarily the Conidae, Cypraeidae, Volutidae, Strombidae, Muricidae, and Olividae, and a few other minor families. All of the species in these groups have commercial value and are eagerly sought after by professional shell collectors. Because of the high market demand, these shell dealers have accumulated detailed locality data on all of these taxa, producing a large body of important biogeographical information that we have used to support our ideas. We also draw upon the range data shown in several recent systematic treatments of these important tropical families; including the collected works of Lorenz (Cypraeidae), Monnier, Limpalaër, Robin, Roux, Tucker and Tenorio (Conidae), Houart (Muricidae), Dekkers and Maxwell (Strombidae), Bail, Limpus, Poppe and Goto (Volutidae), Petuch, Sargent and Sterba (Olividae), and Snyder, Lyons and Vermeij (Fasciolariidae). Throughout this book, the infraspecific taxa at the form level in the Cypraeidae are here treated as subspecies in order to underscore their biogeographical significance. We also do this in order to comply with the recommendations of the Code of the International Commission on Zoological Nomenclature (ICZN) which does not recognize forms as valid taxonomic units.

We hope that the new biogeographic nomenclatural scheme outlined in this book, gleaned from this large data bank, will be used as a baseline for future studies on changing ecosystems and diminished biodiversity. These studies are crucial as our planet enters a new period of global warming and extinction.

Edward J. Petuch, Ph.D. and David P. Berschauer, J.D.

California Sea lions, *Zalophus californianus*, basking in the sun on a rock at "El Arco", also known as Land's End, Cabo San Lucas, Baja California Sur, Mexico. This is the type area of the Lucasian Infraprovince of the Magdalenan Subprovince of the Panamic Province, home to such endemic species as the olive shells *Americoliva grovesi* and *A. violacea*, and the cone shells *Gradiconus dispar*, *G. nybakkeni*, and *G. skoglundae*.

ACKNOWLEDGMENTS

Without the help and interest of the following individuals and organizations, this book would never have been completed. We would like to thank the following people for their invaluable assistance in the production of this book. For their generous donations and for their aiding in the acquisition of the specimens that were used in our illustrations, we thank: **Dr. Felix Lorenz**, Germany; **Damaso Monteiro**, Brazil and Portugal; **Rika Goethaels** and **Fernand De Donder**, Belgium; **John D. Daughenbaugh**, California, USA; **Dr. Anton Oleinik**, Florida Atlantic University, Florida, USA; **John Abbas**, Indonesia; **Kyriakos Papavasileiou**, Greece; **Thierry Vulliet**, Queensland, Australia; **Douglas Shelton**, Alabama, USA; **Darwin Alder**, Texas, USA; **Robert L. Eason, Sr.**, Tennessee, USA; **Marcus Coltro and Jose Coltro**, Brazil, Italy, and USA, and thanks also for allowing us to use selected photographs of specimens from their website; **Gary Smith**, Western Australia, Australia; **A. Kenneth "Kenny" Brown**, Florida, USA; **Olivier Crabos**, Brazil; **Jean-Pierre Barbier**, Philippines; **Eddie Matchett**, Florida, USA; **Clifford Swearingen**, Florida, USA; **Christopher Takahashi**, Hawaii, USA; **Adrian Bishop**, South Australia, Australia; **Kevan and Linda Sunderland**, Florida, USA; **Linda Riley Powers**, Florida, USA; **Tammy Bailey Myers**, Florida, USA; **Louis Rundo**, Ohio, USA; **Luis Vela**, Museo del Mar Mexico, Mexico; **Dr. John and Jennifer Whicher**, England, UK; **Dr. Stephan G. Veldsman**, South Africa; **Werner Massier**, Namibia; **Fabrice Prugnaud**, France; **Pierre Recourt**, Netherlands; **Linda Zylman**, Florida, USA; **Rick Negus**, California, USA; **Frank Swinnen**, Belgium; **Adam Anderson**, Tasmania, Australia; **Remy Devorsine**, Queensland, Australia; **Avis Sousa**, Idaho, USA; **Daniel Notek**, Arizona, USA; **David Waller**, California, USA; **Michael Bruggeman**, Georgia, USA; **Melanie Briskin**, Florida, USA; **Stephen Tressel**, Florida, USA; **Andrew Foster**, Florida, USA; **Donnie Benton**, Florida, USA; **David Wagner**, Florida, USA; **David DeLucia**, Connecticut, USA; **Alain Van't Woudt**, Netherlands; **Antonio Monteiro**, Portugal; **Randy Allamand**, North Carolina, USA; **Leo Ros**, Aruba; **Rob Boot**, Aruba; **Mirko Guardiani**, Italy; **Geraldo Pomponet**, Brazil; **John Ruggero**, New York, USA; **David Preston**, Queensland, Australia; **Dr. Douglas Biggs**, Texas, USA; **Thomas Fair**, Texas, USA; **Jaime Talavera**, Mexico; **Andre Poremski**, Washington, D.C., USA; **Troy Bernier**, Florida, USA; **Dennis Sargent**, Florida, USA; **Roland Houart**, Belgium; **Mark Johnson**, North Carolina, USA; **Randy Bridges**, Arizona, USA; **Charles Powell**, California, USA; **Valda Fraser** and **Mike Fraser**, South Africa. Special thanks are given to **Robert F. Myers**, Florida, USA, for his assistance with the photography of specimens, many of which are used throughout this book, and for his important insights into biogeographical patterns in the Indo-Pacific Super-Region and the Red Sea.

Foreword

From the dawn of civilization, people have been drawn to the exquisite beauty of marine shells, none more so than those found off tropical shores. From ornament to currency, they have been gifted, traded, and collected throughout the world for thousands of years. Today these ancient animals face an uncertain future. Marine mollusks are critical to ecosystem functioning and the biodiversity of benthic communities. Their decline is a forewarning to humanity that degradation of the marine environment will affect us all.

In less than a generation, rising atmospheric carbon dioxide from the combustion of fossil fuels continues to warm the atmosphere at an unprecedented rate, elevating sea surface temperatures (SST) and lowering the pH levels of the oceans. As emissions continue, there will be a profound impact on the physiology and behavior of countless marine taxa. Although its influence on marine mollusks is not yet fully understood, research suggests that thermal stress, particularly among tropical gastropods, raises mortality and compromises populations especially during increasingly frequent marine heatwaves. Furthermore, the transmission of pathogens that have decimated some species of mollusks is exacerbated by the increasing occurrence of El Niño warm-water events, intensified by climate change. Ocean acidification (OA), as well as directly affecting larval survival and shell mineralization of these calcareous animals, also alters predator-prey interactions with as yet unknown consequences for marine ecosystems and the food web. Never has marine conservation planning become more urgent.

Biogeography, the study of the geographical distributions of organisms, is a prime enabler for establishing planning targets in marine conservation on both a regional and global scale for the protection, recovery, and future monitoring of species and ecosystems. The authors of *Tropical Marine Mollusks*, by using mathematical models and methodology derived from endemism rules established from earlier research, and applying these to molluscan fauna of the tropics and warm-water sub-tropics and temperate zones, have been able to define faunal biogeographical regions, subregions, provinces, and subprovinces at a level not previously attempted. From these, evolutionary hot spots around the world are identifiable, critical to establishing conservation priorities. This will enable future studies on biodiversity patterns, especially island endemism, and areas with potentially vulnerable fauna already under threat from coastal development, pollution, and over-fishing, to be targeted for protection.

As a member of the International Union for the Conservation of Nature, actively involved in the assessment of marine mollusks for extinction risk, I applaud this ground-breaking work with its outstanding explanatory maps and images. It will become invaluable, not only to marine ecologists like myself, but also to evolutionary biologists and biogeographers, as well as malacologists and shell collectors. Indeed, any reader interested in the ocean environment will love this work, including those who are just entranced by the sheer beauty of the illustrations and would like to learn more.

Howard Peters, Ph.D.
Department of Environment and Geography
University of York, United Kingdom

CHAPTER 1. Biogeographical Patterns in the Marine Biosphere

For over 160 years, oceanographers, marine biologists, malacologists, and marine ecologists have recognized that the shallow water marine molluscan faunas of the world are distributed in distinct, geographically definable areas. Intensive shell collecting by eighteenth and nineteenth-century naturalists in the English, Spanish, French, Portuguese, and Dutch tropical colonies resulted in the discovery of an overwhelming number of spectacular new species and genera. These new taxa, brought home to Europe by sailors in the late 1600s and early 1700s, were highly prized by wealthy shell collectors and a large, highly-competitive market for rare and desirable species quickly developed. These new taxa supplied the first evidence for the existence of larger and richer molluscan faunas beyond the European and Mediterranean worlds. As more data on new molluscan faunas poured into Europe in the early 1800s and, later in the United States in the mid-1800s and early 1900s, the geographical patterns of these faunas became even better defined. Building on this new pool of information, the European and American malacologists, in recognizing latitudinal distributions, were unknowingly planting the seeds for the development of a completely new branch of marine science: marine biogeography, the study of the spatial distribution of marine organisms and the ecological parameters that determine their distributions.

A formal framework for the study of worldwide biogeography was first proposed in two contemporaneous books, both published in 1856. The first of these, *The Physical Atlas of Natural Phenomena* by Alexander Johnston, divided the world into 25 provinces within a series of nine horizontal, latitudinally-arranged zones that he referred to as "Homoizoic Belts". This geographical classification scheme is generally considered to be the first comprehensive book ever written on worldwide marine biogeography. Johnston's work was based upon earlier discoveries by the "Father of Modern Marine Biology", Edward Forbes, Jr., who recognized a series of "provinces" in the European seas, including the "Arctic, Boreal, Celtic, Lusitanian, Mediterranean, and Black Sea." Forbes also commented on the broad distributional patterns of corals in respect to latitudinal gradations in water temperature, foreshadowing subsequent studies on the physiological and ecological parameters that delineate tropical and subtropical marine faunas. In his *Manual of Conchology*, which was published at the same time as Johnston's work, Samuel P. Woodward, a student of Edward Forbes, was the first worker to formally define a *province*, describing it as "**an area in which 50% of the species are endemic**". Using this "**50% Rule**", Woodward went on to describe and define 25 worldwide marine molluscan faunal provinces and the original book was considered so important that it underwent several editions (including 1856 and 1880 editions). Some of these provinces, such as the Carolinian, Caribbean, South African, Indo-Pacific, Eritrean, Japonic, Californian, and Panamic, are used here in this book, underscoring the brilliance of S.P. Woodward's original pioneer work.

Although only qualitative in nature, Woodward's scheme of 25 separate worldwide molluscan provinces became the standard for marine biogeography throughout the late nineteenth century and the first half of the twentieth century. This view of a 25 province world was overturned in 1953, when Sven Ekman published his highly informative survey of worldwide marine faunas, the classic book *Zoogeography of the Seas*. Although covering all types of marine organisms, not only the Mollusca, Ekman abandoned Woodward's province concept, preferring instead sets of regions and subregions. The area that we here consider to represent four separate faunal provinces (the Carolinian, Caribbean, Brazilian, and Paulinian), Ekman fused together into a single subregion (the "Subtropical American Subregion") of his large "Atlanto-East Pacific Region" (covering these four provinces along with the Californian and Panamic Provinces of the Eastern Pacific). Following this concept of a single giant tropical area in the western Atlantic, Germaine Warmke and R.T. Abbott (in *Caribbean Seashells*, 1961: 319) recognized an expanded Caribbean Province which extended from Cape Hatteras and Bermuda southward through Florida, the Gulf of Mexico, the Caribbean Sea, to Cabo Frio, Brazil. Both Ekman's and Warmke and Abbott's biogeographical schemes are now considered to be overgeneralized and, intrinsically, do not offer the level of resolution needed for the recognition of localized species radiations and island endemism.

The concept of broad, generalized faunal regions was challenged in the 1970s, primarily in the works of James Valentine. Expanding upon Woodward's original marine molluscan province concept, Valentine, in his *Evolutionary Paleoecology of the Marine Biosphere* (1973:337), was the first to offer a quantitative definition of a faunal province. Augmenting Woodward's *50% Rule*, Valentine also applied cluster analysis, multivariant analysis, Jaccard's Coefficient, and other statistical methods to determine the boundaries of provinces and subprovinces around the world. Although retaining three of Woodward's provincial names, the Carolinian, Caribbean, and Patagonian, Valentine also added a fourth province, the Gulf Province (considered here to be a composite of four separate subprovinces within the Gulf component of the Carolinian Province). His analytical studies, although now outdated, were the first to pin down provincial boundaries by using mathematical techniques and Valentine was also the first biogeographer to use plate tectonics and physiological restrictions (limiting factors) to explain provincial distributions.

Several years after Valentine's monumental work, the theoretical ecologist, Geerat Vermeij, took a reductionist viewpoint and defined only two broad provinces for the entire tropical western Atlantic (in his *Biogeography and Adaptation*, 1978). These consisted of a "Tropical Western Atlantic Province" (from Palm Beach, Florida and Bermuda, the southern Gulf of Mexico, the Caribbean Basin, and south to Brazil) and a "Warm-Temperate Northwest Atlantic Province" (Cape Hatteras, North Carolina south to Palm Beach). These broad faunal regions were based primarily on latitudinal patterns of predation on mollusks and not on the actual compositions of the molluscan faunas. Although extremely over-generalized, Vermeij's work inspired the senior author to undertake nine years of intensive field work in eight different countries in South America and the lower Caribbean. From research conducted in these then-unexplored areas, mostly while working on shrimp boats, hundreds of new species and genera were collected, giving new insight into the biogeographical patterns of the lower Caribbean area and South America. The descriptions of all of these important new taxa and their accompanying biogeographical patterns were summarized in one large compendium titled *New Caribbean Molluscan Faunas* (Petuch, 1987). Building on this new taxonomic data, the senior author redefined the provincial arrangement of the tropical western Atlantic, adding a third new province, the Brazilian (Petuch, 1988: 166). In later books, such as *Cenozoic Seas: The View from Eastern North America* (Petuch, 2004) and the *Biogeography and Biodiversity of Western Atlantic Mollusks* (Petuch, 2013), the senior author further subdivided the Brazilian Province into four separate subprovinces. These are re-defined and expanded upon in Chapter 2 of this book.

Modifying the basic definition of a province (as defined by Woodward, Johnston, and Valentine), another well-known marine biogeographer, John Briggs, offered a new and more accurate provincial arrangement in his book *Global Biogeography* (1995). This large and comprehensive work was the first biogeographical text to take into account the geological history of both the continents and world's oceans and their interactions. In an earlier predecessor work, *Marine Zoogeography* (1974), Briggs mainly utilized fish faunas to discern worldwide biogeographical patterns, while in his later work he incorporated biogeographical data taken from studies on many other phyla and classes. This 1995 work, along with that of Valentine (1973), demonstrated that the provinces of other marine organisms, such as echinoderms, crustaceans, and corals, essentially conform to the geographical limits of molluscan faunal provinces. An expanded data bank such as this has provided a much more refined view of the biogeography of entire tropical communities and ecosystems around the world. Interestingly enough, the marine biogeographical classification systems proposed by Briggs, Vermeij, Valentine, Abbott, Ekman, and others all support and corroborate the original ideas that Woodward put forth 164 years ago. Because of his amazing scientific accuracy and foresight, Woodward truly deserves the title of "Father of Molluscan Biogeography". Building on the works of the earlier pioneer scientists, we here present a new and much more detailed overview of the worldwide patterns of molluscan biogeography.

Molluscan Faunal Provinces and Subprovinces

Of the molluscan biogeographers, only Woodward (1856) and Valentine (1973) proposed a formal definition of a province that involved an actual numerical index (the 50% Rule). In 2004, the senior author (Petuch: 21) clarified and constrained Valentine's provincial definition by stating that "*...two adjacent geographical areas can be considered to be separate molluscan provinces if at least 50% of the species-level taxa are endemic to each area. This mutual exclusivity includes the faunas of all possible biotopes within the two provinces.*" In this book we use this more expansive definition of provinciality as the standard of comparison for all the molluscan faunas that fall within the confines of tropical and subtropical seas. Our definition not only takes into account the levels of overall endemism found in a province but also the endemism found in every type of marine habitat within the provincial boundaries. The province-level biogeographical unit can be considered to be a single, giant biological entity or "super-organism" that is evolving separately above the organismal and community levels.

Since this book deals exclusively with the mollusks of warm water seas, we found it necessary to redefine the vernacular terms "tropical" and "subtropical", particularly in reference to the biotic composition of molluscan provinces. Traditionally, tropical seas are defined as low latitude systems that have surface water temperatures that never fall below 20 degrees Celsius. This physical oceanographic definition is here augmented by the addition of a biological component, which consists of suites of high-tropical proxy taxa that can be used as indicators of tropical water conditions. Some of these include the following gastropod families and subfamilies:

Potamididae
Modulidae
Cypraeidae
Cassidae
Ficidae
Strombidae
Rostellariidae
Pisaniidae
Xenophoridae
Fasciolariidae: Peristerniinae
Melongenidae
Turbinellidae: Vasinae
Volutidae: Lyriinae, Cymbiinae, Amoriinae, Meloninae, Plicolivinae, Scaphellinae
Harpidae: Harpinae, Moruminae
Cancellariidae: Cancellariinae, Plesiotritoninae
Olividae: Olivinae
Conidae

If 50% of the species within these families and subfamilies are restricted to a discrete geographical area, and the combined percentages of endemism for all of these higher taxa is greater than 50%, then that area can be considered to be a molluscan faunal province. A quantitative methodology for provincial determination, incorporating mathematical formulas that utilize these percentages of endemism, was proposed by the senior author (Petuch, 2013). We have utilized this methodology for determining all of the provincial boundaries described in this book (see Appendix on p. 303).

Bodies of water that house most, if not all, of these index families and subfamilies as components of resident molluscan assemblages are referred to as **Eutropical Seas** (Petuch, 2004: 21-22; 2013:4). Molluscan provinces that also contain most, if not all, of these taxa are referred to as **Eutropical Provinces** (Petuch, 2004) (see Figure 1.1). Along with these eutropical families, numerous high tropical index gastropod genera can also be used to define eutropical provinces. Some of these include:

• *Turbo*, *Taenioturbo*, *Marmarostoma*, *Lithopoma*, and *Astralium* (all Turbinidae)
• *Modulus, Indomodulus, Trochomodulus, Conomodulus,* and *Laevimodulus* (all Modulidae)
• *Cassis* and *Cypraecassis* (both Cassidae)
• *Turbinella* and *Vasum* (both Turbinellidae)
• *Lambis*, *Harpago*, *Ophioglossolambis*, *Millepes* (all Strombidae)
• *Harpa, Morum, and Oniscidia* (Harpidae)

Environmentally, eutropical areas are demarcated by the presence of **coral reefs** and their associated carbonate sediments, by **Turtle Grass** (*Thalassia*) beds, and by peripheral **mangrove tree forests**. Each of these main eutropical environmental types has its own characteristic associated molluscan fauna, with the same families and genera being found around the world. These classical eutropical molluscan assemblages all derive from the high tropical **Tethys Sea** of the Eocene Epoch (36 million to 52 million years ago), a contiguous belt ocean that surrounded the equatorial region of the Earth. At that time, North and South America were separated by an open strait that existed where Central America is today, and the Arabian Peninsula was flooded and connected what is now the Mediterranean Sea to the Indian Ocean. This worldwide tropical belt ocean was the center of evolution for all the living eutropical marine faunas and floras seen today, including the modern fishes, mollusks, corals, echinoderms, crustaceans, sea grasses, and mangroves. Once Panama and the Arabian Peninsula emerged during the Pliocene (2 million to 5 million years ago), the Pan-Tethyan molluscan fauna was separated into genetically-isolated sections and each of these gave rise to the molluscan provinces described in the following chapters of this book.

At higher latitudes adjacent to the eutropical areas are cooler "subtropical" and warm-temperate seas, where water temperatures seasonally fall below 20 degrees Celsius but not below 10 degrees Celsius. These bodies of water, referred to here as **Paratropical Seas**, house only some of the classic tropical index families and genera and lack coral reefs, carbonate platforms, Turtle Grass beds, and extensive mangrove forests. Molluscan provinces that occur within these cooler oceanic climates are referred to as **Paratropical Provinces** (Petuch, 2004) and they sometimes form a mirror-image pair (north and south) on either side of the eutropical provinces (see Figure 1.2). Paratropical provinces typically contain mixed faunas, with Tethyan-derived tropical components co-existing with boreal and antiboreal-derived cold water components. In the Recent Americas, a classic example of a paratropical province is seen in the Californian Province of the Eastern Pacific. Although the water temperatures are below 20 degrees year round and the area is devoid of coral reefs, Turtle Grass beds, and mangrove forests, the Californian Province does contain a few tropical molluscan elements such as the cowrie genus *Neobernaya*, the cone shell *Californiconus californicus*, and the potamidid *Cerithidea*. These cold-tolerant Tethyan derivatives co-exist with classic cold water North Pacific genera such as the borsoniid *Ophiodermella*, the pseudomelatomid *Megasurcula*, the naticid *Euspira*, and large species radiations of the muricids *Paciocinebrina* and *Pteropurpura* and the abalone *Haliotis* (see Chapter 3). As pointed out by Vermeij (1978), the cold-loving (cryophilic) offshoots of tropical families are usually generalist feeders with weak shell architecture, indicating that their invasion into higher and lower latitudes may be the result of avoidance of competition and predation. Another classic example of a paratropical province is seen in the cold water Flindersian Province of southern Australia. Here, a huge radiation of cold-tolerant cowries has evolved, including multiple species clusters in the genera *Zoila*, *Umbilia*, *Notocypraea*, and *Austrocypraea* (see Chapter 9).

Geographically widespread provinces typically contain areas with multiple endemic species radiations and localized evolutionary centers. These special endemic organisms are almost always species that are nonvagile, with direct development larvae that do not have planktotrophic stages. Nonvagile organisms such as these are highly susceptible to genetic isolation and rapidly undergo speciation, particularly along peripheral areas where seasonal climate changes are more extreme. Because of this, multiple areas within a province may develop their own characteristic faunas and become distinctive enough to be separable from the rest of the province. This evolutionary pattern is frequently seen in provinces that contain island-covered shallow platforms or large, complex archipelagos of islands. Geographically definable faunal subdivisions within a province, containing localized centers of allopatric speciation (speciation by genetic isolation) with at least a 25% level of endemism, are referred to as **Subprovinces** (Petuch, 2013:6). Some of the geographically smaller provinces, such as the Verdesian, Namibian, Marquesan, and Rapauian, do not have any recognizable subprovinces. Other larger provinces, such as the Carolinian and the Guinean, each have 5 distinct subprovinces, while the Panamic and Lemurian Province have 6. The Caribbean Province, with its multiple archipelagos of islands, scattered wide banks, and varied continental shorelines has 7 distinct subprovinces, the most of any single province in the world. Maps of the world showing the distribution of the marine molluscan provinces are shown here in Figures 1.1 and 1.2.

After applying both quantitative and qualitative analyses to the worldwide tropical marine gastropod fauna (using the mathematical models and methodology outlined in Petuch, 2013: 1-7; see Appendix at p. 303), we here present a new and more detailed classification scheme for molluscan biogeographical patterns. Using the 50% Rule for province definition and the 25% Rule for subprovince definition, we now recognize that the molluscan faunas of the eutropical and paratropical seas are contained in 2 Super-Regions, 9 Regions, 9 Subregions, 24 Provinces, and 68 Subprovinces. These will be discussed in detail in the following chapters of this book and they are arranged here by hierarchical order:

Western Atlantic Region

Northwestern Atlantic Paratropical Subregion

Carolinian Molluscan Province
- Georgian Subprovince
- Floridian Subprovince
- Suwannean Subprovince
- Texan Subprovince
- Yucatanean Subprovince

Northwestern Atlantic Tropical Subregion
Caribbean Molluscan Province
- Bermudan Subprovince
- Bahamian Subprovince
- Antillean Subprovince
- Nicaraguan Subprovince
- Venezuelan Subprovince
- Grenadian Subprovince
- Surinamian Subprovince

Southwestern Atlantic Tropical Subregion
Brazilian Molluscan Province
- Cearaian Subprovince
- Bahian Subprovince

Southwestern Atlantic Paratropical Subregion
Paulinian Molluscan Province
- Janeiran Subprovince
- Uruguayan Subprovince

Eastern Atlantic Region

Northeastern Atlantic Paratropical Subregion
Mediterranean Molluscan Province
- Ionian Subprovince
- Algerian Subprovince
- Canarian Subprovince
- Madeiran Subprovince

Eastern Atlantic Tropical Subregion
Verdesian Molluscan Province
Guinean Molluscan Province
- West Saharan Subprovince
- Senegalian Subprovince
- Biafran Subprovince
- Angolan Subprovince
- Helenean Subprovince

Southeastern Atlantic Paratropical Subregion
Namibian Molluscan Province

Eastern Pacific Region

Northeastern Pacific Paratropical Subregion
Californian Molluscan Province
- Diegan Subprovince
- Cedrosian Subprovince
- Guadalupean Subprovince

Eastern Pacific Tropical Subregion
Panamic Molluscan Province
- Magdalenan Subprovince
- Cortezian Subprovince
- Jaliscoan Subprovince
- Cocosian Subprovince

Chiriquian Subprovince
Ecuadorian Subprovince

Indo-Pacific Super-Region

Central Pacific Region

Hawaiian Molluscan Province
Marquesan Molluscan Province
Rapanuian Molluscan Province
Polynesian Molluscan Province
Tahitian Subprovince
Micronesian Subprovince

Western Pacific Tropical Region

Indo-Malaysian Molluscan Province
Indonesian Subprovince
Melanesian Subprovince
Neocaledonian Subprovince
Philippinian Subprovince
Japonic Molluscan Province
Shikokuan Subprovince
Ryukyuan Subprovince
South China Subprovince

Indian Tropical Region

Lemurian Molluscan Province
Andamanian Subprovince
Bengalian Subprovince
Malabaran Subprovince
Mascarenean Subprovince
Madagascan Subprovince
Mozambican Subprovince
Eritrean Molluscan Province
Aqaban Subprovince
Dahlakian Subprovince
Omanian Subprovince
Somalian Subprovince

Australian Super-Region

North Australian Tropical Region

Solanderian Molluscan Province
Moretonian Subprovince
Cairnsian Subprovince
Coralian Subprovince
Dampierian Molluscan Province
Exmouthian Subprovince
Carpentarian Subprovince

South Australian Paratropical Region

Peronian Molluscan Province
Macquarian Subprovince

Maugean Molluscan Province
- Victorian Subprovince
- Tasmanian Subprovince

Flindersian Molluscan Province
- Adelaidean Subprovince
- Euclean Subprovince
- Perthian Subprovince

Southern Africa Region

South African Molluscan Province
- Capean Subprovince
- Transkeian Subprovince
- Natalean Subprovince

Of the 24 provinces listed here, 15 are considered to be Eutropical Provinces, containing the full complement of tropical index families, subfamilies, and genera and housing coral reefs, Turtle Grass beds, and mangrove forests. These are:

Eutropical Provinces (with a complete complement of Tethyan Taxa)

Caribbean Province
Brazilian Province
Verdesian Province
Guinean Province
Panamic Province
Hawaiian Province
Marquesan Province
Rapanuian Province
Polynesian Province
Indo-Malaysian Province
Japonic Province
Lemurian Province
Eritrean Province
Solanderian Province
Dampierian Province

The remaining 9 provinces are here considered to be Paratropical Provinces, in that they often lack coral reefs, Turtle Grass beds, and mangrove forests, and have only a partial complement of tropical families, subfamilies, and genera. These are:

Paratropical Provinces (with only a partial complement of Tethyan Taxa)

Carolinian Province (with the exception of the Floridian and Yucatanean Subprovinces)
Paulinian Province
Mediterranean Province
Namibian Province
Californian Province
Peronian Province
Maugean Province

Flindersian Province
South African Province

In several areas, such as the southern Caribbean Sea and the Western Pacific, shallow neritic (inner continental shelf) mollusks from peripheral cooler water provinces move into deeper outer neritic and upper bathyal water (outer continental shelf and upper continental slope) if they extend their ranges into eutropical provinces. These wide-ranging deep water species are here referred to as **Panbathyal Taxa**, and the bathymetric range of these organisms is controlled by the depth of the thermocline in tropical seas. A classic example of the submergence of panbathyal taxa is seen in the faunal inter-relationship of the Shikokuan and Rykyuan Subprovinces of the Japonic Province and the Philippinian, Indonesian, and Melanesian Subprovinces of the Indo-Malaysian Province. In the cooler waters of Japan and the Ryukyu Islands, the cone shells *Afonsoconus kinoshitai*, *Graphiconus armadillo, Graphiconus kuroharai*, and *Fusiconus ichinoseana* all occur in depths of 100 m or less and are frequently collected by local fishing boats. South of this area, however, these same species submerge into depths of 300-500 m or more, following the cold water temperatures below the thermocline and dispersing as far south as New Caledonia (see Figure 6.23 for more examples of panbathyal taxa). A similar pattern of panbathyal submergence is seen in the Grenadian Subprovince of the southern Caribbean Province. Here, species such as the cone shells *Sandericonus sorenseni* and *Sandericonus perprotractus* and the strombinid *Cotonopsis lindae* were originally described from the cold, upwelling waters off Barbados, in 100 m depth or less, but are now known to occur all along the Lesser Antilles in depths of 200-300 m (more examples are shown on Figure 2.33). Panbathyal faunas are also known from the Lemurian Province of the Indian Ocean.

Infraprovinces and Provinciatones

Highly localized "hot spots" of evolution and biodiversity often appear within subprovincial areas, particularly in isolated enclosed bays or on isolated offshore islands. These small centers of intense speciation, here referred to as **Infraprovinces**, are of special interest because of their limited geographical size and vulnerability to environmental degradation and extinction. Infraprovinces increase the overall biodiversity of their resident subprovinces and may, in turn, act as the "seeds" for new, future subprovinces. The largest number of infraprovinces is seen in the Caribbean and Mediterranean Seas (15 in the Caribbean and 7 in the Mediterranean), both of which contain thousands of isolated islands and smaller enclosed basins. Some island infraprovinces, such as the Noronhan Infraprovince of Brazil (Fernando de Noronha Island) and the Brandonian Infraprovince of Mauritius (St. Brandon's Shoals), harbor Miocene-Pliocene relictual genera, such as the tonnid *Malea noronhensis* on Fernando de Noronha and the ficid *Ficus dandrimonti* on St. Brandon's. Small, deep, isolated basins within enclosed seas can also produce infraprovincial "hot spots", such as the Felipean Infraprovince in northernmost Gulf of California. Here, Pleistocene relictual taxa from the Californian Province, such as the muricid *Forreria* (*F. corteziana*) and the bursid *Crossata* (*C. sonorana*) have managed to survive in the deep water petroleum seep environments of the thermally-isolated deep Wagner Basin. We here list the 52 best-known and best-studied infraprovinces from the world's oceans and these are discussed in detail in the following chapters.

WESTERN ATLANTIC REGION
Carolinian Province
Palm Beach Infraprovince (Palm Beach coast, Florida)
Chokoloskean Infraprovince (Ten Thousand Islands, Florida)
Apalachicolan Infraprovince (Florida Panhandle to Mississippi)

Caribbean Province
Biminian Infraprovince (Bimini Chain, Bahamas)
Eleutheran Infraprovince (Eleuthera Is., Bahamas)
Abacoan Infraprovince (Abaco Islands, Bahamas)
Belizean Infraprovince (Great Barrier Reef and Atolls of Belize)
Cuban Infraprovince (southern coast of Cuba)
Jamaican Infraprovince (Jamaica)
Hispaniolan Infraprovince (Haiti and the Dominican Republic)
Puerto Rican Infraprovince (Puerto Rico)
Caymanian Infraprovince (Cayman Islands)
Blasian Infraprovince (San Blas Islands, Panama)
Colombian Infraprovince (Colombian coast south of the Magdalena River)
Martiniquean Infraprovince (Martinique)
Roquesian Infraprovince (Los Roques Atoll, Venezuela)
Aruban Infraprovince (Aruba, Curacao, and Bonaire Islands)
Barbadan Infraprovince (Barbados)

Brazilian Province
Noronhan Infraprovince (Fernando de Noronha Islands)
Itaparican Infraprovince (Todos os Santos Bay, Bahia State)
Abrolhosian Infraprovince (Abrolhos Archipelago and Abrolhos Platform)
Trindadean Infraprovince (Trindade and Martim Vaz Archipelago)

Paulinian Province
Macaen Infraprovince (Cabo Frio area)

EASTERN ATLANTIC REGION
Mediterranean Province
Aegean Infraprovince (Aegean Sea)
Levantine Infraprovince (Turkey to Egypt)
Adriatic Infraprovince (Adriatic Sea)
Libyan Infraprovince (Libya and Tunisia)
Sicilian Infraprovince (northern Sicily and Sardinia)
Alboranian Infraprovince (Alboran Sea)

Guinean Province
Gorean Infraprovince (Cape Verde Peninsula, Senegal)
Gambian Infraprovince (Gambia)

Tomean Infraprovince (Sao Tome e Principe)
Luandan Infraprovince (northern Angola)
Namibe Infraprovince (southern Angola)

EASTERN PACIFIC REGION
Panamic Province
Lucasian Infraprovince (Cabo San Lucas area, Mexico)
Felipean Infraprovince (Wagner Basin, northernmost Gulf of California, Mexico)
Clippertonian Infraprovince (Clipperton Island)
Galapagan Infraprovince (Galapagos Islands)

CENTRAL PACIFIC REGION
Rapanuian Province
Salas-Gomesian Infraprovince (Salas y Gomez Island)

Polynesian Province
Tuamotuan Infraprovince (Tuamotu Archipelago)
Samoan Infraprovince (Samoa)
Kwajaleinian Infraprovince (Kwajalein Atoll)

WESTERN PACIFIC TROPICAL REGION
Indo-Malaysian Province
Sundan Infraprovince (southern Java and Sunda Islands, Indonesia)
Solomonian Infraprovince (Solomon Islands)
Norfolkian Infraprovince (Norfolk Ridge and Norfolk Island)
Suluan Infraprovince (Sulu Sea, southern Philippines and northern Malaysia)
Vietnamese Infraprovince (Vietnam)

INDIAN TROPICAL REGION
Lemurian Province
Brandonian Infraprovince (St. Brandon's Shoals, Mauritius)
Toliaran Infraprovince (southern Madagascar)
Masiran Infraprovince (Al Masirah Island, Oman)

NORTH AUSTRALIAN TROPICAL REGION
Solanderian Province
Howeian Infraprovince (Lord Howe Island, Tasman Sea)

Dampierian Province
Ashmorean Infraprovince (Ashmore Reefs, NW Australia)

As more research and field work are conducted in tropical seas around the world, there is no doubt that many more infraprovinces will be discovered by future workers. These highly localized areas of special evolution are in need of recognition and protection, as they are particularly vulnerable to over-collecting and habitat destruction due to human development.

Depending on the patterns and configurations of the oceanic currents, the boundaries between two provinces are often blurred, represented by wide, poorly-defined areas of faunal overlap. These transition zones characteristically contain sympatric faunal elements from both parent provinces, often co-existing in unusual and novel molluscan assemblages. This type of provincial overlap zone is referred to as a **Provinciatone** (Petuch, 2004:21; 2013:7) and it is analogous to the smaller-scale ecotone (an area of overlap between two ecosystems; Odum, 1971). Provinciatones characteristically contain three types of faunal components: species from a lower latitude eutropical or paratropical province; species from a higher latitude paratropical or cold-temperate province; and **provinciatonal endemics**, species that are restricted to the boundaries of the provinciatone. A classic example of a provinciatone is seen in the Platensian Provinciatone of the Paulinian Province. Here, in a faunal transitional area extending from Uruguay to northern Argentina, tropical genera such as the cone shell *Lamniconus* (*L. carcellesi*), the marginellid *Volvarina* (*V. warreni*), and the terebrid *Duplicaria* (*D. gemmulata*) co-exist with subantarctic taxa such as the volutes *Adelomelon ancilla*, *Odontocymbiola subnodosa*, and *Provocator corderoi*, and the provinciatonal endemics *Buccinanops duartei* (Nassariidae) and *Olivancillaria duartei* (Olividae). A similar pattern is seen in the Northern Californian Provinciatone, where a broad transition zone extends from southern Oregon to Point Conception, California. Along this stretch of coastline, some cold-tolerant species from the Californian Province co-exist with warm-tolerant species from the Oregonian Province, along with the provinciatonal endemic *Tegula montereyi* (Figure 4.5 L). Only three major provinciatones are known from the world's oceans. These include the **Platensian Provinciatone** (Uruguay-northern Argentina; Chapter 2), the **Namaquan Provinciatone** (southern Namibia-northwestern South Africa; Chapter 3), and the **Northern Californian Provinciatone** (southern Oregon-Point Conception, California; Chapter 4).

Geographical Heterochrony

Research conducted by the senior author along northern South America showed that the marine ecosystems offshore of Venezuela and Colombia contained many gastropod genera that were thought to have become extinct by the late Pliocene (Petuch, 1979; 1981; 1982a; 1982b; 1987; 2013:8). These "living fossil" genera were known from the paleontological record of the West Indies, Panama, and northern South America but were not known, up to that time, to have survived into the present. These small areas of prehistoric left-overs are referred to as **Relict Pockets** (Petuch, 1982; 2013:8-9). Some of the relictual taxa found in the Colombian relict pocket included the cancellariid *Aphera* (first-known living Caribbean species, *A. lindae*), the muricid *Lindapterys* (*L. lindae*; previously known only as a fossil from the early Miocene Chipola Formation), the borsoniid *Paraborsonia* (first-known living species, *P. lindae*), and living representatives of archaic groups such as the *Tenorioconus consobrinus* complex (*T. granarius*; Conidae), *Muracypraea henekeni* complex (*M. tristensis*; Cypraeidae), *Sconsia laevigata* complex (*S. lindae*; Cassidae), *Cancellomorum dominguense* complex (*C. lindae*; Harpidae), the *Ficus pilsbryi* complex (*F. lindae*; Ficidae), the *Panamurex gatunensis* complex (*P. petuchi*, Muricidae), and many others (some shown here on Figures 2.29 and 2.30). The presence of "living fossil" molluscan assemblages, in a geographically-small relict pocket that is surrounded by larger and more modern molluscan faunas, demonstrates that rates of community evolution are not constant across an entire province. The faunas of some biogeographical areas evolve at a much slower rate and often retain large numbers of archaic and relictual taxa. The presence of these small ancient faunas within a larger, more recently-evolved fauna is referred to as **Geographical Heterochrony**, where the rates of evolution and extinction vary over a biogeographical area, often reaching evolutionary stasis and retaining much of the original ancient community structure (Petuch, 1982a; 1988; 2013: 9).

Subsequent to the senior author's initial research in South America, more relict pockets were discovered and recognized, the most important being from the area around Roatan Island, Honduras (Petuch, 1980; 1981; 1987). Here, several gastropod genera that were known only from the Plio-Pleistocene fossil record of southern Florida (Caloosahatchee and Bermont Formations) were found to be extant in the Bay Islands. Some of these included the large cerithiid *Cerithioclava* (*C. garciai*), the pleioptygmatid giant miter *Pleioptygma* (*P. helenae*), and the small cone shell *Cariboconus* (several species, with the only-known fossil species coming from the Bermont Formation of the Everglades area). The archaic nature of the Roatan relict pocket demonstrates that the entire area is geographically heterochronous and actually represented a left-over of an early Pleistocene fauna from southern Florida, one that must have migrated southward during the cold glacial intervals and found a refuge in the warmer western Caribbean. Since the original research on these two relict pockets, more examples of geographical heterochrony have been found around the world, including the Yucatan Peninsula of Mexico, where late Pleistocene (Fort Thompson Formation) Florida species such as *Melongena (Rexmela) bispinosa* and *Turbinella wheeleri* have survived, and also areas along northeastern Brazil and off Barbados (discussed in the next chapter). The Indonesian and Philippines areas of the Indo-Malaysian Province also appear to contain relict pockets and archaic faunas.

Figure 1.1 Map of the world's oceans, showing the areal distributions of the **Eutropical Molluscan Provinces**: the **Caribbean Province** (green), the **Brazilian Province** (dark blue), the **Verdesian Province** (lime green), the **Guinean Province** (brown), the **Panamic Province** (orange), the **Hawaiian Province** (gold), the **Marquesan Province** (purple), the **Rapanuian Province** (light yellow), the **Polynesian Province** (aquamarine), the **Indo-Malaysian Province** (yellow), the **Japonic Province** (forest green), the **Lemurian Province** (light rose), the **Eritrean Province** (burgundy), the **Solanderian Province** (red), and the **Dampierian Province** (light blue).

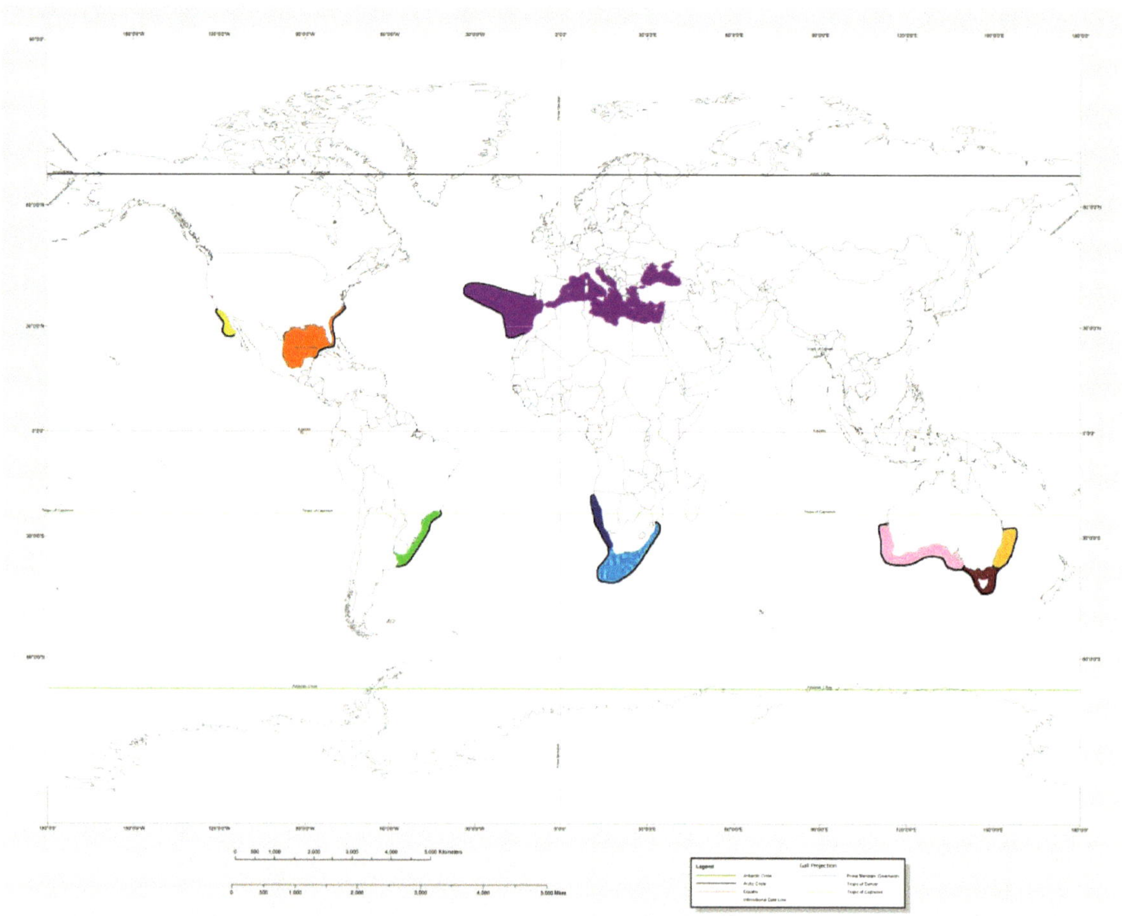

Figure 1.2 Map of the world's oceans, showing the areal distributions of the **Paratropical Molluscan Provinces**: the **Carolinian Province** (orange), the **Paulinian Province** (green), the **Mediterranean Province** (purple), the **Namibian Province** (dark blue), the **Californian Province** (yellow), the **Peronian Province** (gold), the **Maugean Province** (burgundy), the **Flindersian Province** (light rose), and the **South African Province** (light blue).

CHAPTER 2. Western Atlantic Region

The Western Atlantic Region spans the entire eastern coasts of North and South America, from Labrador and Arctic Canada south to Patagonia and the Strait of Magellan. This vast area encompasses a wide range of water temperatures and substrate types and comprises four distinct subregions that are relevant to the scope of this book: the **Northwestern Atlantic Paratropical Subregion**, the **Northwestern Atlantic Tropical Subregion**, the **Southwestern Atlantic Tropical Subregion**, and the **Southwestern Atlantic Paratropical Subregion**. These four subregions also encompass four warm-temperate and tropical molluscan provinces and 17 distinct subprovinces. The four paratropical and tropical provinces include, from north to south: the **Carolinian** (warm temperate); **Caribbean** and **Brazilian** (both tropical); **Paulinian** (warm temperate), and their spatial arrangement is shown on Figure 2.1. This core of warm water provinces is bounded on the north by the subarctic Labradoran Province and the cold-temperate Acadian and Virginian Provinces, and in the south by the cold-temperate Patagonian Province and the subantarctic Magellanic Province. The tropical and warm temperate (paratropical) areas of the Western Atlantic Region are discussed in detail in the following sections and the important index gastropods of the provinces and subprovinces are shown under their respective biogeographical units. (For details on the faunal subdivisions of the cold temperate Virginian Province, see Petuch, Myers, and Berschauer, 2015: 58-59.)

Faunistically, the Western Atlantic Region is defined by the combined distributions of several ubiquitous index species, which range from North Carolina to central Brazil. Some of these, like *Charonia variegata*, *Proadusta surinamensis*, *Oniscidia dennisoni*, *Stephanoconus regius*, and *Chelyconus testudinarius*, are known to have long-lived planktonic larvae and can travel great distances by riding the oceanic currents. The conid genus *Jaspidiconus* is completely restricted to this region and is one of the classic regional index taxa. Unlike other conidean genera, this group of small cones has evolved species radiations in all four of the molluscan provinces, with at least 80 species occurring in the area extending from North Carolina to São Paulo, Brazil. Of these, 55 species are shown on the figures in the Iconography at the end of this chapter. Some of the primary index gastropods of the Western Atlantic Region are shown here on Figure 2.6.

Northwestern Atlantic Paratropical Subregion

The Northwestern Atlantic Paratropical Subregion encompasses a single major biogeographical unit, the Carolinian Molluscan Province. This area, which extends from Cape Hatteras, North Carolina, to the Dry Tortugas, Florida Keys, and the entire Gulf of Mexico contains two main suites of endemic taxa: one along the southeastern United States and the other within the Gulf of Mexico. During the warm climatic times of the late Pliocene and early Pleistocene, the Florida Peninsula was submerged as a shallow bank and there was complete genetic flow between the mollusks of the Atlantic Ocean and Gulf of Mexico (Petuch, 2004). The peninsula, and its surrounding platform, was submerged at least 20 times during the Pleistocene, leading to a distinct

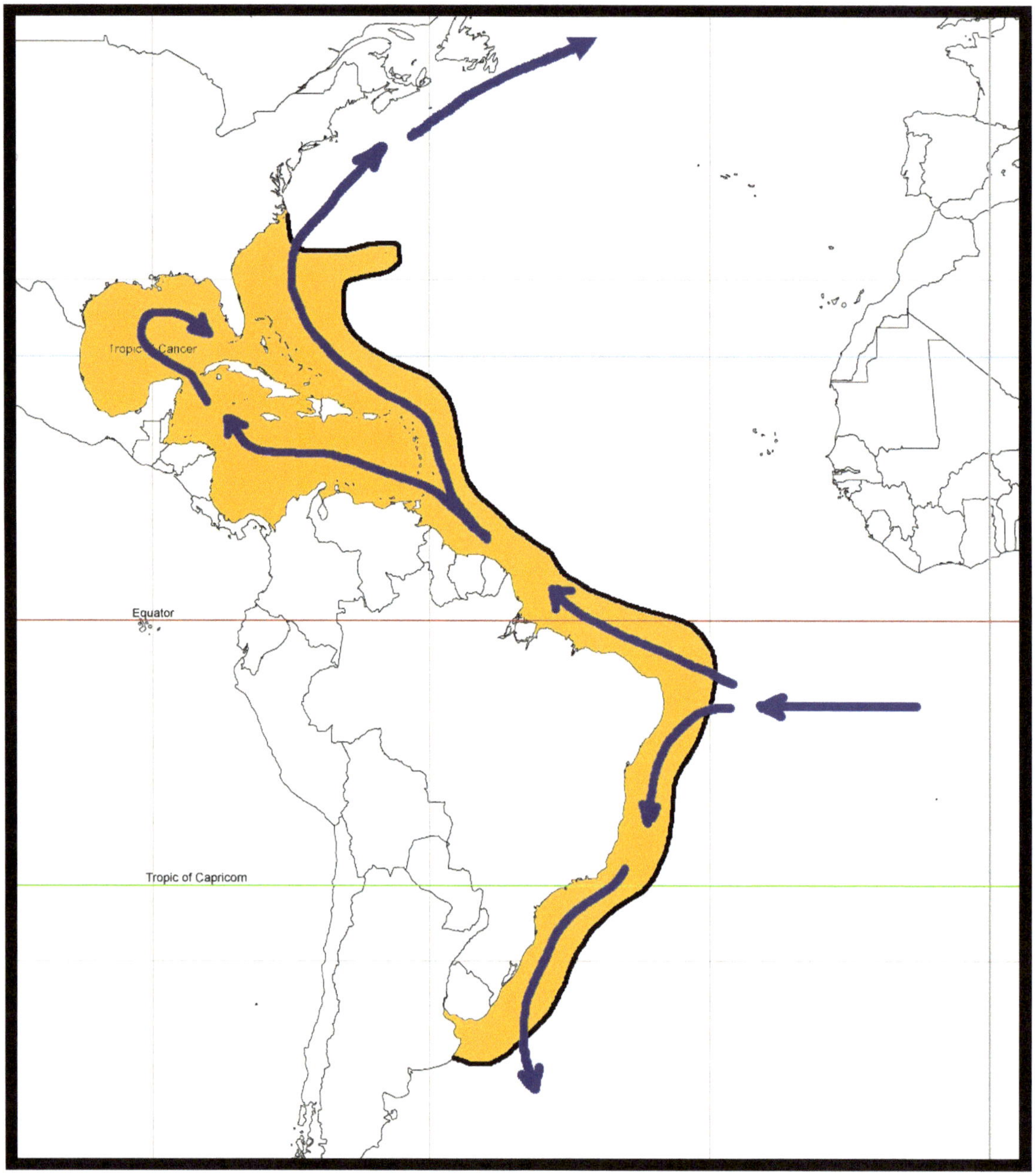

Figure 2.1 Map of the **Western Atlantic Region**, showing the areal extents of the combined tropical and paratropical subregions (gold) and the primary warm water currents (blue arrows).

homogeneity in the resident molluscan faunas on both sides of Florida. These periods of submergence were interrupted by periods of emergence, when the entire peninsula and the Florida Platform became dry land and Florida acted as a barrier to dispersal of the mollusks in both oceanic systems. During these times of emergence, some of which lasted from 500,000 to 800,000 years, the mollusks of the oceanographically-isolated Gulf of Mexico began to diverge from their Atlantic siblings and evolved into new endemic species. Two especially long periods of sea level lows appear to have had the greatest impact on molluscan speciation, with one occurring during the Calabrian-Ionian Age boundary and the other during the Ionian-Tarantian Age boundary. These times of extended genetic separation led to the evolution of sets of twin cognate species and many examples are readily demonstrated in the fossil record of southern Florida (see Petuch, 2004; 2013 for detailed discussions of these speciation patterns). Some of these include mid-Pleistocene bivalves such as the scallop *Argopecten anteamplicostatus*, which gave rise to the Recent *Argopecten taylorae* (Gulf of Mexico) and *Argopecten concentricus* (eastern coast of Florida) and the mid-Pleistocene giant cockle *Dinocardium hazeli*, which gave rise to the Recent *Dinocardium vanhyningi* (Gulf of Mexico) and *Dinocardium robustum* (eastern coast of Florida). With the final emergence of the Floridian Peninsula during the last major sea level drop at the end of the Late Pleistocene, the molluscan faunas of the Gulf of Mexico and Atlantic Ocean were effectively isolated from each other.

By the beginning of the late Pleistocene, the malacofaunas of the Gulf had evolved their own characteristic endemic species radiations and, based upon climate, had become sequestered into three distinct subprovincial areas. Likewise, the malacofaunas of the open Atlantic coast had become sequestered into two distinct subprovinces, with the southern one having become almost tropical in nature, complete with coral reef complexes, Turtle Grass beds, and mangrove forests. This tropical area, the Floridian Subprovince (the Florida Keys), acted as a barrier to genetic flow between the warm temperate faunas of the Georgian Subprovince (Atlantic coast) and the Suwannean Subprovince (eastern Gulf coast). In the Recent, this bimodality is readily apparent in the gastropod fauna, with examples being seen in the potamidid genus *Cerithidea* (with the Gulf *C. turrita* and the Atlantic *C. costata*), in the littorinid genus *Littoraria* (with the Gulf *L. irrorata sayi* and the Atlantic *L. irrorata*), in the conid genera *Gradiconus* (with the Gulf *G. anabathrum* and the Atlantic *G. philippii*) and *Jaspidiconus* (with the Gulf *J. stearnsi* and the Atlantic *J. pfleugeri*), in the olivid genus *Americoliva* (with the Gulf *A. nivosa choctaw* and the Atlantic *A. nivosa clenchi*), and in the busyconid genus *Fulguropsis* (with the Gulf *F. pyruloides* and the Atlantic *F. rachelcarsonae*: see Petuch, Myers, and Berschauer, 2015: 142-143; Petuch and Berschauer, 2019). The evolution of sibling species cognate pairs on either side of Florida is similar to what occurred between the western Atlantic Ocean and Eastern Pacific Ocean after the closing of the Isthmus of Panama (see Chapter 4).

Carolinian Molluscan Province

The Carolinian Molluscan Province, named for North and South Carolina is, faunistically and ecologically, one of the most diverse areas in the Western Atlantic Region. Encompassing the southeastern coast of the United States from Cape Hatteras, North Carolina to the Florida Keys, and the entire Gulf of Mexico to Cabo Catoche, Yucatan Peninsula, the Carolinian Province contains five distinct subprovinces, each having over 25% endemism at the species level. Within the geographical range of these five subdivisions, the province encompasses a wide variety of habitats, including areas where water temperatures drop to non-tropical ranges in the winter (such as the Carolinas and Georgia, the entire northern coast of the Gulf of Mexico, and the coast of Texas) to high-tropical areas where mangroves and coral reefs flourish (such as the Florida Keys and the Yucatan Peninsula of Mexico). Because of the extensive areas of cooler water temperatures during the winter months, the Carolinian Province is classified as a paratropical area.

Following the extinction event at the end of the Miocene (5.3 million years ago, the "Messinian Event"; see Petuch and Drolshagen, 2010: 12, 20; Petuch, 2012), eastern North America and the Gulf of Mexico areas evolved a new, highly endemic molluscan fauna and produced a completely new faunal region, the **Caloosahatchian Paleoprovince** (named for the Caloosahatchee Formation of Florida; see Petuch, 1982; 1997; 2004: 42-48). Some of the major Pliocene and early Pleistocene Caloosahatchian index gastropods included the busyconid genera *Busycon*, *Sinistrofulgur*, *Lindafulgur*, *Busycoarctum*, *Busycotypus*, and *Fulguropsis* (see Petuch, Meyers, and Berschauer, 2015), the volutid genera *Aurinia*, *Rehderia*, and *Scaphella (sensu stricto)*, the fasciolariid genera *Cinctura*, *Heilprinia*, and *Triplofusus*, and the melongenid subgenus *Rexmela*. These endemic Caloosahatchian genera survived the multiple glacial events and periodic extinctions of the Pleistocene "Ice Ages" and have persisted on into the Recent as components of the Carolinian Province (examples shown here on Figures 2.7-2.17). Over the past 1.8 million years, the Carolinian Province has evolved to be an impoverished version of the older and more diverse Caloosahatchian Paleoprovince, with many of the classic older genera having died out by the mid-Pleistocene (Petuch, 2004). Almost 75% of the Carolinian mollusks are endemic to the provincial boundaries, making this a very strong provincial unit (Petuch, 2013: 15-20). Some of the more important widespread Carolinian Province endemic species include the muricids *Hexaplex fulvescens* and *Pteropurpura bequaerti*, the giant cowrie *Macrocypraea (Lorenzicypraea) cervus*, the fasciolariids *Triplofusus papillosus* and *Cinctura hunteria*, the volute *Clenchina gouldiana*, and the cone shells *Dauciconus amphiurgus*, *Dalliconus mcgintyi*, *Kohniconus delessertii*, and *Lindaconus atlanticus* (all shown here on Figure 2.7). The five Carolinian subprovinces include: the **Georgian**, **Floridian**, **Suwannean**, **Texan**, and **Yucatanean**, and these are described in the following sections.

Underscoring their oceanographic isolation during the Pleistocene, the five subprovinces of the Carolinian Province have often evolved swarms of endemic taxa, resulting in the formation of unique species complexes. This is particularly obvious in the evolutionary patterns of the common bay scallop (the *Argopecten irradians* complex) which has evolved a cluster of four different but closely-related taxa: with *Argopecten concentricus* ranging from Cape Hatteras to Fort Pierce, Florida (Georgian Subprovince); *Argopecten taylorae* ranging from Biscayne Bay, Florida, around the Florida Keys, and northward along western Florida to Mobile Bay, Alabama (Floridian and Suwannean Subprovinces); and *Argopecten amplicostatus* ranging from the Mississippi River Delta, along Texas, to the Yucatan Peninsula of Mexico (Texan and Yucatanean Subprovinces) (see Petuch, 2013: 17). The closely-related *Argopecten irradians* is a cold water species, endemic to the Virginian Molluscan Province (outside the scope of this book), and ranges from Cape Cod to Cape Hatteras. A similar pattern is seen in the four species in the busyconid genus *Sinistrofulgur*: with *S. laeostomum* ranging from southern New Jersey to northeastern Florida (northern Georgian

Subprovince); *S. sinistrum* ranging from northeastern Florida, around the Florida Keys, and northward to the Mississippi River Mouth (southern Georgian, Floridian, and Suwannean Subprovinces); *S. pulleyi* ranging from the Mississippi River, along the Texas coast, and to Veracruz, Mexico (Texan Subprovince); and *S. perversum* ranging along the Yucatan Peninsula of Mexico (Yucatanean Subprovince) (see Petuch, Myers, and Berschauer, 2015, for busyconid speciation patterns). This speciation pattern is also seen in the olivid genus *Americoliva*, with *A. nivosa clenchi* in the Georgian Subprovince, *A. nivosa bollingi* in the Floridian Subprovince, *A. nivosa choctaw* in the Suwannean Subprovince, and *A. nivosa maya* in the Yucatanean Subprovince (see Petuch and Berschauer, 2019, for an overview of the *A. nivosa* species complex).

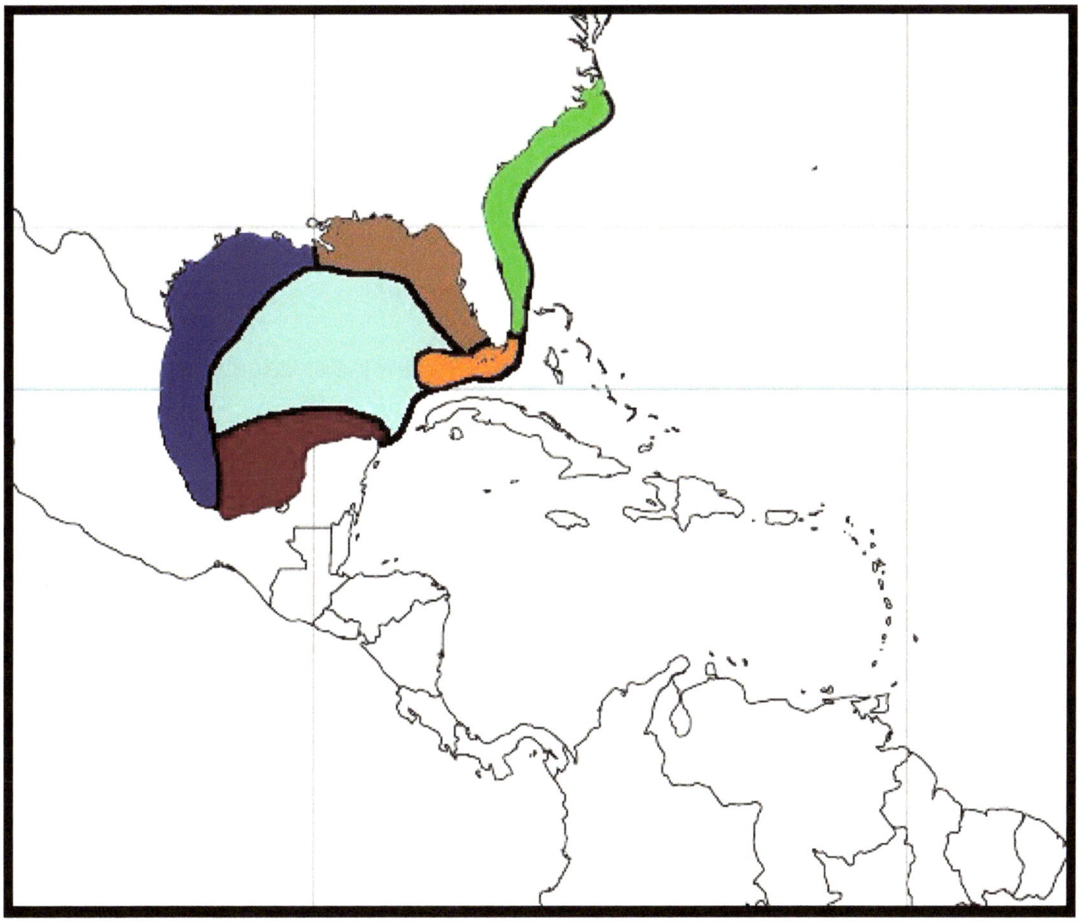

Figure 2.2 Map of the Carolinian Molluscan Province, showing the areal extents of its subprovinces: the **Georgian Subprovince** (green), the **Floridian Subprovince** (orange), the **Suwannean Subprovince** (brown), the **Texan Subprovince** (blue), and the **Yucatanean Subprovince** (burgundy). The pale blue color demarcates the central abyssal plain of the Sigsbee Deep, which contains its own endemic bathyal and abyssal faunas.

The Carolinian Province is also noteworthy in containing a remarkable radiation of the family Modulidae, the largest known from anywhere in the world. At least eight species in three genera are now known, including: *Modulus pacei*, *Modulus kaicherae*, *Trochomodulus calusa foxhalli*, and the relictual *Conomodulus lindae* (only-known living member of the Miocene genus *Conomodulus*), all of which occur in the Georgian Subprovince; *Trochomodulus calusa* which is restricted to the Floridian Subprovince;

Modulus floridanus which is restricted to the Suwannean Subprovince; and two undescribed *Modulus* species that are found only in the Yucatanean Subprovince (some shown here on Figures 2.8, 2.9, and 2.11). Many of the primary index gastropods for the Carolinian Province are shown here on Figure 2.7.

Georgian Subprovince

Named for the State of Georgia, which lies at the geometric center of the subprovince, the Georgian Subprovince ranges from Cape Hatteras, North Carolina, south to Palm Beach County, Florida. For the most part, the subprovince is physiographically dominated by large stretches of open sandy beaches and extensive brackish water coastal lagoon systems; all the result of the numerous rivers and braided streams that feed into the coastal areas from the Appalachian Mountains. The water temperatures of the near-shore areas of the Georgian Subprovince fluctuate greatly over the year, becoming only temperate in the winter and warming to near subtropical conditions in the summer. These extremes in water temperature, coupled with fluctuating salinities in the coastal lagoons, are too harsh for most neritic mollusks, resulting in an impoverished paratropical fauna.

Offshore areas of the Georgian Subprovince, along the edge of the continental shelf, are continuously bathed in the tropical and subtropical waters of the northward-flowing Gulf Stream and actually support Caribbean-type marine communities. These warm water conditions have allowed large coral bioherms ("patch reefs") to become established at the edge of the neritic zone and to extend all the way northward to off Cape Hatteras. On these offshore coral assemblages, many classic Caribbean reef-dwelling gastropods have become established, including the strombid *Macrostrombus costatus*, the harpid *Oniscidia dennisoni* (see Petuch, 1972), the cypraeids *Luria cinerea* and *Naria acicularis*, and the conids *Chelyconus ermineus* and the wide-ranging species *Jaspidiconus mindanus*. The offshore star coral (*Solenastera hyades* and *Siderastrea siderea*) patch reefs house a bizarre composite malacofauna, one where these Caribbean tropical reef mollusks co-exist with classic Carolinian hard bottom-dwelling species such as the cypraeid *Macrocypraea (Lorenzicypraea) cervus*, the muricid *Hexaplex fulvescens*, and the pisaniid *Hesperisternia multangulus* (see Petuch, 1972; 1974; 2013: 35-37).

Within the Georgian Subprovincial area, large areas of open sand sea floors typically are present between the tropical offshore coral bioherms and the warm temperate coastal communities. These deep sublittoral-neritic zone habitats, in 20-50 m depths, cover huge areas of the local continental shelf and support a distinctive ecosystem that is composed of, and supported by, immense beds of scallops. The dominant species within these pectinid shoals is the large "Carolina Calico Scallop", *Argopecten gibbus carolinensis*, an endemic bivalve that forms the basis for a large commercial fishery. These scallop shoals extend from off Cape Hatteras, North Carolina all the way to Fort Pierce, Florida, and contain the largest biomass of any of the known Carolinian Province benthonic ecosystems. The *Argopecten gibbus carolinensis* beds also house many gastropod species that are found nowhere else and these appear to have evolved to live exclusively in association with scallop shoals. Some of these scallop bed gastropods include the fasciolariids *Cinctura keatonorum* and *Polygona williamlyonsi*, the muricid *Vokesimurex morrisoni*, the volutid *Rehderia georgiana*, and the olivid *Americoliva nivosa clenchi* (Petuch, 2013: 34-35) (shown here on Figure 2.8).

The deep water areas (200-300 m depths) of the Georgian Subprovince, along the outer edge of the continental shelf and upper continental slope, also house a distinctive and highly endemic molluscan fauna. Outside of the influence of the warm Gulf Stream, these outer neritic and bathyal zone areas are, oceanographically, cold temperate in nature and support Arctic (Boreal) faunas. Many classic North American boreal species, such as the aporrhaid *Arrhoges occidentalis* (shown here on Figure 2.8), the buccinid *Buccinum abyssorum*, and the arcticid bivalve *Arctica islandica*, range all the way south to off South Carolina and Georgia, but only far offshore, in depths of over 250 m along the upper continental slope. The water temperature of these bathyal environments approximates that of the shallow water areas of the high Arctic, allowing these northern mollusks to migrate southward and become established within the faunal limits of the Georgian Subprovince. Some of the key index species for the Georgian Subprovince are shown here on Figure 2.8, in the Iconography at the end of this chapter.

Along the southernmost end of the Georgian Subprovince, from Fort Pierce to Dania Beach in southeastern Florida, the **Palm Beach Infraprovince** stands out as an area of special interest within the Atlantic coastal section of the Carolinian Province. Named for Palm Beach County, Florida (Petuch, 2013), the geometric center of the infraprovince, this evolutionary "hot spot" contains an unusually large number of coral reef-associated endemic mollusks, including species such as the cypraeid *Proadusta surinamensis* and the cone shell *Atlanticonus granulatus* (both of which are relatively common in coral rubble piles in 20 m depth). The offshore coral reef systems, which are perpetually bathed in the warm Gulf Stream waters, provide the ideal habitat for a host of endemic reef species, including the modulid *Modulus kaicherae*, the muricids *Favartia goldbergi*, *Pygmaepterys richardbinghami*, *Dermomurex glicksteini*, and *Coralliophila pacei*, and the conids *Dauciconus glicksteini*, *Kellyconus binghamae*, *Tuckericonus flamingo*, and *Jaspidiconus vanhyningi*. Some of the more important Palm Beach infraprovincial endemics are shown here on Figure 2.9. The coastline of the Palm Beach Infraprovince also contains large expanses of tropical brackish water lagoons and tidal creeks, all remnants of the ancient **Seminole Lagoon System**, a once-larger estuarine system that was present farther inland during the late Pleistocene (see Petuch and Roberts, 2007 for details on the late Pleistocene lagoon systems, and Petuch, 2004: 250 for illustrations of fossil specimens of living Palm Beach species). These coastal lagoons, such as the extensive Lake Worth Lagoon, house endemic Palm Beach taxa such as the muricid *Stramonita buchecki*, the modulids *Modulus pacei* and *Trochomodulus calusa foxhalli*, the elongated cerithiid *Cerithium lutosum lindae*, the large and inflated melongenid *Melongena (Rexmela) corona winnerae*, and the bivalves

Mercenaria harti and *Semele donovani* (some shown here on Figure 2.9). Many of the gastropods found in the Palm Beach Infraprovince are also prominent organisms within the Floridian Subprovince to the south, such as the olivid *Americoliva nivosa bollingi* and the bubble shell *Haminoea taylorae*, demonstrating that the Palm Beach Infraprovince represents the northernmost limit of the overall Florida Keys fauna.

The Gulf of Mexico Fauna

Comprising over two-thirds of the entire Carolinian Province, the Gulf of Mexico is essentially an inland sea, connecting to the Caribbean Basin in the south (through the Yucatan Straits) and with the Atlantic Ocean in the southeast (through the Florida Straits). Faunistically, this enclosed basin houses four separate subprovinces of the Carolinian Province: the **Floridian** (Florida Keys archipelago) and **Yucatanean** (Yucatan Peninsula and Campeche Banks) in the south; the **Suwannean** (western and northwestern Florida) in the east, and the **Texan** (Texas coast) in the west. Along the northern Gulf coast, the Mississippi River Delta acts as an ecological barrier, with its immense fresh water effluent separating and isolating the shallow marine faunas of the Suwannean and Texan Subprovinces. Because of the harsh and variable oceanographic conditions of the northern and western Gulf coastal areas, only a few hardy intertidal endemic gastropods are known to be widespread throughout the Gulf of Mexico. Two of these include species that are associated with oyster banks, such as the muricid *Stramonita canaliculata* (in the Suwannean, Texan, and Yucatanean Subprovinces) and the small pisaniid *Solenosteira cancellaria* (in the Suwannean and Texan Subprovinces; Figure 2.12). The deeper water areas of the Gulf of Mexico (deep neritic and upper bathyal zones), on the other hand, are much more faunally homogenous and contain a widespread Gulf endemic fauna. Some of these endemics include the dwarf Carrier Shell *Xenophora microdiscus*, the conids *Conasprelloides cancellata* (endemic to the Gulf; often confused with similar species from northern South America), *Dalliconus armiger*, *Gradiconus largillierti*, and *Gradiconus optabilis*, the fasciolariids *Heilprinia timessus* and *Fasciolaria bullisi*, the muricid *Paziella nuttingi*, the cancellariid *Cancellaria richardpetiti*, and the volutid *Clenchina robusta*. Most of these, and other Gulf endemics, are shown here on Figure 2.12.

With the cooler winter water temperatures in both the east and west and with the year-round near-tropical water temperatures in the south, the four subprovinces of the Gulf of Mexico have evolved a number of species complexes composed of geographically-isolated sibling species and subspecies. One of these Gulf species complexes is seen in the busyconid genus *Fulguropsis*: where *F. keysensis* is restricted to the Floridian Subprovince; *F. pyruloides* is restricted to the Suwannean Subprovince; *F. galvestonense* and *F. texanum* are restricted to the Texan Subprovince; and *F. spiratum* is restricted to the Yucatanean Subprovince. Another of these complexes is seen in the olivid genus *Americoliva*, where *A. sayana sarasotaensis* is confined to the Suwannean Subprovince and *A. sayana texana* is confined to the Texan Subprovince. The true *A. sayana sayana* is restricted to the Georgian Subprovince of the open Atlantic coast and ranges from North Carolina south to Boca Raton, Florida. No *A.*

sayana or *sayana*-type olives are found in the Florida Keys and the Floridian Subprovince, demonstrating that the Atlantic Ocean Georgian Subprovince *A. sayana sayana* is geographically widely-separated from its Gulf of Mexico sibling *A. sayana sarasotaensis*. Another example is seen in the conid genus *Gradiconus*: where the Floridian Subprovince houses three endemic species (*G. burryae*, *G. tranthami*, and *G. mazzolii*); the Suwannean Subprovince houses three endemic species (*G. anabathrum*, *G. largillierti*, and *G. optabilis*); and the Yucatanean Subprovince houses two endemic species (*G. maya* and *G. sennottorum*). The marine climate of the Texan Subprovince is too extreme and no *Gradiconus* species are present along the Texas coast. Because of all of these isolating factors, the Gulf of Mexico fauna is now seen as much more species-rich and diverse than was originally thought.

Of special interest within the deep central areas of the Gulf of Mexico is a remarkable molluscan fauna that is associated with cold petroleum seeps that emanate from deep-earth hydrocarbon reservoirs. These oil and methane seeps occur in depths ranging from 350 m to 1200 m all along the Sigsbee Escarpment, which borders the northern edge of the Sigsbee Deep south of the coasts of southeastern Texas, Louisiana, and Mississippi. The seeping petroleum supports huge mats of the methanogenic bacterium *Beggiatoa* and also beds of the deep water mussel *Bathymodiolus childressi*, which often forms aggregations that carpet the sea floor for kilometers. This deep water mussel, which is attached to the substrate by byssal threads, occurs together with the vesicomyid clams *Calyptogenia ponderosa* and *Vesicomya chordata* and the thyasiroidean clam *Thyasira*. All of these bivalves, attached or free-living, thrive in this hydrocarbon-rich environment as a result of harboring symbiotic methanogenic bacteria, which allow them to metabolize methane and petroleum bi-products. The *Bathymodiolus* beds provide a solid substrate for a variety of rare and seldom-seen deep water mollusks, including the giant limid bivalve *Acesta bullisi* and the cancellariid gastropod *Cancellaria rosewateri*. Most noteworthy of the mussel-associated organisms is the neritoidean gastropod, *Bathynerita naticoidea*, which is now known to belong to the family Phenacolepadidae (shown here on Figure 2.12M). These small nerites occur abundantly on the shells of the sessile *Bathymodiolus* mussels, where they graze on *Beggiatoa* bacterial films (see Clarke, 1989 for an overview of *Bathymodiolus* and *Bathynerita* community).

Floridian Subprovince

Named for the Florida Keys island chain, which forms the nucleus of this biogeographical unit, the Floridian Subprovince is essentially an outlier of the Caribbean Province that contains a large component of Carolinian taxa. The subprovince extends from Miami and Biscayne Bay, Miami-Dade County, southward across the entire Florida Keys archipelago to the Dry Tortugas and Dry Rocks, and northward to Cape Sable and the Shark River, Monroe County. The entire subprovincial area contains tropical oceanographic conditions and supports coral reef systems composed of over 45 different species of scleractinians (stony corals) and 35 species of octocorallians (sea whips, sea fans, sea plumes, and soft corals). During the late Pleistocene cold times and sea level drops, the warmer Florida Keys acted as a refugium for tropical and warm-temperate mollusks that had evolved in areas farther north along the Floridian Peninsula (Petuch and Myers, 2014). Many of these late Pleistocene relicts are still extant and are represented by a large component of endemic Keys species, some of which include the calliostomatid *Calliostoma adelae*, the modulid *Trochomodulus calusa*, the muricids *Favartia pacei*, *Murexiella kalafuti*, *Dermomurex pacei*, *Coralliophila kalafuti*, and *Murexiella caitlinae*, the melongenid *Melongena (Rexmela) bicolor*, the volutids *Scaphella junonia elizabethae*, *Clenchina dohrni*, *C. florida*, and *Rehderia schmitti*, the olivids *Americoliva nivosa bollingi*, *A. recourti*, and *A. matchetti*, the cancellariid *Cancellaria adelae*, and the conids *Gradiconus burryae*, *Jaspidiconus acutimarginatus* (recently re-discovered and recognized as a Florida Keys endemic species; see Eason, 2018), and *Jaspidiconus pealii* (most shown here on Figures 2.10 and 2.11). These unique Floridian taxa occur together with widespread Caribbean species such as *Aliger gigas* and *Atlanticonus granulatus* and widespread Carolinian species such as *Triplofusus papillosus* and *Macrocypraea (Lorenzicypraea) cervus*, forming a combined malacofauna that is the richest known from the eastern coast of the United States (with over 1500 known species; see Petuch and Myers, 2014 for illustrations and lists of the mollusks found in the numerous Keys ecosystems).

The warm water oceanographic conditions of the Floridian Subprovince typically support a rich variety of tropical marine environments, including Turtle Grass (*Thalassia*) beds, extensive sponge bioherms ("sponge reefs"), mangrove forests, true zonated coral reefs, and coral bioherms. Within this wide variety of habitats, a number of gastropod species radiations have taken place, producing swarms of endemic taxa. Of special interest is a species radiation of the conid genus *Gradiconus*, with three sibling species having evolved: *G. burryae* which is restricted to Turtle Grass beds, *G. mazzolii* which is restricted to sponge bioherms, and *G. tranthami* which is restricted to the offshore coral reefs (see Petuch and Myers, 2014 for photographs of the living animals and their radular teeth). These endemic Floridian cones are shown on Figure 2.10. Another radiation of special interest is seen in the octocoral-feeding ovulid genus *Cyphoma*, where three endemic Keys species, *C. sedlaki*, *C. rhomba*, and *C. alleneae* (shown on Figures 2.10 and 2.11) occur together with the more widespread Caribbean and Carolinian *C. gibbosum*, *C. signatum*, and *C. mcgintyi*. These six *Cyphoma* species, along with *Pseudocyphoma intermedium*, the deep reef-dwelling *Cyphoma gibbulum*, *C. aureocinctum*, and *Calcarolvula piragua*, and several small *Simnialena* and *Pseudosimnia* species, form the largest single fauna of ovulid gastropods known from anywhere in the Atlantic Ocean.

Suwannean Subprovince

Named for the Suwannee River of northwestern Florida, the Suwannean Subprovince is confined to the eastern Gulf of Mexico, where its boundaries extend from Cape Sable and the Shark River, Monroe County, along the Ten Thousand Islands, northward up the entire west coast of the Floridian Peninsula, and westward along the Florida Panhandle to the Mississippi River Delta (Petuch, 2004). Because of cooler water conditions during the winter months (minimally warm temperate), the western coast of Florida houses a relatively impoverished molluscan fauna. These seasonal harsh environmental conditions have allowed only a small number of endemic mollusks to flourish in the shallow water areas along western Florida, and some of these cold-tolerant gastropods include the conids *Jaspidiconus stearnsi* and *Gradiconus anabathrum*, the muricids *Eupleura tampaense* and *Vokesinotus perrugatus*, the busyconid *Fulguropsis pyruloides*, and the small terebrid *Strioterebrum vinosum*. Deeper offshore areas (20-50 m depths) within the Suwannean Subprovince offer a more stable marine climate and support a much richer fauna of endemic gastropods. Some of these neritic zone endemics include the ovulids *Cyphoma lindae* and *Cyphoma robustior*, the muricids *Murexiella taylorae*, *Favartia lindae*, and *Vokesimurex lindajoyceae*, the fasciolariid *Aristofusus stegeri*, the olivid *Americoliva sunderlandi*, and the volutes *Aurinia dubia* and *Caricellopsis matchetti*. Most of these are shown here on Figure 2.13.

Of special interest within the Suwannean Subprovince is the area along the outermost edge of the continental shelf (80-200 m depths) off western Florida, from south of Apalachicola to north of the Dry Tortugas. This long, narrow strip of deep water sea floor is carpeted by a thick growth of the red coralline alga *Porolithon*, which occurs as small interlocking branches and clumps (rhodoliths). These immense deep water red algal (dasycladecean) beds support a highly endemic molluscan fauna replete with many seldom-seen species, such as the muricids *Chicoreus rachelcarsonae* and *Phyllonotus whymani*, the conid *Dauciconus aureonimbosus*, the busyconid *Lindafulgur lyonsi*, the fasciolariid *Cinctura tortugana*, the large spotted terebrid *Cinguloterebra lindae*, and the pisaniid *Hesperisternia harasewychi*. All of these coralline algae-dwelling deep water endemic species have taken on a bright pink, red, or orange color, matching the bright pink and orange substrate created by the dasycladacean corallines. Some examples of the offshore Suwannean algal bed endemics are shown here on Figures 2.13 and 2.14.

The southernmost end of the Suwannean Subprovince occurs within the Ten Thousand Islands area of southwestern Florida, from Marco Island (Cape Romano) south to the Shark River mouth (near Cape Sable) and encompasses a localized evolutionary "hot spot" referred to here as the **Chokoloskean Infraprovince** (named for Chokoloskee Island, Collier County, the geometric center of the infraprovince). This stretch of coastline is primarily composed of closely-packed mangrove islands and supports large biohermal reeflike structures composed entirely of the vermetid gastropod *Petaloconchus nigricans* (referred to locally as "Worm Shell Reefs"; shown here on p. 352) (Petuch and Myers, 2014:156-167). Some of these are acres in size and form true deterministic reef systems that provide a stable environment for a number of local endemics (some shown on Figure 2.14). These include the turritellid *Vermicularia fargoi owensi*, the pisaniid *Gemophos tinctus pacei* (which feeds on the living *Vermetus* shells), the moon snail *Naticarius verae* (which lives in sandy pockets on the worm shell reefs), and the endemic estuarine venerid bivalve *Mercenaria browni*. The Chokoloskean Infraprovince also represents a faunal transition zone between the Suwannean and Floridian Subprovinces and is the only known locality where the Florida Keys endemic busyconid *Fulguropsis keysensis* occurs together with the Florida west coast *Fulguropsis pyruloides*.

Another area of special interest within the Suwannean Subprovince is in the "Big Bend" and eastern Panhandle region of northern Florida, and this area also demarcates an evolutionary "hot spot". Referred to here as the **Apalachicolan Infraprovince** (named for the Apalachicola River Delta and the city of Apalachicola), this infraprovince ranges from near Pensacola, Florida, to Steinhatchee, Florida, and is continuously bathed in the warmer waters of the Gulf Loop Current. This warm water current forms just north of the Yucatan Strait and moves straight northward until it impinges upon the western Florida Panhandle. From there, it flows eastward and then southward along the Florida west coast, keeping the area warmer than it should be during the cold winter months. The oceanographic sequestering and isolation has allowed an endemic fauna to evolve all along the Florida Panhandle coast. Some of the more important intertidal Apalachicolan endemics include the pisaniid *Hesperisternia grandana*, the large melongenid subspecies *Melongena (Rexmela) corona johnstonei*, and the olivid *Americoliva nivosa choctaw*. The offshore areas of the Apalachicolan Infraprovince also house an interesting molluscan fauna, which includes large endemic scaphelline volutes such as *Aurinia kieneri*, the ribbed *Aurinia kieneri ethelae*, and *Scaphella junonia johnstoneae* (shown here on Figure 2.14), and the large deep water cerithiids *Cerithium chara* and *Cerithium lymani*.

Texan Subprovince

Extending from the Mississippi River Delta, westward and southward along the Texas and Mexico coasts as far as Veracruz, Mexico, the Texan Subprovince (named for the state of Texas) is the single largest subprovince of the Carolinian Province. Because of the cold winter weather, massive influxes of fresh water during the rainy season, and the extreme summer heating, the coastal environments of the Texan Subprovince fluctuate greatly in water temperature and salinity. These harsh and variable conditions cause the Texan and eastern Mexican coastal lagoons and barrier islands to support only an impoverished molluscan fauna, the least species-rich in the entire Gulf of Mexico. Unlike all of the other Gulf Carolinian subprovincial areas, no cone shells (Conidae) are known from shallow water anywhere along the Texas coast. The melongenid genus *Rexmela*, which is so prominent in the Floridian, Suwannean, and Yucatanean Subprovinces, is also completely absent in the Texan Subprovince, apparently excluded by the very cold winter marine climate. Outside of the influence of the warm Gulf Loop Current, the Texas coast is oceanographically isolated from the rest of the neighboring subprovinces and has evolved its own unique molluscan fauna. The enclosed Texan coastal lagoon systems are now known to harbor several interesting, ecologically-hardy endemic species, some of which include the muricid *Stramonita alderi*, the olivid *Americoliva sayana texana*, the potamidid *Cerithidea hegewischii*, the large busyconid *Sinistrofulgur pulleyi* (all shown on Figure 2.15), and the venerid bivalve *Mercenaria texana*. Deeper areas directly offshore, in depths of 5 to 100 m also support an interesting fauna of Texan neritic endemics, including the busyconids *Fulguropsis texanus* and *Fulguropsis plagosus galvestonensis*, the fasciolariid *Cinctura lilium*, and the conid *Dalliconus sauros* (these are all illustrated on Figure 2.15). The strikingly-patterned volutid *Scaphella junonia curryi* (Figure 2.15 K), which is primarily a Bay of Campeche and eastern Mexican species, does range as far north as Corpus Christi, Texas, and can be considered to be part of this Texan offshore malacofauna. Besides these few endemic taxa, the majority of the Texan molluscan fauna is composed of physiologically-plastic, widespread Carolinian species, such as the giant fasciolariid *Triplofusus papillosus*, the muricid *Hexaplex fulvescens*, and the large cypraeid *Macrocypraea (Lorenzicypraea) cervus*.

Of special interest within the boundaries of the Texan Subprovince are two small, shallow banks that lie 90 kms and 150 kms south of Galveston, at the edge of the Texas-Louisiana Shelf. These small, isolated shallow platforms, the East and West Flower Garden Banks, are far enough away from the mainland that they have stable year-round warm water conditions, allowing for coral reefs to flourish. These subtropical temperatures, along with clear, sediment-free water, have produced the conditions for the development of the northernmost coral reefs within the Gulf of Mexico. Only around 24 species of stony corals are known from the Flower Garden Reefs, and this faunal impoverishment reflects the marginal oceanographic conditions of the northwestern Gulf area, especially during the winter months. Besides the normal complement of widespread Carolinian index taxa, these Texan reefs also house an impoverished Caribbean-type gastropod fauna, composed of widespread taxa such as the cypraeids *Macrocypraea zebra*, *Luria cinerea*, and *Naria acicularis*, the strombids *Aliger gigas* and *Aliger gallus*, the cassids *Cassis flammea* and *Cypraecassis testiculus*, the muricid *Siratus consuela*, and the conids *Jaspidiconus* cf. *mindanus* and *Chelyconus testudinarius* (an unusual solid orange or golden color form). For a detailed overview of the Flower Garden Banks marine faunas, see Bright and Pequegnat, 1974.

Yucatanean Subprovince

Like the Florida Keys and the Floridian Subprovince, the Yucatanean Subprovince (named for the Yucatan Peninsula of Mexico) contains near-tropical oceanographic conditions, with water temperatures being warm enough, year round, for extensive coral reefs to flourish. Extending from near Veracruz, Mexico to Isla Contoy (Contoy Light) and Cabo Catoche, Quintana Roo, Mexico, and encompassing the entire offshore Campeche Bank, the Yucatanean Subprovince is the second-largest subprovincial unit within the Gulf of Mexico. Having remained as an "island" of tropical oceanic conditions that is surrounded by cooler water areas, the Yucatanean area has provided a distinctive set of carbonate environments that has allowed for the evolution of a highly endemic molluscan fauna (examples of some of these endemics are shown here on Figures 2.16 and 2.17). These endemic Yucatan taxa occur together with widespread Carolinian index taxa such as *Triplofusus papillosus*, *Hexaplex fulvescens*, and *Macrocypraea (Lorenzicypraea) cervus* and also several classic Caribbean Province index taxa such as the strombid *Aliger gigas*, the melongenid

Melongena melongena (ranging to Veracruz), and the turbinellid *Turbinella angulata*, forming a composite molluscan assemblage that is unique within the Gulf of Mexico and the Tropical Western Atlantic.

The endemic component of the Yucatanean fauna is very impressive, with overall levels of endemism approaching 35% (Petuch, 2013: 69-74). Some of the more conspicuous and prominent endemic gastropods include the busyconids *Lindafulgur candelabrum*, *Busycoarctum coarctatum*, *Fulguropsis spiratum*, and *Sinistrofulgur perversum*, the pisaniid *Engina dicksoni* the turbinellid *Siphovasum latiriforme*, the conids *Lindaconus therriaulti* and *Gradiconus sennottorum*, the fasciolariids *Cinctura (Hollisteria) branhamae* and *Aristofusus couei*, the olivids *Americoliva nivosa maya* and *Americoliva contoyensis*, the large marginellid *Prunum labiatum*, the large scaphelline volutids *Scaphella junonia butleri*, *Scaphella junonia stimpsonorum*, *Aurinia macginnorum*, and *Caricellopsis contoyensis*, and the terebrid *Cinguloterebra stegeri*. The tropical coastal lagoons of the Yucatan Peninsula also house a rich and highly endemic gastropod fauna, most of which are associated with extensive Turtle Grass beds (*Thalassia testudinum*) and Red Mangrove forests *(Rhizophora mangle*) that dominate the lagoonal environments. Some of these shallow water tropical lagoonal endemics include the melongenid *Melongena (Rexmela) bispinosa* and its form or subspecies *Melongena (Rexmela) bispinosa martiniana*, the conids *Gradiconus maya*, *Jaspidiconus chaac*, and *Jaspidiconus ixchel*, and the large muricid *Phyllonotus mexicanus* (some shown on Figures 2.16 and 2.17).

Of special interest along northeastern Yucatan, particularly in the areas of Laguna Yalahua and Isla Holbox, is the large endemic turbinellid, *Turbinella wheeleri* (Figure 2.16L).This impressive species, one of the largest gastropods found along the Yucatan Peninsula, was originally described as a late Pleistocene fossil from the Fort Thompson shell beds in the Everglades region of southern Florida. Here, fossil specimens of this remarkable turbinellid are found together with fossil specimens of the melongenids *Melongena (Rexmela) bispinosa*, *M. (Rexmela) corona*, and *Melongena melongena* (see Petuch, 1994; Petuch, 2004; Petuch and Roberts, 2007), demonstrating that all four species were present on the Florida Peninsula 150,000 years ago and that they lived together in a unique ecosystem unlike anything seen today (all four species are abundant in the Coffee Mill Hammock Member of the Fort Thompson Formation; see Petuch, 2004: 247-253; Petuch and Roberts, 2007: 180-185). During the late Pleistocene, oceanic temperatures became much colder in Florida and the northern Gulf of Mexico and *Melongena melongena*, *Melongena (Rexmela) bispinosa*, and *Turbinella wheeleri* became regionally extinct (extirpated), with only the cold-tolerant *Melongena (Rexmela) corona* having survived on the Florida Peninsula. Today, the high-tropical *Melongena melongena* lives along northern Yucatan and ranges throughout the Caribbean Basin and along northern South America. The once-Floridian *Melongena (Rexmela) bispinosa* and *Turbinella wheeleri*, on the other hand, are now known to be relictual species, having survived the cold-climate late Pleistocene extinction event by migrating into a refuge in the warm water areas along the eastern Yucatan Peninsula.

Northwestern Atlantic Tropical Subregion

The Northwestern Atlantic Tropical Subregion, like the previous paratropical subregion, encompasses only a single major biogeographical unit, the Caribbean Molluscan Province. Extending from Bermuda and the Bahamas in the north, throughout the Caribbean Basin, and along the northern coast of South America to the Amazon River Mouth, this tropical province is biogeographically complex, encompassing seven distinct subprovincial areas. Unlike the previous paratropical subregion, the Northwestern Atlantic Tropical Subregion incorporates multiple island archipelagoes and offshore carbonate banks, creating the genetic isolation that results in the formation of numerous infraprovincial "evolutionary hot spots". Continuously bathed in the northward-flowing North Equatorial Current and Antilles Current, this tropical subregion contains the warmest water found in the entire western Atlantic Ocean. These hyper-tropical oceanographic conditions have allowed for massive coral reef development in many areas of the Caribbean Province, especially the Bahamas Banks, the numerous carbonate banks off Honduras and Nicaragua, the Great Barrier Reef and atolls along Belize, and the banks and atolls off Venezuela (especially the Los Roques Atoll). The coastal areas of central and northern South America also house immense regions of mangrove forests and sea grass-filled estuarine lagoons, which support a rich molluscan fauna and numerous infraprovincial areas.

Caribbean Molluscan Province

Named for the Caribbean Sea, which is the geographic center of this major faunal unit, the Caribbean Province spans a wide area; from the isolated island of Bermuda, to the Bahama Banks, the Greater Antilles Island Arc, the Lesser Antilles, the Central Caribbean islands and banks, and the entire coast of Central and South America, from Belize to the Mouth of the Amazon River (Figure 2.3). Because of the great diversity of habitats within the provincial boundaries, large numbers of geographically-restricted endemic mollusks have evolved within seven broad areas. Faunal analyses have shown that these areas represent separate subprovinces, each with over 30% endemism at the species level (Petuch, 2013). These include: the **Bermudan Subprovince**, the **Bahamian Subprovince**, the **Antillean Subprovince**, the **Nicaraguan Subprovince**, the **Venezuelan Subprovince**, the **Grenadian Subprovince**, and the **Surinamian Subprovince**. The geographical extents and faunal compositions of the subprovinces are discussed in more detail throughout this chapter. Based on the main marine habitat types, the Caribbean subprovinces fall into two broad categories: the **continental mainland type**, typified by continental coastlines that are dominated by terrigenous sediments and organic-rich coastal lagoons and have muddy water conditions (the Nicaraguan, Venezuelan, and Surinamian Subprovinces); and the **offshore island type**, typified by island arcs and isolated bank systems that are dominated by carbonate sediments and clean, open-oceanic water conditions (the Bermudan, Bahamian, Antillean, and Grenadian Subprovinces). Some of the classic wide-ranging Caribbean Province index species include the giant strombid *Aliger gigas*, the cowries *Luria cinerea*, *Naria acicularis*, and *Macrocypraea zebra*, the modulid *Modulus modulus*, the olivids *Americoliva reticularis*, *Aericoliva bifscaiata*, and *Cariboliva scripta*, and the cone shells *Dauciconus daucus*, *Gladioconus mus*, and *Atlanticonus granulatus*. All of these index species, and others, are illustrated here on Figure 2.18.

The molluscan fauna of the Caribbean Province derives from the Miocene and Pliocene-aged Gatunian Paleoprovince, which spanned both sides of the then-open Bolivar Straits, across what is now the Isthmus of Panama (Petuch, 2004). During the late Pliocene, this narrow connection between the Atlantic and Pacific Oceans began to close off, effectively isolating the Caribbean Basin and setting the stage for the evolution of several distinctive local gastropod species radiations. Some of these included species swarms of the conid genus *Tenorioconus*, the cypraeid genus *Muracypraea*, and the olivid genus *Americoliva*. In the Recent Caribbean, *Muracypraea* is confined to the Venezuelan Subprovince and *Tenorioconus* is confined to the Grenadian and Nicaraguan Subprovinces, while *Americoliva* has evolved entire suites of endemic species throughout all seven subprovinces (and also in the Carolinian, Brazilian, and Paulinian Provinces). Although *Muracypraea* was present on the Eastern Pacific side of the Gatunian Paleoprovince (which extended from southern California to Peru; Petuch, 2004), it became extinct there by the early Pleistocene and the genus survives, as four relictual taxa, only in the southern Caribbean. The conid genus *Tenorioconus* was also represented in the Eastern Pacific by several species during the Pliocene, but now is present only as a single species, *Tenorioconus archon* (see Chapter 4). In the Recent Caribbean, *Tenorioconus* has undergone another species explosion, with as many as 18 species and subspecies being known from Honduras to Venezuela and the Lesser Antilles.

Bermudan Subprovince

The Bermudan Subprovince, named for Bermuda (Petuch, 2013), is the smallest of the subprovincial units of the Caribbean Province. Encompassing only Bermuda and the slopes of the Bermuda Seamount, this subprovincial area is faunally impoverished, containing only a small number of widespread Caribbean taxa and a small number of endemic species. During the late Pleistocene, the oceanic temperatures around Bermuda were below tropical levels and much of the local warm water molluscan fauna became extinct. Since the Holocene, the Gulf Stream current now bathes the island system in warm water and the oceanic climate has again become subtropical. Many widespread Caribbean species, especially those with long-lived planktotrophic larvae, have colonized the island and have become re-established on the shallow carbonate sand shoals and coral reefs. Due to the genetic isolation and stressed ecological conditions during the late Pleistocene, a small number of endemic species and subspecies evolved in the shallow platforms and reefs that surround Bermuda. Some of these include the intertidal cerithiid *Cerithium lutosum bermudae*, the olivid *Americoliva nivosa* (*A. jenseni* is a synonym; see Petuch and Berschauer, 2020), and the conid *Jaspidiconus bermudensis* (shown here on Figure 2.19). The deep neritic and bathyal areas on the sides of the Bermuda Seamount offered more stable environmental conditions during the climatologically-chaotic late Pleistocene and acted as a refugium for several endemic

Bermudan genera, including the fasciolariid *Lightbournus* (*L. russjenseni*) and the conid *Bermudaconus* (*B. lightbourni*). This same deep water area also supported the evolution of a unique Bermudan bathyal fauna, with endemic species such as the slit shells *Perotrochus insularis* and *Entemnotrochus bermudensis* and the winged muricid *Timbellus lightbourni*. Some of the endemic Bermudan gastropods are shown here on Figure 2.19.

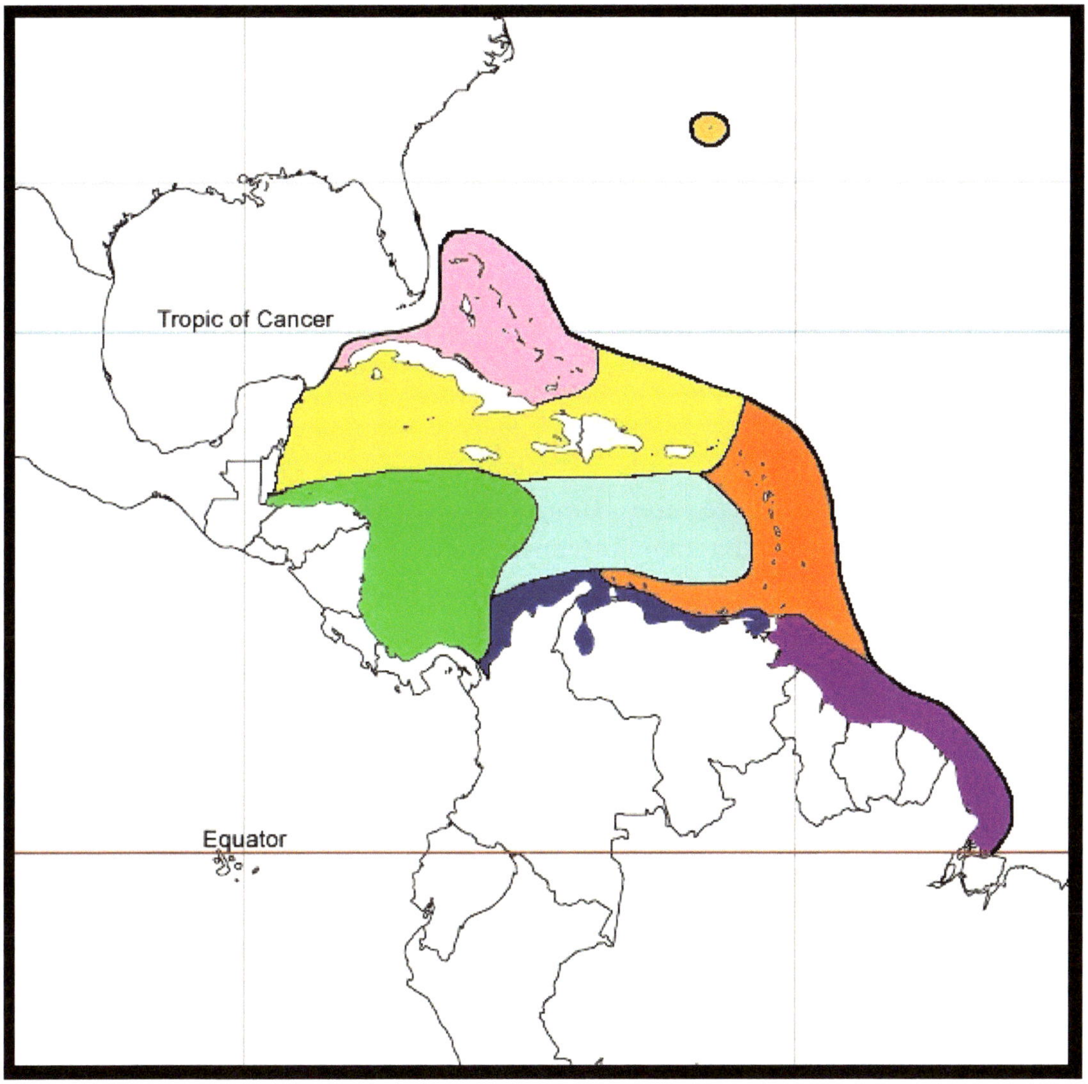

Figure 2.3 Map of the Caribbean Molluscan Province, showing the areal extents of its subprovinces: the **Bermudan Subprovince** (gold), the **Bahamian Subprovince** (light rose), the **Antillean Subprovince** (yellow), the **Nicaraguan Subprovince** (green), the **Venezuelan Subprovince** (dark blue), the **Grenadian Subprovince** (orange), and the **Surinamian Subprovince** (purple). The pale blue area demarcates the abyssal plains of the deep Colombian and Venezuelan Basins, which contain endemic bathyal and abyssal faunas.

Bahamian Subprovince

Encompassing the Little and Great Bahama Banks and the Turks and Caicos Islands, the Bahamian Subprovince (named for the Bahamas; Petuch, 2013) is composed entirely of broad, shallow carbonate banks covered with underwater dunes of unconsolidated oolite pellets, extensive coral reef tracts, and oolitic limestone islands. The underwater dunes act as ecological barriers, preventing reef-dwelling species from crossing the submarine deserts and have genetically isolated species with direct development. This has resulted in the evolution of large species swarms in several families of gastropods, especially the Muricidae and Conidae. Some of the typical widespread Bahamian index species include the muricids *Murexiella deynzerorum* and *Murexsul zylmanae*, and the relictual modulid *Laevimodulus honkeroroum* (the only-known living member of the Mio-Pliocene genus *Laevimodulus*), the lyriine volute *Enaeta cylleniformis*, and the conids *Tuckericonus flavescens*, *Dauciconus bahamensis*, *Purpuriconus stanfieldi*, and *Magelliconus jacarusoi*. These, and other widespread Bahamian endemic species, are shown here on Figure 2.20. The still-unexplored northern coast of Cuba and Cay Sal Bank also appear to belong in the Bahamian Subprovince, sharing many taxa with the Bahamas Banks, such as the conid *Purpuriconus arangoi* and the cerithiid *Fastigiella carinata*. For a review of the Bahamian *Purpuriconus* species radiation, see Petuch (2000).

Because of the intervening immense underwater sand dune deserts, the isolated main reef complexes around the Bahamas have formed several evolutionary "hot spots" (infraprovinces), each with its own endemic fauna. The three main evolutionary centers include the **Biminian Infraprovince**, the **Eleutheran Infraprovince**, and the **Abacoan Infraprovince.** The **Biminian Infraprovince** extends along the western edge of the Great Bahamas Bank, from North Bimini Island south to Victory Cay and the entire Bimini Chain of islands. Numerous endemic gastropods have evolved within this isolated island chain, including the cerithiid *Cerithium biminiense*, the muricid *Murexsul sunderlandi* (also found on Cay Sal), the volute *Enaeta lindae*, the terebrid *Strioterebrum biminiensis*, and a large radiation of endemic cone shells containing species such as *Purpuriconus donnae*, *Purpuriconus richardbinghami*, *Magelliconus zylmanae*, *Jaspidiconus herndli*, and *Cariboconus sahlbergi*. These and other Biminian endemic species are shown here on Figure 2.21. The **Eleutheran Infraprovince** occurs along the eastern edge of the Great Bahamas Bank, and encompasses Eleuthera Island, Cat Island, and San Salvador Island. Numerous endemic gastropods have evolved around these elongated, reef-edged islands, including the muricid *Chicoreus dunni*, the marginellid *Volvarina jimcordyi*, and the cone shells *Purpuriconus abbotti*, *Jaspidiconus marcusi*, and *Jaspidiconus exumaensis* (all shown here on Figure 2.22). The **Abacoan Infraprovince** is centered within the Abaco Islands, along the eastern edge of the Little Bahamas Bank. This small area, which is separated from the Great Bahamas Bank and Eleutheran Infraprovince by the very deep and wide Northeast Providence Channel, has evolved its own distinctive fauna, which includes the muricid *Murexsul honkeri*, the fasciolariid *Polygona paulae*, and the conids *Purpuriconus jucundus*, *Jaspidiconus branhamae*, and *Gradiconus honkerorum* (the only *Gradiconus* species known from the Bahamas). These are illustrated on Figure 2.22.

Of special interest along the western Bahamas is a highly endemic, geographically-restricted molluscan fauna that is found at 400 m depth along the base of the Bimini Wall, off the Bimini Chain of islands (see Petuch, 2002; Oleinik,

Petuch, and Aley, 2012). This narrow shelf, averaging less than a kilometer in width, is covered with dense thickets of hydrocorals, fan sponges, and crinoids, and supports a unique assemblage of gastropods that includes Bimini Wall endemics such as the conid *Lindaconus lindae*, the cystiscid *Persicula biminiensis*, the buccinid *Chickcharnea fragilis*, the nassariids *Antillophos freemani* and *Antillophos bahamasensis*, the volutid *Scaphella biminiensis*, the olivid *Americoliva bahamasensis*, and the turrid *Polystira starretti*. In some places, accumulations of dead *Lindaconus lindae* are so dense that they literally form a solid pavement on the sea floor (the "*Conus* Pavement"; see Petuch, 2002). These dead cone shells, in turn, act as the substrate for hydrocoral attachment, allowing dense forests of delicately-branched cnidarians to build large biohermal structures on what is predominantly a soft sediment sea floor area. Some of the Bimini Wall endemics are shown here on Figure 2.21.

Antillean Subprovince

Named for the Greater Antilles island arc (Petuch, 2013), the Antillean Subprovince extends from Belize (the entire Great Barrier Reef of Belize and offshore atolls) to Cuba, Hispaniola, Jamaica, and Puerto Rico, eastward to the Virgin Islands. The large islands that make up this subprovince are considered to be "continental islands", those that have long coastlines composed of volcanic rocks and terrigenous sediments that were carried to the coast by large river systems. Because of their igneous rock-derived shorelines and beaches, these continental islands offer a greater variety of habitats than do the isolated coral cays and carbonate banks of the Bahamas. The Greater Antilles islands are separated from each other by deep channels, and these oceanic barriers effectively isolate most of the mollusks that lack planktotrophic larvae. Because of this genetic isolation, each island has evolved its own set of endemic species and its own characteristic molluscan fauna. Within the subprovincial limits, the olivid genus *Americoliva* has evolved clusters of endemic taxa, including the elongated *Americoliva antillensis* and a swarm of subspecies of the small *Americoliva broderipii*, with endemic subspecies occurring on Hispaniola, Puerto Rico, and Jamaica. The conid genera *Magelliconus*, *Purpuriconus*, and *Jaspidiconus* also evolved clusters of endemic species on Cuba, the Cayman Islands, Hispaniola, Jamaica, Puerto Rico, and the Virgin Islands. Some of these are shown here on Figure 2.23.

Unlike all of the other Caribbean subprovinces, the Antillean has developed the largest number of infraprovinces, with at least five being distributed along the largest islands and one along the coast of Belize and its offshore atolls. The only mainland evolutionary hot spot, the **Belizean Infraprovince**, encompasses both the Great Barrier Reef of Belize and the offshore Glover's Atoll, Lighthouse Reef Atoll, and Turneffe Atoll and also the Banco Chinchorro Atoll of extreme southernmost Quintana Roo State, Mexico (the only four true atoll reefs in the Western Caribbean). These reef complexes house interesting endemic species including conids such as *Purpuriconus belizeanus* and the Banco Chinchorro endemics *Cariboconus deynzerorum* and *Jaspidiconus chinchorroensis*, the muricid *Dermomurex coonsorum*

(Glover's Atoll), and the small olivid *Americoliva mooreana*, which lives in deep water (300 m) at the base of the barrier reef system. On the opposite side of the Yucatan Strait, the **Cuban Infraprovince** encompasses the southern coastline of Cuba and the Isla de la Juventud and adjacent small island groups such as the Jardines de la Reina. Of particular interest along this coast is the endemic cowrie *Macrocypraea (Lorenzicypraea) cervus lindseyi*, a dwarf subspecies (or possibly a full species) of the Carolinian index taxon *Macrocypraea (Lorenzicypraea) cervus* (Figure 2.23I, J). This unusual cowrie is found mostly within the Golfo de Guacanayabo and may represent a relictual population of the cooler water *M. (Lorenzicypraea) cervus* that has survived in this isolated area since the Pleistocene.

Besides Cuba and Belize, the Greater Antilles islands of Jamaica, Hispaniola, and Puerto Rico each have their own evolutionary hot spots and endemic species radiations, primarily of the gastropod families Conidae, Olividae, and Muricidae. The **Jamaican Infraprovince**, confined to the island of Jamaica, contains a number of highly-restricted taxa, including the olivid *Americoliva broderipi jamaicensis*, the conid *Magelliconus explorator*, and the very rarely-seen colubrariid *Cumia sunderlandi* (Figure 2.24). The larger island of Hispaniola houses the **Hispaniolan Infraprovince**, which harbors endemic taxa like the conids *Purpuriconus cardinalis* (Figure 2.23, primarily from Haiti) and *Jaspidiconus lindapowersae* and the olivids *Americoliva broderipi* (Dominican Republic) and *Americoliva broderipi zombia* (Haiti). The **Puerto Rican Infraprovince**, which includes the main island of Puerto Rico and the adjacent small islands of Mona, Culebra, and Vieques, has also evolved several endemic species radiations of cone shells and olives. Some of these include conids in the genera *Magelliconus* (*M. mayaguensis* and *M. speciosissimus*), and *Jaspidiconus*, (*J. anaglypticus*, *J. boriqua*, and *J. culebranus*), and the olivid *Americoliva broderipi* unnamed subspecies (examples shown here on Figure 2.24). Although they are not continental islands, the coral cays of the Cayman Islands (collectively the **Caymanian Infraprovince**) also house a small endemic species radiation of cone shells, including *Magelliconus sphaecelatus* and *Jaspidiconus janapatriceae* (Figure 2.24).

Nicaraguan Subprovince

Extending from Guatemala (near Puerto Barrios) to the Golfo de Uraba, on the Panama-Colombia border, the Nicaraguan Subprovince (named for Nicaragua, the geographic center) includes not only the coastal areas but all of the large, shallow carbonate banks scattered throughout the central Caribbean Basin, particularly off the coasts of Honduras and Nicaragua. Like isolated islands, many of these shallow platforms (1-20 m deep) and drowned former atolls are separated from each other, and from the mainland, by deep channels, preventing the dispersal of species with direct-development larvae. Because of this genetic isolation, many of these submerged carbonate platforms, such as Gorda Bank, Serranilla Bank, Rosalind Bank, Bajo Nuevo Bank, and Misteriosa Bank have evolved their own endemic faunas, in many ways mimicking the terrestrial speciation patterns seen in the Galapagos Islands. During Pleistocene sea level low stands, many of these banks, particularly those on the

Nicaraguan Rise, were connected as a single land mass and the marine faunas were distributed along the coastlines. When sea level rose during Pleistocene warm interglacial times (and in the Holocene), these homogeneous shallow water coastal faunas were cut into isolated populations by the deep intervening channels and formed distinct genetically-separate populations on each of the banks and coral cay archipelagos (this process was described graphically by Petuch, 1981). These isolates later speciated to form large radiations, particularly in the gastropod families Volutidae and Conidae. The central Caribbean basin remained warm enough, during even the coldest times in the Pleistocene, to act as a refugium for several groups of regionally extirpated gastropods. Of particular interest are relictual species of Caloosahatchean Paleoprovince genera that have managed to survive into the Recent only in the Nicaraguan Subprovince, such as the giant miter *Pleioptygma* (*P. helenae*; Figure 2.25E) and the cerithiid *Cerithioclava* (*C. garciai*).

The Nicaraguan Subprovince also harbors one of the richest cone shell (Conidae) faunas found anywhere in the western Atlantic, with at least 35 species in total, and with over 20 being endemic to the subprovince. The genetic isolation produced by the offshore banks and islands has led to the evolution of species swarms within the genera *Purpuriconus*, *Jaspidiconus*, *Gradiconus*, and *Cariboconus*. Some of these include *Purpuriconus kukulcan*, *Jaspidiconus allamandi*, *J. masinoi*, *J. roatanensis*, *J. sargenti*, *Gradiconus paschalli*, *G. garciai*, *Cariboconus magnottei*, and *Atlanticonus ritae*, and these are shown here on Figures 2.25 and 2.26 at the end of this chapter. Other interesting species include the dwarf volute *Enaeta bessei*, and the conids *Dauciconus sunderlandi*, *Gradiconus aureopunctatus*, and *Tenorioconus harlandi* (the westernmost Caribbean species of its genus). The Nicaraguan Subprovince is also remarkable in that it contains the single largest species radiation of the volutid genus *Voluta* (see Petuch, 1981; Berschauer, 2019). To date, at least 10 species are known from the coasts of Honduras, Nicaragua, Costa Rica, and Panama and from the myriad of offshore banks and island chains. Most of these are restricted to small geographical areas, often on only one or two offshore banks. Some of these, such as *Voluta polypleura*, *V. demarcoi*, *V. kotorai*, *V. morrisoni*, *V. retemirabila*, *V. sunderlandi*, and *V. hilli* are shown here on Figure 2.27. Unlike the Bahamian and Antillean Subprovince, the Nicaragua Subprovince houses one of the richest muricid faunas in the western Atlantic and also one that contains numerous large species. Some of these include impressive taxa such as *Vokesimurex samui*, *Vokesimurex garciai*, and *Chicoreus bullisi*. The muddy coastal waters also support unusual and distinctive molluscan assemblages (see Petuch, 1998), containing characteristic species such as the ficid *Ficus villai*, the olivids *Agaronia hilli* and *Agaronia leonardhilli* (the only species of *Agaronia* known from the Caribbean Sea), the marginellid *Prunum sunderlandorum*, and the pseudomelatomid *Hindsiclava tippetti*. These are all illustrated on Figures 2.25 and 2.26.

The Caribbean coast of Panama contains one of the most distinctive sets of biotopes found in the entire area. Comprised of a series of coral cays (the San Blas Islands) and reef-like structures made up of fused platforms and nodules of pink coralline algal (rhodoliths), this area stands as an "island" of carbonate environments within a region dominated by muddy coastal lagoons and river mouths. Because of its ecological isolation from the carbonate environments to the northeast and northwest, the San Blas Islands have evolved their own evolutionary "hot spot", referred to here as the **Blasian Infraprovince**. These shallow water coralline algal reef areas support a rich array of endemic species including the fasciolariid *Hemipolygona cuna*, the conids *Atlanticonus glenni* and *Tenorioconus panamicus*, the volutes *Voluta ernesti* and *Voluta lindae*, the olivid *Americoliva ernesti*, and the marginellid *Prunum leonardhilli*. The deeper water areas offshore of the islands and reefs also house a highly endemic fauna, containing rarely-seen species such as the muricids *Chicoreus hilli* and *Murexiella edwardpauli*, the pseudomelatomid *Knefastia hilli*, the volutid *Voluta virescens lacertina*, and conids such as *Gradiconus ernesti* and *Jaspidiconus kellyae* (all shown here on Figure 2.28). On open carbonate sea floors, these endemic species are often associated with large beds of the Blasian endemic turritellids, *Torcula howardpetersi* (Figure 2.28M) and *Torcula marianopsis*, which form a detritivore trophic base for an entire endemic deep water community (see Petuch, 1991; Petuch and Berschauer, 2020b). This infraprovince was originally referred to as the "Blasian Subregion" by the senior author (Petuch, 1991).

Venezuelan Subprovince

Extending from the Golfo de Uraba, on the Panama-Colombia border, to the mouth of the Orinoco River, the Venezuelan Subprovince is almost entirely a coastal biogeographical unit. With the exceptions of a few isolated islands off northern Colombia and the Monjas Islands off the Goajira Peninsula, this subprovince is dominated by extensive coastal lagoon systems and rocky headlands. Because of the continuous trade winds blowing away the surface water, almost the entire area is under the influence of cold water upwelling systems, creating plankton-rich, high-productivity oceanic conditions. Because of this, no extensive coral reef systems have developed anywhere between Panama and the Cumana Peninsula of Venezuela. In some areas, particularly off the Goajira Peninsula of Colombia, the cold, nutrient rich waters support huge aggregations of sponges, which literally carpet the sea floor at depths of around 30-100 m. This atypical Caribbean environment houses a highly endemic fauna replete with Miocene-Pliocene relictual gastropods, including members of the genera such as the cypraeid *Muracypraea*, the borsoniid *Paraborsonia*, the strombinid *Strombina*, the muricid *Panamurex*, and the olivid *Amalda*. These once-widespread Caribbean taxa are now extinct in other adjacent areas but have managed to survive only within the subprovincial boundaries, creating an area of "living-fossils" referred to as a **Secondary Relict Pocket** (see Petuch, 1982).

The area extending from the Goajira Peninsula to the Golfo Triste, and including the enclosed Golfo de Venezuela, harbors a very rich fauna of cone shells, with a high percentage of endemicity. This fauna is dominated by species radiations of the genera *Gradiconus* and *Conasprelloides* and includes endemic taxa such as *Gradiconus gibsonsmithorum*, *G. paulae*, *G. parascalaris*, and *G. paraguana* and *Conasprelloides tristensis*, *C. venezuelanus*, *C. finkli*, and *C. penchaszadehi*. These occur along with species of other conid genera such as *Tenorioconus granarius*, *Tenorioconus sanguineus*, *Attenuiconus poulosi*, and *A. honkeri* (examples shown here on Figure 2.29). This rich conid fauna occurs along with a distinctive molluscan assemblage that includes the endemic olivoideans *Americoliva porcea*, *Americoliva figura*, *Eburna glabrata*, and *Amalda tankervillii*, the muricids *Panamurex velero* and *Panamurex petuchi*, the buccinid *Truncaria lindae*, and three ecologically-exclusive cypraeids of the relict genus *Muracypraea*; *M. mus* (which lives in intertidal Turtle Grass beds and is often exposed at low tide; see Petuch, 1976), *M. bicornis* (which lives on sponge beds in 3-30 m depths), and *M. tristensis* (which lives in offshore deep water areas of 50-100 m depth). All of these, along with relictual taxa such as *Strombina pumilio* and *Paraborsonia lindae*, give the Golfo de Venezuela a characteristic "prehistoric" appearance, resembling the faunas of the Gatunian Paleoprovince during the Miocene and Pliocene (Petuch, 1982; 1987; 2004).

The Colombian coast between the Golfo de Uraba and the Goajira Peninsula, including the Golfo de Morrosquillo, the Cartagena lagoon systems, and the Santa Marta bays, is dominated by muddy, organic-rich sea floors. These environments derive from the effluent of several main river systems that drain the northern Andes Mountains, including the Atrato in the south and the Magdalena in the north. Because of the river (fluvial) input, the sediments just offshore are generally composed of sand, mud, and clays and a large component of plant debris, including leaves, branches, and tree trunks. These organic-rich reducing environments support a highly restricted molluscan fauna with a high level of endemicity, referred to here as the **Colombian Infraprovince**. The senior author worked on shrimp boats all along this coast while undertaking his doctoral research (with the Vikingos de Colombia shrimp fisheries company, out of Cartagena), and was able to observe, first hand, a large number of previously-undescribed mollusks, all of which are restricted to this evolutionary hot spot (Petuch, 1987). Some of these remarkable discoveries included a new southern subspecies of the relictual cypraeid *Muracypraea bicornis* (*M. bicornis donmoorei*, shown here on Figure 2.30), the first northern South American shallow water ficid, *Ficus lindae*, a new shallow water (2-20 m depths) cassid of the normally-deep water genus *Sconsia* (*S. lindae*), the large muricid *Vokesimurex bayeri*, the second-known Caribbean species of the harpid genus *Oniscidia* (*O. lindae*), several new species of olivids including *Americoliva goajira* and *Amalda williamsoni*, and new conids such as *Jaspidiconus tayrona* and *Dauciconus vikingorum* (shown here on Figure 2.30). The molluscan assemblages of the organic-rich Colombian offshore area also contain other large gastropods such as the volutes *Voluta virescens* and *Caricellopsis evelynae*, the cone shell *Lindaconus phlogopus*, the muricid *Vokesimurex messorius*, and the large marginellid *Prunum poulosi* (some shown here on Figure 2.30).

Grenadian Subprovince

Extending from Aruba to Barbados and northward through the Lesser Antilles to Anguilla, the Grenadian Subprovince (named for the island nation of Grenada) is comprised entirely of offshore island chains. Geomorphologically, these vary greatly, ranging from low, eroded volcanic islands (like Aruba, Curacao, and Barbados), to chains of coral cays (like Las Aves and Los Testigos) and coral atolls (like Los Roques Atoll, off the Venezuelan coast), to high volcanic islands (like Grenada, St. Vincent, St. Lucia, and Martinique). Being isolated from continental oceanic conditions, the islands of the Grenadian Subprovince are surrounded by extensive coral reef growth and their adjacent sea floors are dominated by a mixture of carbonate and eroded volcanic sediments. This wide variety of habitats and oceanographic isolation has resulted in the evolution of a very rich and highly endemic molluscan fauna, with distinct species and subspecies being found on separate islands and archipelagos. Of primary interest within the Grenadian Subprovince is the presence of a large species swarm of the conid genus *Tenorioconus*, with at least 15 species and subspecies being known from the scattered island chains and archipelagoes. Some of these include *Tenorioconus cedonulli* (St. Vincent), *T. insularis* (St. Lucia), *T. dominicanus* (Dominica), *T. domincanus grenadensis* (southern Grenadines), *T. martinicanus* (northern Grenadines), *T. trinitarius* (Trinidad and Margarita Island), *T. mappa* (Tobago), *T. caracanus* (Cumana Peninsula, Venezuela), *T. sanguineus* (Venezuelan coast) (illustrated here on Figure 2.32), and several new and undescribed species from the Aves, Testigos, Orchila, Tortuga, and Blanquilla Islands off Venezuela. Other *Tenorioconus* species are found on island groups along northern South America, such as Aruba, Bonaire, and Los Roques Atoll, and these are listed under their respective island groups later in this chapter. The subprovince characteristically houses a large number of geographically-restricted taxa, with most being restricted to only one or two islands. Some of these include the turbinellids *Globivasum globulus* and *Globivasum whicheri*, the lyriine volutes *Mitraelyria beaui*, *M. archeri*, and *M. sabaensis*, the olivid *Americoliva olivacea*, and the conid *Jaspidiconus berschaueri* (some shown here on Figures 2.31 and 2 .32).

The deep water areas (200-300 m) between, and surrounding, the isolated volcanic islands of the Lesser Antilles also contain a highly endemic molluscan fauna. Many of these upper bathyal species were first encountered in deep water dredgings around the island of Barbados in the late 1970s, as part of intensive faunal surveys undertaken by Dr. Finn Sander, Director of the McGill University Marine Laboratory at St. James. Although originally thought to be endemic to the Barbados Seamount, many of these bathyal taxa were later collected off other islands, such as Martinique, Guadeloupe, and Curacão, demonstrating that they had much wider distributions within the subprovince. Some of these Grenadian panbathyal gastropods include the pleurotomariid *Perotrochus sunderlandorum*, the relict strombinid *Cotonopsis lindae*, the ranellid *Sassia (Cymatiella) lewisi*, the muricid *Lindapterys sanderi*, the cone shell *Attenuiconus maryleeae*, and a radiation of the deep water conid genus *Sandericonus* including *S. sanderi*, *S. hunti*, *S. perprotractus*, and *S. sorenseni* (see Figure 2.33). This deep water molluscan assemblage ranges all along the Lesser Antilles, from as far as Anguilla in the north to Grenada in the south. A new *Sandericonus* fauna, containing species such as *S. lamyi*, *S. desiradensis*, *S. laurenti*, and *S. oualeiriensis*, was also recently discovered in deep water within a limited area off Guadeloupe Island (Rabiller and Richard, 2019).

Because of the geographical isolation due to intervening deep water channels and basins, several areas within the Grenadian Subprovince have become evolutionary hot spots, forming a series of infraprovinces. One of these, the **Martiniquean Infraprovince** (named for the island of Martinique and also including the islands of Guadeloupe and Dominica), houses several important endemic reef-dwelling gastropods, including the winged murex *Timbellus phyllopterus* and the cone shells *Dauciconus norai*, *Dauciconus boui*, *Poremskiconus colombi*, and *Poremskiconus hennequini* (illustrated here on Figure 2.34). Off the coast of Venezuela, the isolated coral reef complex of the Los Roques Atoll also supports an evolutionary hot spot, referred to here as the **Roquesian Infraprovince**. This large atoll (over 400 sq. kms.), comprises over 300 islands and hundreds of separate reef systems and houses a number of interesting endemic gastropods. Some of the more important Roquesian endemics include the endemic colubrariid genus and species *Roquesia lindae*, the cone shell *Tenorioconus duffyi* and the muricid *Murexsul duffyi* (which all live on living coral reefs), and the olivid *Americoliva fulgurator bullata* and the strombinid *Strombina francesae*, which live on the immense expanses of carbonate sand between the reefs and islands (see Figure 2.34). For details on the Roquesian Infraprovince, see Petuch (1992).

The Dutch West Indian island of Aruba contains the **Aruban Infraprovince**, the single richest evolutionary hot spot found anywhere in the southern Caribbean. Geologically, Aruba and its surrounding coastal areas are an extension of the Venezuelan continental shelf and contain nutrient rich, continental-type oceanic conditions. This island of nutrient-rich water stands out in stark contrast to the sterile, oligotrophic open oceanic water of the neighboring islands of Curacao and Bonaire. Because of the higher nutrient input, larger plankton resources, and more food resources, the Aruban gastropod fauna is one of the richest in the entire Western Atlantic, containing a large number of endemic taxa. Primary among these are the muricid *Murexiella hilli*, the olivids *Americoliva fulgurator* and *Eburna balteata*, and an exceptionally large fauna of conids, including the endemic genus *Arubaconus* (*A. hieroglyphus*), an endemic species radiation of the genus *Tenorioconus*, including *T. curassaviensis*, *T. monicae*, and *T. rosi*, and several other endemic species such as *Perplexiconus wendrosi*, *Jaspidiconus vantwouldti*, *J. booti*, and *Lindaconus baylei arubaensis*. On the opposite side of the Grenadian Subprovince from Aruba, the island of Barbados is now known to house its own evolutionary hot spot, the **Barbadan Infraprovince**. Technically not part of the volcanic island chain that makes up the Lesser Antilles, Barbados is, instead, a Pacific Ocean seamount that was dragged into the Caribbean Basin along with the Caribbean Tectonic Plate during the Late Cretaceous. Today, the Barbados Seamount is surrounded by very deep water and stands by itself outside the West Indian Arc. This oceanographic isolation has led to the development of a rich deep water gastropod fauna with a high degree of endemism. Preliminary sampling of these still-unexplored molluscan assemblages has yielded numerous new and interesting gastropods, including the mitrid *Mitra hilli*, the olivid *Americoliva barbadensis*, the conid *Dalliconus coletteae*, and the large drilliid *Clathrodrillia petuchi* (see Figure 2.36).

Surinamian Subprovince

Extending from the mouth of the Orinoco River, Venezuela, to the mouth of the Amazon River, Brazil, the Surinamian Subprovince (named for the country of Suriname) is the farthest-south biogeographical component of the Caribbean Province. Besides Suriname and the Orinoco River area, this biogeographical unit also encompasses the muddy, mangrove-lined coasts of Guyana, French Guiana, and Amapa State, Brazil. Throughout this area, the coastline is flooded with an immense amount of fresh water effluent from large river systems like the Oyapok, Maroni, Courantine, Cuyuni, and Essequibo and hundreds of smaller steams and rivers. As a result of the input of the myriad coastal rivers, the offshore sea floors are composed of organic-rich muds, with few rocky outcrops and almost no coral growth. The entire subprovince acts as an ecological barrier for most of the classic coral reef-dwelling Caribbean gastropods, resulting in a very impoverished molluscan fauna.

The Surinamian brackish water lagoons and coastal mangrove forests support an interesting gastropod fauna that contains a large number of neritids, such as *Vitta zebra* and *Neritina virginea brasiliana*, and *Vitta piratica*. These occur along with large predatory gastropods such as the melongenids *Melongena melongena* and *Pugilina tupiniquim* and the muricids *Thaisella trinitatensis* and *Thaisella coronatus*. The offshore areas, with their rich organic mud substrate, support immense numbers of penaeid shrimp, which form the basis of an important regional fishery. The commercial shrimpers frequently capture larger mollusks in their nets and they are the sole source of study specimens from this area. Among the more typical species that are collected include the muricids *Poirieria oregonia*, *Siratus springeri*, *Siratus thompsoni*, and *Phyllonotus guyanensis*, the fasciolariid *Fasciolaria guyanensis*, the turbinellid *Turbinella laevigata rianae*, and the conids *Conasprelloides brunneobandatus*, *C. guyanensis*, and *Kellyconus rachelae* (some shown here on Figure 2.36).

Southwestern Atlantic Tropical Subregion

As in the previous two subregions, the Southwestern Atlantic Tropical Subregion encompasses only a single major biogeographical unit, the Brazilian Molluscan Province. This area, which extends from the mouth of the Amazon River, around the "nose" of Brazil, and south to Cabo Frio in central Rio de Janeiro State, contains the second-richest molluscan fauna found in the entire Western Atlantic (Petuch, 2013). The mollusks of this southern tropical subregion derive from the malacofaunas of the Pliocene Gatunian Paleoprovince (see Petuch, 1982; 2004), which extended from southern Brazil, across the then-open Isthmus of Panama, and into the eastern Pacific; in essence spanning two separate oceans. Because of this ancient Pacific relationship, the living mollusks of the Southwestern Atlantic Tropical Subregion and the Brazilian Province have closer affinities to the Panamic Molluscan Province (see Chapter 4) of western Central America than they do to those of the Caribbean Province to the north.

During the Miocene, Pliocene, and early Pleistocene, the Amazon River region of northern Brazil was flooded with an immense marine embayment known as the **Amazon Paleosea** (see Petuch, 1988; 2004: 41, 51-52). Extending from the foothills of the Andes Mountains in the west to the mouth of the present-day Amazon River in the east, the narrow, fjord-like Amazon Paleosea housed a highly endemic and bizarre marine fauna, unlike anything else seen in the tropical western Atlantic. The strange estuarine areas at the western end of the sea formed the habitat for a large species radiation of *Neritina* nerites, large *Tympanotonus*-like ornate potamidids (undescribed genera), and spiny melongenids, along with large walrus and narwhal-like tusked sirenian mammals and giant crocodilians (all contained in the Piraban Subprovince of the Miocene Baitoan Paleoprovince and the Juruaian Subprovince of the Pliocene Gatunian Paleoprovince). By the mid-Pleistocene, the immense influx of sediments from the erosion of the Andes filled in the entire Amazon Paleosea, destroying all the previous aquatic environments and faunas. By the late Pleistocene, the Amazon River began to form within the depression of the older Amazon Paleosea and its fresh water effluent separated the surviving Gatunian Paleoprovince faunas into a northern Caribbean fauna and a southern Brazilian fauna. Since the Holocene, the amount of fresh water flowing into the Atlantic has increased dramatically and now acts as an ecological and physiological barrier between the two molluscan provinces.

Brazilian Molluscan Province

Named for the country of Brazil (Petuch, 1988; 2004), this tropical province extends from the Amazon River mouth south to the Cabo Frio area in Rio de Janeiro State. Malacologically, the Brazilian Province contains the full complement of classic tropical index families, including the Conidae, Cypraeidae, Strombidae, and Modulidae. Having evolved from the surviving remnants of the Gatunian Paleoprovince, a large part of the Recent Brazilian malacofauna shows a stronger affinity to the Panamic Province of eastern Central America than it does to the neighboring Caribbean Province. Several classic Panamic gastropod genera, such as the tonnid *Malea* (shown here on Figure 2.39), the buccinid *Northia*, and the pisaniid *Caducifer* occur along the Brazilian coast and several classic Brazilian gastropod species more closely resemble their Panamic congeners than they do their Caribbean neighbors to the north. Some of these cognates include the cypraeid *Macrocypraea dissimilis* (close to the Panamic *M. cervinetta*), the ovulid *Cyphoma macumba* (close to the Panamic *C. emarginatum*), the strombid *Titanostrombus goliath* (close to the Panamic *T. galeatus*), the fasciolariid *Pustulatirus ogum* (close to the Panamic *P. mediamericana*), and the harpid *Morum bayeri* (close to the Panamic *Morum tuberculosum*) (some shown on Figures 2.37 and 2.40). The Brasilian Province contains two subprovinces, the **Cearaian** and **Bahian**, and three distinct infraprovinces.

The Brazilian molluscan fauna also contains a large number of endemic genera, including the large marginellid *Bullata*, the fasciolariids *Aurantilaria* and *Goniofusus*, and the conid genera *Artemidiconus*, *Brasiliconus*, and *Coltroconus* (with the entire genus being confined to the Abrohos Platform). These are illustrated here on

Figures 2.37, 2.38, 2.40, and 2.42. The widespread western Atlantic conid genus *Jaspidiconus* and the southern Caribbean-Brazilian genus *Poremskiconus* have also undergone explosive evolution within the Brazilian Province, producing large clusters of sibling species along northern Brazil, the area around Bahia de Todos os Santos of Bahia State, and across the Abrolhos Platform and Abrolhos Archipelago. Some of these are shown on Figures 2.37-2.42. Other classic Brazilian Province index species include the abalone *Haliotis aurantia*, the large modulid *Modulus bayeri*, the large turbinellids *Turbinella laevigata* and *Vasum cassiforme*, the volutids *Voluta ebraea* and *Plicoliva zelindae*, and the large ornate muricid *Siratus tenuivaricosus*.

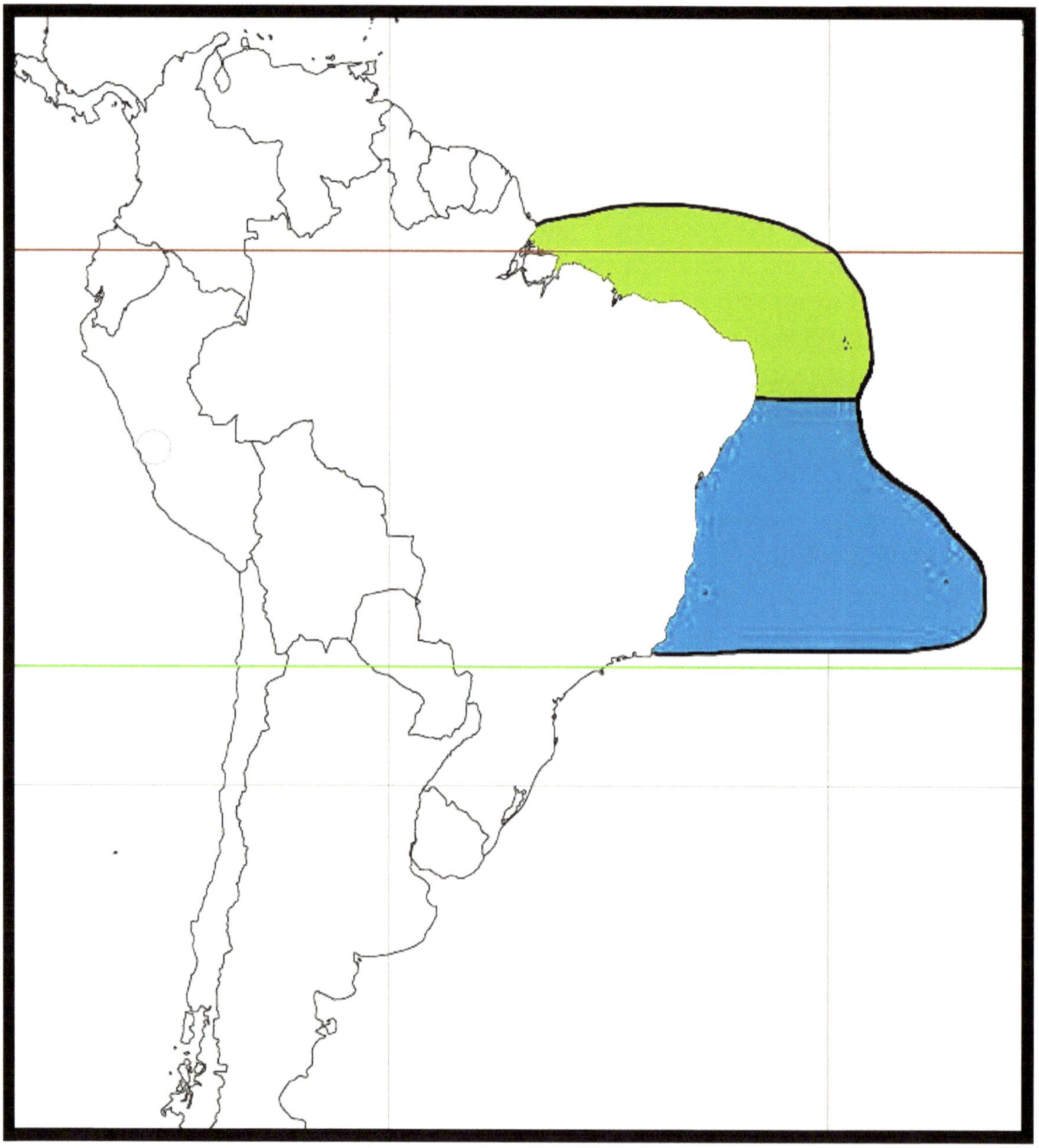

Figure 2.4 Map of the Brazilian Molluscan Province, showing the areal extents of its subprovinces: the **Cearaian Subprovince** (lime green) and **Bahian Subprovince** (blue).

Cearaian Subprovince

Named for the State of Ceará in northern Brazil, this subprovince encompasses the area extending from the mouth of the Amazon River (Pará State) south around the "nose" of Brazil to near Aracajú, Sergipe State. Ecologically, the shallow water areas of the Cearaian Subprovince are dominated by extensive "reefs" made up of red coralline algae (fused algal rhodoliths), primarily of the algae *Porolithon*, *Sporolithon*, *Lithophyllum*, *Lithoporella*, and *Popolithon*, and also *Homotrema* encrusting foraminifera and embedded vermetid gastropods (*Thylaeodus* species). These distinctive and unique reef complexes often form immense wall-like structures (*i.e.* the "recife" off Recife, Pernambuco State) that extend for hundreds of kilometers. The offshore island of Atol das Rocas (the only true atoll in the South Atlantic) is also composed almost entirely of these coralline algae rhodolith reefs, making it the only-known algal atoll in the world. These unusual algal reefs also house highly endemic and geographically-isolated molluscan faunas, especially in the areas off Maranhão and Rio Grande do Norte States (Rio do Fogo). Large species radiations of the conid genera *Jaspidiconus* and *Poremskiconus* occur throughout the subprovince and some of these are illustrated here on Figures 2.38 and 2.39. The offshore carbonate sediment sea floors also house a distinct gastropod fauna with a high level of endemism, and typically include species such as the olivid *Americoliva circinata jorioi*, the marginellid *Bullata lilacina*, the harpids *Morum mariaodeteae* and *Oniscidia matthewsi* (see Petuch and Berschuaer, 2020a for a review of the Brazilian *Morum* species), and the large conid *Conasprelloides hazinorum* (all shown here on Figures 2.38 and 2.39).

Of special interest within the Cearaian Subprovince is the Fernando de Noronha Archipelago off northeastern Brazil (Pernambuco State), which encompasses the main landmass of Ilha Fernando de Noronha, along with Ilhas Rata, Sela Gineta, Rasa, do Meio, São Jose, and 15 other smaller islands. This isolated volcanic archipelago has evolved a molluscan fauna with a high level of endemism, especially in the intertidal and supratidal areas, producing an evolutionary hot spot referred to here as the **Noronhan Infraprovince**. The Noronhan endemic periwinkle *Echinolittorina vermeiji*, the endemic nerite *Nerita (Ritena) chlorostoma deturpensis*, and the endemic limpet *Collisella noronhensis* are abundant in the high intertidal areas along rocky cliffs and headlands. The tide pools on exposed rocky platforms and the shallow subtidal rock outcrops also support an interesting gastropod fauna that includes the Noronhan endemic volute *Enaeta leonardhilli*, the endemic fasciolariid *Polygona vermeiji*, and the Noronhan Grinning Tun *Malea noronhensis*. This last-mentioned species, the only living *Malea* in the Atlantic Ocean, has also been collected off the Atol das Rocas and is apparently endemic to the Brazilian offshore island systems. These infraprovincial endemics are illustrated on Figure 2.39.

Bahian Subprovince

Named for the State of Bahia, this southern subprovince of the Brazilian Province extends from Aracajú, Sergipe State south to the Cabo Frio area of Rio de

Janeiro State (from Farol de São Thome to Arraial do Cabo). Unlike the coralline algae-dominated environments of the Cearaian Subprovince, the Bahian ecosystems are dominated by bioherms containing sponges, brown algae, endemic scleractinian corals such as *Mussismilia braziliensis*, *M. hispida*, *M. harttii*, and *Favia leptophylla*, and endemic hydrocorals such as *Millepora braziliensis* and *M. nitida* (Leao *et al*., 2003). These reefal environments of the Bahian Subprovince house a diverse gastropod fauna that exhibits a very high percentage of regional endemism. Some of these endemics include the muricids *Murexiella iemanja*, *M. leonardhilli*, and *Pygmaepterys oxossi*, the fasciolariids *Polygona bayeri* and *Pustulatirus ogum*, the harpid *Morum berschaueri*, the giant marginellid *Bullata bullata*, the olive shell *Americoliva circinata*, the strange olive-like volute *Plicoliva zelindae*, the nutmeg shell *Cancellaria petuchi*, and the cone shells *Dauciconus riosi* and *D. worki*. The Bahian Subprovince also houses the largest species radiations of the conid genera *Jaspidiconus* and *Poremskiconus* found anywhere in the western Atlantic, with at least 12 known species of *Jaspidiconus* and at least 8 known species of *Poremskiconus* being found on the shallow reef systems off Sergipe, Bahia, and Espirito Santo States (some shown here on Figures 2.40, 2.41, and 2.42).

The immense Todo os Santos Bay, which contains the capital city of Salvador and over 30 smaller cities, is the largest single bay along the entire eastern coast of Brazil. This complex estuary represents the flooded valley of the Rio Paraguacu and contains numerous smaller ancillary bays and several large hyposaline lagoons. At the mouth of the bay, the large Itaparica Island essentially cuts off the western half of the bay from the Atlantic Ocean, producing quiet lagoonal conditions. This large island, environmentally intermediate between lagoonal and open oceanic conditions is now known to be an evolutionary hot spot and is here referred to as the **Itaparican Infraprovince**. Of particular interest on Itaparica Island is a small local endemic radiation of the conid genus *Jaspidiconus*, including the pustulose *J. ogum* and *J. henckesi* and the smooth *J. marinae*. These small cones are restricted to very small intertidal areas around the island and are found nowhere else. Deeper within Todos os Santos Bay and behind Itaparica Island, another endemic cone, *Jaspidiconus pomponeti*, has been found to inhabit only one small ancillary bay (Itapajipe). These Itaparican endemic cones are shown here on Figure 2.41.

Evolutionarily, the single most important area within the Bahian Subprovince is the Abrolhos Platform off southern Bahia State (Petuch, 1979). This large, shallow rectangular shelf houses hundreds of coral reef complexes and also the Abrolhos Archipelago, a caldera-like circle of five high volcanic islands. The Abrolhos Platform reef complexes also represent the farthest-south major coral reefs in the South Atlantic and they demarcate the southernmost edge of the tropical western Atlantic. This coralline area also represents a major evolutionary hot spot which is here referred to as the **Abrolhosian Infraprovince**. The infraprovincial area also extends to the nearby Minerva Seamount, Sulfur Bank, and Royal Charlotte Bank, guyot-like flat-topped seamounts that rise to within 100 m of the sea surface and which are capped with carbonate sediments. Of primary interest within the Abrolhosian Infraprovince is the conid genus *Coltroconus*, comprising at least six tiny species, and which is known only from the Abrolhos Platform and neighboring seamounts (all six known species are shown on Figure 2.42). Some of the *Coltroconus* species are endemic to individual seamounts, with *C. bianchii* being endemic to the Minerva Seamount, *C. henriquei* being endemic to the Royal Charlotte Bank, and with *C. schirrmeisteri* being endemic to the Sulfur Bank. The Abrolhosian Infraprovince also houses the endemic harpid *Morum damasoi* and a species radiation of the conid genus *Poremskiconus*, with at least seven species being known from the Platform area. Some of these are shown on Figure 2.42. The basaltic volcanic islands of the offshore Abrolhos Archipelago also have an interesting rocky intertidal and tide pool molluscan fauna that is known to contain several endemic taxa. Some of the most prominent of these isolated endemics are the lottiid limpet *Collisella abrolhosensis*, the small volute *Plicoliva zelindae oceanica*, and a possible new dwarf subspecies of *Macrocypraea dissimilis* that resembles a tiny version of the Trindadean endemic cowrie *M. mammoth* (discussed next).

More than 1,100 km east of the coastline of Espirito Santo State, at the southernmost limit of the Bahian Subprovince, the Trindade and Martim Vaz Archipelago represents an extreme offshore outpost of the Brazilian Province fauna. Being so far from the coastline, this small group of islands has experienced long periods of genetic isolation and has evolved its own characteristic molluscan fauna with a fairly high level of endemism. Now known to be an isolated hot spot of evolution, we here refer to the area of this remote island group as the **Trindadean Infraprovince**, for Trindade Island, the largest component of an archipelago that contains five large islands and numerous small islets and emergent rocks. Several important and interesting macrogastropods are now known to be index taxa for the infraprovince, some of these include the lottiid limpet *Collisella marcusi*, the large cowrie *Macrocypraea mammoth* (shown here on Figure 2.39L), the nerite *Nerita (Ritena) chlorostoma trindadeensis*, and the large cone shell *Dauciconus jorioi*.

Southwestern Atlantic Paratropical Subregion

Like its mirror-image paratropical subregion in the North Atlantic, the Southwestern Atlantic Paratropical Subregion contains only a single biogeographical unit, the Paulinian Molluscan Province. Also similar to the Carolinian Province, the biogeographical units of this paratropical subregion contain faunas that are a mixture of tropical, warm temperate, and temperate organisms. South of the area extending from Farol de São Thome to Arraial do Cabo (the Cabo Frio region) in northern Rio de Janeiro State, the water temperatures become much cooler, often dropping below tropical conditions during the winter (see Castelao and Barth, 2006). Because of these seasonal low temperatures, all extensive coral reef growth ceases south of Arraial do Cabo and only three hardy scleractinian species, (*Madracis decactis*, *Astrangia rathbuni*, and *Mussismilia hispida*) are known from São Paulo, Paraná, and Santa Catarina States. These eurythermal corals are found only in isolated bays and nearshore islands where water temperatures stay warmer during the winter months (Leao, Kikuchi, and Testa, 2003). South of the island city of Florianópolis, Santa Catarina State, these last hardy corals disappear and this marks the beginning of a broad provinciatonal transition zone that extends to Uruguay, the Rio de la Plata, and northern Argentina. The molluscan faunas of this paratropical subregion reflect this temperature gradient.

Paulinian Molluscan Province

Although originally described as a subprovince of the Brazilian Province (Petuch, 2013: 163-175), recent, more-detailed analyses of the faunas along the southern Brazilian coast now show that the level of endemism in the key index families reaches levels of 50% or more. This high percentage of endemism demonstrates that the area from Farol de São Thome, Rio de Janeiro State south to northern Argentina represents a full, distinct province all to itself. Named for São Paulo State, this biogeographical unit is here referred to as the Paulinian Province. More detailed faunal analyses also show that the Paulinian Province is a much more complicated biogeographical unit than originally thought, comprising two distinct subprovinces (the **Janeiran** and the **Uruguayan**) and a broad faunal transition area between the Uruguayan Subprovince and the cold water Patagonian Province (the Platensian Provinciatone).

The northern boundary of the Paulinian Province is easily delineated by the abrupt disappearance of classic Brazilian Province index gastropods such as the ovulid *Cyphoma macumba*, the strombid *Titanostrombus goliath*, the turbinellids *Vasum cassiforme* and *Turbinella laevigata*, and the fasciolariid *Aurantilaria aurantiaca*. South of the Cabo Frio region (Farol de São Thome to Arraial do Cabo), these index taxa, and most of the other classic Brazilian Province species, are replaced by a completely different malacofauna with a high level of endemism. Within this paratropical province, the large species radiations of the conid genera *Jaspidiconus* and *Poremskiconus* and the fasciolariid genus *Goniofusus*, which dominate the neritic ecosystems of the tropical Brazilian Province to the north, are absent and are replaced by radiations of the warm-temperate conid genus *Lamniconus* and the fasciolariid genus *Apertifusus*. The Paulinian Province also houses a much richer volutid fauna, comprising species in the genera *Odontocymbiola*, *Zidona*, *Pachycymbiola*, and *Adelomelon*. While volutes are small and relatively rare along the Bahian Subprovince to the north, the Volutidae is represented within the Paulinian Province by many very large and ecologically-prominent taxa. The sandy coastal beaches and lagoons also house a completely different malacofauna, with the nassariid genus *Buccinanops* and the olivid genus *Olivancillaria* dominating the intertidal and shallow subtidal areas (examples of all of these are shown here on Figures 2.43-2.46).

Janeiran Subprovince

The Janeiran Subprovince, named for Rio de Janeiro State, contains the richest molluscan fauna of the entire Paulinian Province. Extending from Farol de São Tome, northern Rio de Janeiro State, south to Florianópolis, Santa Catarina State, this subprovincial area includes several large nearshore islands and island groups such as Ilha Grande, Ilhabela, Ilha de Itacuruca, Ilha da Gipoia, Ilha de Buzios, Ilha Santa

Catarina, and hundreds of smaller islands, along with numerous large bays and lagoons, such as Baia de Caraguatatuba and Baia de Ubatuba. The enclosed bays and protected island groups of this environmentally-diverse area have warmer water conditions than the areas farther south and this has allowed three species of scleractinian corals and two hydrocorals to flourish within isolated pockets along the coast. These marginal coral bioherms also support a highly endemic fauna of gastropods, comparable in species-richness to the fauna of the Abrolhos Platform (Abrolhosian Infraprovince). The Janeiran Subprovince islands and protected bays typically house distinctive paratropical gastropods such as the ovulids *Cyphoma versicolor* and *C. guerrinii*, the cypraeid *Naria acicularis marcuscoltroi*, the strombid *Strombus worki* (often found on algae-covered gravel beds), the olive shell

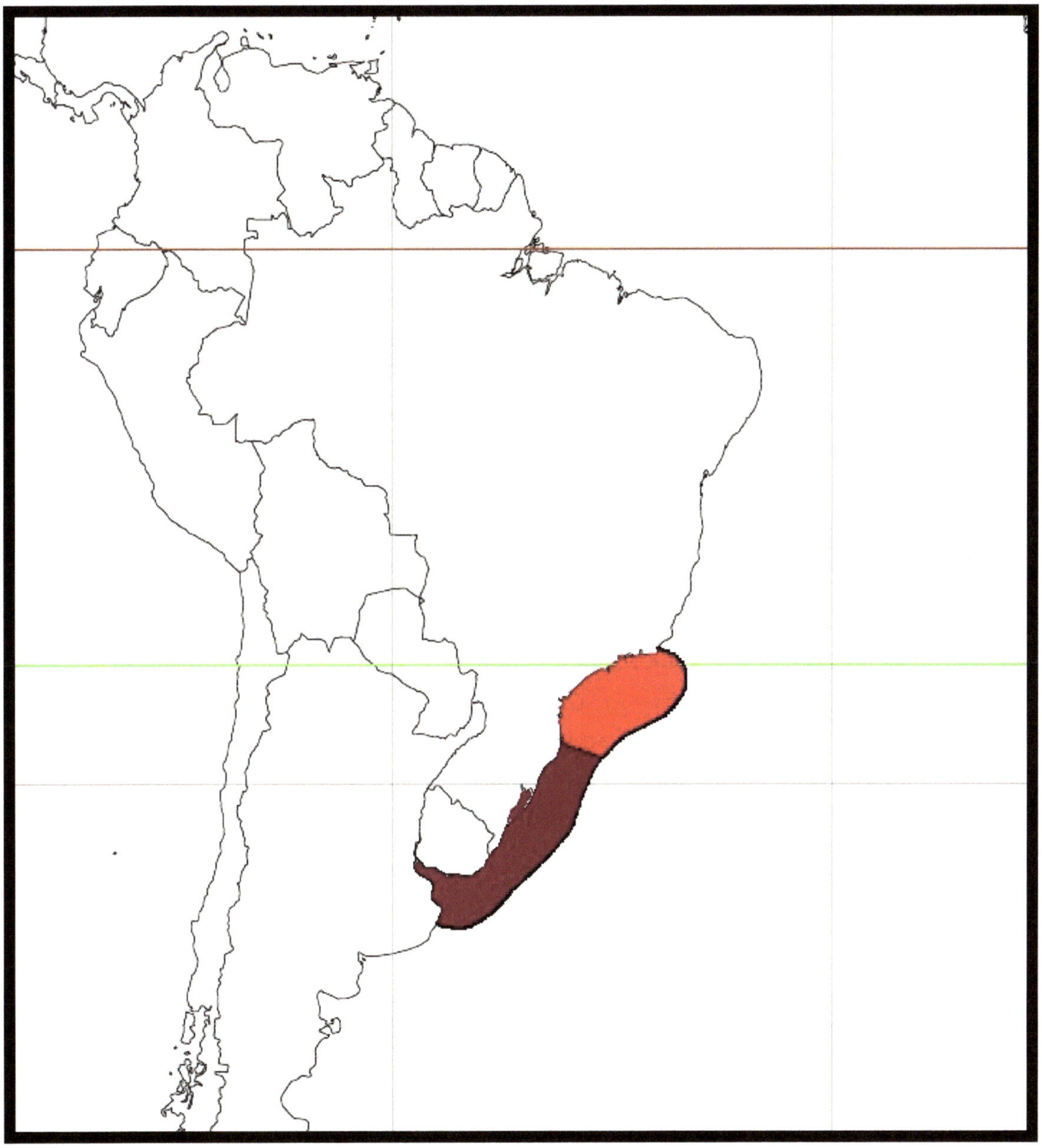

Figure 2.5 Map of the Paulinian Molluscan Province, showing the areal extents of its subprovinces: the **Janeiran Subprovince** (red) and the **Uruguayan Subprovince** (burgundy).

Americoliva circinata tostesi, the fasciolariid *Leucozonia granulilabris*, and the conid *Jaspidiconus simonei* (only northern Rio de Janeiro State). Farther offshore, in the cooler and deeper waters, a large fauna of endemic cone shells occurs, some of which include *Lamniconus clerii*, *L. tostesi*, *L. xanthocinctus*, *L. patriceae*, *L. lemniscatus*, *L. clenchi*, *Sandericonus carioca*, *Conasprelloides capricorni*, and *Dalliconus edpetuchi*. These large and impressive cone shells live together with the volutes *Odontocymbiola cleryana*, *O. americana*, *O. simulatrix*, and *O. simulatrix nana*, the fasciolariid *Apertifusus mariaodetae*, the ranellid *Ranella barcellosi*, and the muricid *Trophon pelseneeri* (all shown here on Figures 2.44 and 2.45).

Of special interest within the Janeiran Subprovince is the area off the Cabo Frio region, from Arraial do Cabo northward to Farol de São Tome. This northernmost end of the Paulinian Province houses a special subtropical area known as the "Coral Oasis" (Leao, Kikuchi, and Testa, 2003), where at least seven species of scleractinian corals and two hydrocorals form scattered small patch reefs. This "Coral Oasis" also represents an evolutionary hot spot containing numerous highly-localized endemic gastropod species radiations and is here referred to as the **Macaen Infraprovince** (for Macaé, Rio de Janeiro State). Some of the more important restricted species that live offshore of this geographically-small area include the volutes *Odontocymbiola saotomensis* and *O. macaensis*, the fasciolariid *Goniofusus damasoi*, and the cone shell *Lamniconus petestimpsoni* (all shown here on Figure 2.44). Probably the most noteworthy endemic species from this area is the large charoniid *Charonia marylenae*, which is endemic to the Cabo Frio region (Figure 2.43B). This large and impressive newly-described species, belonging to the worldwide paratropical *Charonia lampas* species complex, appears to represent a relictual population of charoniids that once had a much wider distribution along South America. A possible late Pliocene ancestor of *C. marylenae* is known from fragments collected in the Mare Formation of Venezuela (Weisbord, 1962; 630, plate 25). This demonstrates that the *C. lampas* complex was present in the Pliocene Caribbean Basin (Gatunian Paleoprovince) for at least 3 million years and it is possible that this ancestral species (which was later named *Charonia weisbordi*; see Gibson-Smith, 1976), extended down the entire eastern coast of South America during the Pliocene. Since that time, the *C. lampas* complex has died out everywhere else in the western Atlantic except for the relictual *C. marylenae*, which has managed to survive along Cabo Frio.

Uruguayan Subprovince

Named for the country of Uruguay (Republica de Uruguay del Este), this southern subprovince of the Paulinian Province extends from Florianópolis, Santa Catarina State southward to Punta del Este, Uruguay and encompasses Rio Grande do Sul State of Brazil and Rocha and Maldonado Departments of Uruguay. The water temperatures of the Uruguayan Subprovince are generally too cold to support both coral growth and tropical gastropods and the classic Janeiran shallow water ecosystems are absent south of Florianópolis. Particularly noticeable is the abrupt disappearance of gastropods in the families Cypraeidae, Ovulidae, and Strombidae and also the olivid genus *Americoliva*, all of which are faunal components of typical Janeiran neritic ecosystems. The Uruguayan Subprovince, however, does share many other taxa with the southern part of the Janeiran Subprovince (off Santa Catarina State), including the large volutes *Pachycymbiola brasiliana*, *Adelomelon beckii*, *Zidona dufresnei*, and *Odontocymbiola pescalia*, the large muricid *Coronium acanthodes*, the fasciolariid *Apertifusus frenguellii*, and a species radiation of the olivid genus *Olivancillaria*, including *O. contortiplicata*, *O. urceus*, *O. vesica*, *O. auricularia*, *O. carcellesi*, *O. brasiliana*, *O. deshayesiana*, and *O. orbignyi*. This demonstrates that the molluscan fauna of the colder water southern area is still representative of the Paulinian Province.

The area extending from the southernmost tip of Rio Grande do Sul State, Brazil, across the southeastern coast of Uruguay (including the offshore islands of Isla Gorriti and Isla de Lobos), and into the northern side of the Rio de la Plata embayment, also houses an impoverished but distinctive molluscan fauna. Of special interest here are endemic species such as the olivid *Olivancillaria teaguei*, the nassariid *Buccinanops duartei*, the muricid *Hanetia rushii*, the volutes *Adelomelon martensi* and *A. barrattinii*, and the cone shell *Lamniconus carcellesi* (some illustrated here on Figure 2.46). Ranging from Rio Grande do Sul State, Brazil south to San Clemente del Tuyu, Argentina, *Lamniconus carcellesi* is the farthest-south dwelling species of cone shell in the entire Atlantic Ocean. The area extending from Buenas Aires, Argentina to the southern coast of the Rio de la Plata embayment, and south to the city of Mar del Plata, Argentina, represents a broad faunal transition zone, here referred to as the **Platensian Provinciatone** (named for the Rio de la Plata). South of Mar del Plata, the northward-flowing Falklands / Malvinas Current creates subantarctic oceanic conditions and the warm temperate Uruguyan endemic gastropods abruptly disappear. Between the Rio de la Plata and Mar del Plata, this provinciatonal area houses an odd mixture of paratropical and Patagonian mollusks. Here, cold-adapted members of normally-tropical groups such as the marginellid *Volvarina warreni* and the terebrid *Duplicaria gemmulata* occur together with the Patagonian-subantarctic volutes *Odontocymbiola subnodosa*, *Adelomelon ancilla*, and *Provocator corderoi* (shown here on Figure 2.46). In the northern half of the provinciatone, the cone shell *Lamniconus carcellesi* also occurs along with the marginellid, the terebrid, and the volutes, forming a bizarre local molluscan assemblage, unlike any other known from the Americas.

ICONOGRAPHY OF GASTROPODS OF THE WESTERN ATLANTIC REGION

(Principal Index Gastropods are shown on Figures 2.6 to 2.46)

Figure 2.6 Widespread Index Gastropods of the Western Atlantic Region, occurring in the Carolinian, Caribbean, and Brazilian Provinces.

A= *Pseudocyphoma intermedium* (Sowerby I, 1828), length 26.5 mm, 10 m depth, off Cabo Rojo, Puerto Rico; **B=** *Proadusta surinamensis* (Perry, 1811), length 26.8 mm, 50 m depth off Fortaleza, Ceará, Brazil; **C=** *Charonia variegata* (Lamarck, 1816), length 205 mm, 10 m depth off Gonave Island, Haiti; **D=** *Cypraecassis testiculus* (Linnaeus, 1758), length 33 mm, 5 m depth on Pickles Reef, Plantation Key, Florida Keys, Florida; **E=** *Aliger gallus* (Linnaeus, 1758), length 123 mm, 10 m depth off Gonave Island, Haiti; **F=** *Macrostrombus costatus* (Gmelin, 1791), length 114 mm, 1 m depth off Missouri Key, Florida Keys, Florida; **G=** *Phyllonotus oculatus* (Reeve, 1845), length 83 mm, 10 m depth off St. James, Barbados; **H=** *Oniscidia dennisoni* (Reeve, 1842), length 36 mm, on shell rubble and sponges, 80 m depth off Cabo la Vela, Goajira Peninsula, Colombia; **I=** *Chelyconus testudinarius* (Hwass, 1792), length 65 mm, 2 m depth off Malmok, Aruba; **J=** *Stephanoconus regius* (Gmelin, 1791), length 50 mm, 2 m depth on Pickles Reef, Plantation Key, Florida Keys, Florida; **K=** *Ficus pellucida* Deshayes, 1856, length 33 mm, 200 m depth off Key West, Florida; **L=** *Jaspidiconus mindanus* (Hwass, 1792), length 35 mm, in sand under coral, 10 m depth off La Vauclin, Martinique Island.

Figure 2.7 Widespread Carolinian Province Mollusks.
A= *Macrocypraea (Lorenzicypraea) cervus* (Linnaeus, 1771), length 101 mm, in a crab trap set in 10 m depth off Rabbit Key, Ten Thousand Islands, Florida; **B**= *Hexaplex fulvescens* (Sowerby I, 1834), length 144 mm, 20 m depth off Panama City, Florida; **C**= *Pteropurpura bequaerti* (Clench and Farfante, 1945), length 36 mm, dredged from 250 m depth off Tampa, Florida; **D**= *Cinctura hunteria* (Perry, 1811), length 48 mm, low tide off Turtle Key, Ten Thousand Islands, Florida; **E**= *Triplofusus papillosus* (Sowerby I, 1825), length 312 mm, low tide on Siesta Key, Sarasota, Florida (= *giganteus* Kiener, 1840); **F**= *Clenchina gouldiana* (Dall, 1887), length 54 mm, 250 m depth off St. Augustine, Florida; **G**= *Scaphella junonia* (Lamarck, 1804), length 107 mm, 20 m depth in crab traps, off Pavilion Key, Ten Thousand Islands, Florida; **H**= *Dalliconus mcgintyi* (Pilsbry, 1955), length 68 mm, 250 m depth off Tampa, Florida; **I**= *Dauciconus amphiurgus* (Dall, 1889), length 31 mm, 20 m depth off Panama City, Florida; **J**= *Kohniconus delessertii* (Recluz, 1843), length 53 mm, 50 m depth off Cape Canaveral, Florida; **K**= *Lindaconus atlanticus* (Clench, 1942), length 51 mm, low tide, Pavilion Key, Ten Thousand Islands, Florida.

Figure 2.8 Index Gastropods of the Georgian Subprovince, Carolinian Province.
A= *Conomodulus lindae* (Petuch, 1987), height 11 mm, 70 m depth off Palm Beach Island, Palm Beach, Florida; **B**= *Vokesimurex morrisoni* Petuch and Sargent, 2011, length 41 mm, 100 m depth off Fort Pierce, Florida; **C**= *Cinctura keatonorum* Petuch, 2013, length 92 mm, 50 m depth off Cape Canaveral, Florida; **D**= *Fulguropsis rachelcarsonae* Petuch, Myers, and Berschauer, 2016, length 96 mm, Singer Island, Lake Worth Lagoon, Palm Beach, Florida; **E**= *Liomesus stimpsoni* Dall, 1889, length 43 mm, trawled from 200 m depth off Tybee Island, Georgia; **F**= *Prunum canilla* (Dall, 1927), length 16 mm, trawled from 200 m depth off Cape Fear, North Carolina; **G**= *Rehderia georgiana* (Clench, 1946), length 80 mm, dredged from 250 m depth off Cape Canaveral, Florida; **H**= *Gradiconus philippii* (Kiener, 1847), length 35 mm, 30 m depth off Beaufort, North Carolina; **I**= *Jaspidiconus pfluegeri* Petuch, 2004, length 17 mm, 5 m depth off Singer Island, Lake Worth Lagoon, Palm Beach, Florida (ranges to the central Florida Keys); **J**= *Strioterebrum onslowensis* Petuch, 1974, length 32 mm, 20 m depth off Amelia Island, Florida (ranges from North Carolina to the northern Florida Keys); **K**= *Arrhoges occidentalis* (Beck, 1836), length 37 mm, 35 m depth off Great Cranberry Island, Frenchmen's Bay, Maine (ranges from Arctic Canada to 250 m depth off South Carolina); **L**= *Americoliva nivosa clenchi* Petuch and Berschauer, 2019, length 57 mm, trawled from 25 m depth in an *Argopecten gibbus carolinensis* scallop bed in Onslow Bay, south of Beaufort, Carteret County, North Carolina; **M** = *Tuckericonus flamingo* (Petuch, 1980) holotype length 19.4 mm, trawled at 70 m depth off Dania Beach, Broward County, Florida.

Figure 2.9 Index Gastropods of the Palm Beach Infraprovince, Carolinian Province.
A= *Modulus kaicherae* Petuch, 1987, diameter 12 mm, 75 m depth off Palm Beach Island, Palm Beach County, Florida; **B=** *Modulus pacei* Petuch, 1987, diameter 14.5 mm, 1 m depth on rocks, off Peanut Island, Lake Worth Lagoon, Palm Beach County, Florida; **C=** *Coralliophila pacei* Petuch, 1987, length 31 mm, 20 m depth off Palm Beach Island, Palm Beach County, Florida; **D=** *Favartia goldbergi* Petuch and Sargent, 2011, length 10.5 mm, 75 m depth off Palm Beach Island, Palm Beach County, Florida; **E=** *Stramonita buchecki* Petuch, 2013, length 30 mm, on oysters at low tide, Singer Island, Lake Worth Lagoon, Palm Beach County, Florida; **F=** *Melongena (Rexmela) corona winnerae* (Petuch, 2004), length 147 mm, low tide on sand flats, Singer Island, Lake Worth Lagoon, Palm Beach County, Florida; **G=** *Gemophos filistriatus* Vermeij, 2006, length 20.4 mm, 75 m depth off Palm Beach Island, Palm Beach County, Florida; **H=** *Dauciconus glicksteini* (Petuch, 1987), length 23 mm, 75 m depth off Palm Beach Island, Florida; **I=** *Jaspidiconus vanhyningi* (Rehder, 1944), length 16 mm, 5 m depth off Pompano Beach, Broward County, Florida (also found in the northern Florida Keys); **J=** *Kellyconus binghamae* (Petuch, 1987), length 14 mm, 75 m depth off Palm Beach Island, Palm Beach County, Florida; **K=** *Cerithium lutosum lindae* Petuch, 1987, length 15 mm, low tide under rocks, Peanut Island, Lake Worth Lagoon, Palm Beach County, Florida; **L=** *Trochomodulus calusa foxhalli* (Petuch and Myers, 2014), diameter 13 mm, in Shoal Grass bed, low tide off Singer Island, Lake Worth Lagoon, Palm Beach County, Florida.

Figure 2.10 Index Gastropods of the Floridian Subprovince, Carolinian Province.
A= *Cyphoma sedlaki* Cate, 1976, length 13.5 mm, 3 m depth off Sombrero Light, Florida Keys, Florida; **B=** *Murexiella kalafuti* Petuch, 1987, length 15 mm, 30 m depth off Dry Tortugas, Florida Keys, Florida; **C=** *Melongena* (*Rexmela) bicolor* (Say, 1827), length 33 mm, low tide on Middle Torch Key, Florida Keys; **D=** *Fulguropsis keysensis* Petuch, 2013, length 58 mm, 2 m depth off Middle Torch Key, Florida Keys, Florida; **E=** *Clenchina florida* (Clench and Aguayo, 1940), length 57 mm, 300 m depth off Key West, Florida Keys, Florida; **F=** *Rehderia schmitti* (Bartsch, 1931), length 82 mm, 300 m depth north of the Dry Tortugas, Florida Keys, Florida; **G=** *Scaphella junonia elizabethae* Petuch and Sargent, 2011, length 110 mm, 10 m depth off Garden Island, Dry Tortugas, Florida Keys, Florida; **H=** *Gradiconus burryae* (Clench, 1942), length 43 mm, near Turtle Grass bed, Missouri Key, Florida Keys, Florida; **I=** *Gradiconus mazzolii* Petuch and Sargent, 2011, length 21 mm, on sponge bed off Middle Torch Key, Florida Keys, Florida; **J=** *Gradiconus tranthami* (Petuch, 1995), length 26 mm, 5 m depth off Pickles Reef, Plantation Key, Florida Keys, Florida (pink color forms from the Dry Tortugas were named *tortuganus* and *antoni*; now considered synonyms of *tranthami*); **K=** *Jaspidiconus pealii* (Green, 1830), length 13.2 mm, in Turtle Grass, off Middle Torch Key, Florida Keys, Florida; **L=** *Pusula juyingae* Petuch and Sargent, 2011, length 15 mm, 20 m depth off Elliott Key, northern Florida Keys, Florida (often incorrectly referred to as *P. costispunctata*, which is a similar Panamic Province species).

Figure 2.11 Index Gastropods of the Floridian Subprovince, Carolinian Province.
A= *Dibaphimitra florida* (Gould, 1850), length 55 mm, 5 m depth off Plantation Key, Florida Keys, Florida (the Nicaraguan Subprovince *D. janetae* is often incorrectly synonymized with this Keys endemic, which differs from its western Caribbean congener in having a more inflated shell with a purple shell interior and purple protoconch; see Figure 2.26G); **B**= *Cyphoma alleneae* Cate, 1973, length 33.5 mm, 2 m depth off Missouri Key, Florida Keys, Florida; **C**= *Trochomodulus calusa* (Petuch, 1988), diameter 12 mm, on Turtle Grass blades in 1 m depth off Middle Torch Key, Florida Keys, Florida; **D**= *Cyphoma rhomba* Cate, 1978, length 21 mm, 5 m depth on Pickles Reef, Plantation Key, Florida Keys, Florida; **E**= *Dermomurex pacei* Petuch, 1987, length 17 mm, 3 m depth off Pickles Reef, Plantation Key, Florida Keys, Florida; **F**= *Favartia pacei* Petuch, 1988, length 16 mm, 1 m depth off Missouri Key, Florida Keys, Florida; **G**= *Clenchina dohrni* (Sowerby III, 1903), length 52 mm, 300 m depth off Key West, Florida Keys, Florida; **H**= *Americoliva recourti* Petuch and Myers, 2014, length 22 mm, 150 m depth south of Lower Matecumbe Key, Florida Keys, Florida; **I**= *Cancellaria adelae* Pilsbry, 1940, length 39 mm, 5 m depth on Pickles Reef, Plantation Key, Florida Keys, Florida; **J**= *Jaspidiconus fluviamaris* Petuch and Sargent, 2011, length 16 mm, 5 m depth south of Garden Island, Dry Tortugas Islands, Florida Keys, Florida; **K**= *Calliostoma adelae* Schwengel, 1951, diameter 17.4 mm, on sponges, 1 m depth off Middle Torch Key, Florida Keys, Florida; **L**= *Americoliva matchetti* Petuch and Myers, 2014, length 19 mm, 200 m depth east of Elliott Key, northern Florida Keys, Florida; **M**= *Jaspidiconus acutimarginatus* (Sowerby II, 1866), length 19 mm, 2 m depth in sand, off Missouri Key, Florida Keys, Florida; **N**= *Murexiella caitlinae* Petuch & Myers, 2014, holotype length 30 mm, 1.5 m depth off Missouri Key, Florida Keys, Florida.

Figure 2.12 Widespread Gulf of Mexico Index Gastropods.
A= *Clenchina robusta* (Dall, 1889), length 80 mm, trawled from 400 m depth in the Florida Straits, south of Key West, Monroe County, Florida; **B=** *Heilprinia timessa* (Dall, 1889), length 84 mm, trawled from 100 m depth off Panama City, Florida; **C=** *Gradiconus largillierti* (Kiener, 1848), length 43 mm, 20 m depth off Panama City, Florida; **D=** *Fasciolaria bullisi* Lyons, 1972, length 100 mm, trawled from 250 m depth west of Tampa, Florida; **E=** *Conasprelloides cancellatus* (Hwass, 1792), length 35 mm, trawled from 150 m depth south of Grande Isle, Louisiana; **F=** *Neverita delessertiana* (Recluz, 1843), height 32 mm, low tide on sand, Round Key, Ten Thousand Islands, Collier County, Florida; **G=** *Paziella nuttingi* (Dall, 1896), length 42 mm, 150 m depth north of the Dry Tortugas, Florida; **H=** *Solenosteira cancellaria* (Conrad, 1846), length 24 mm, on oyster banks at low tide, Cedar Key, Florida; **I=** *Cancellaria richardpetiti* Petuch, 1987, length 47 mm, trawled from 200 m depth south of Apalachicola, Florida; **J=** *Gradiconus optabilis* (A. Adams, 1845), length 30 mm, dredged from 200 m depth north of the Dry Tortugas, Florida Keys (a rare "lost species"; *G. selectus* may be the same); **K=** *Dalliconus rainesae* (McGinty, 1953), length 23 mm, trawled from 200 m depth north of the Campeche Banks, Yucatan, Mexico; **L=** *Xenophora microdiscus* Petuch, 1994, width 17 mm, trawled from 150 m depth, 150 km west of Tampa, Florida (originally named as a late Pleistocene fossil from southern Florida, but now known to be living in deep water within the Gulf of Mexico); **M=** *Bathynerita naticoidea* Clarke, 1989, width 9.4 mm, found on a *Bathymodiolus* deep water mussel, from an oil seep in 580 m depth due south of Marsh Island, Mississippi River Delta, Louisiana.

Figure 2.13 Index Gastropods of the Suwannean Subprovince, Carolinian Province.
A= *Chicoreus rachelcarsonae* Petuch, 1987, length 40 mm, 200 m depth due west of Tampa, Florida; **B**= *Favartia lindae* Petuch, 1987, length 18 mm, 100 m depth off Tampa, Florida; **C**= *Vokesimurex lindajoyceae* (Petuch, 1987), length 32.4 mm, trawled from 200 m depth due west of Tampa, Florida; **D**= *Murexiella taylorae* Petuch, 1987, length 15 mm, 100 m depth off Tampa, Florida; **E**= *Phyllonotus whymani* Petuch and Sargent, 2011, length 43 mm, in a crab trap, 100 m depth north of the Dry Tortugas, Florida; **F**= *Fulguropsis pyruloides* (Say, 1822), length 77 mm, low tide off Rabbit Key, Ten Thousand Islands, Collier County, Florida; **G**= *Lindafulgur lyonsi* (Petuch, 1987), length 110 mm, on coralline algal rubble sea floor, 160 m depth off Tampa, Florida; **H**= *Aurinia dubia* (Broderip, 1827), length 135 mm, trawled from 250 m depth off Sarasota, Florida; **I**= *Americoliva sunderlandi* (Petuch, 1987), length 22 mm, trawled from 80 m depth west of Tampa, Florida; **J**= *Gradiconus anabathrum* (Crosse, 1865), length 43 mm, low tide, Marco Island, Collier County, Florida; **K**= *Jaspidiconus stearnsi* (Conrad, 1869), length 20 mm, low tide, Marco Island, Collier County, Florida; **L**= *Hesperisternia harasewychi* (Petuch, 1987), length 24.8 mm, trawled from 150 m depth north of the Dry Tortugas, Florida.

Figure 2.14 Index Gastropods of the Chokoloskean and Apalachicolan Infraprovinces, Carolinian Province. **(Chokoloskean Infraprovince) A=** *Naticarius verae* (Rehder, 1947), width 28 mm, low tide, Marco Island, Florida; **B=** *Gemophos tinctus pacei* Petuch and Sargent, 2011, length 25.4 mm, low tide on a worm shell reef, Turtle Key, Ten Thousand Islands, Florida; **C=** *Melongena (Rexmela) corona* (Gmelin, 1791) form *perspinosa* (Pilsbry and Vanatta, 1934), length 60 mm, low tide on oysters, Chokoloskee Island, Ten Thousand Islands, Florida; **(Apalachicolan Infraprovince) D=** *Hesperisternia grandana* (Abbott, 1986), length 27 mm, 1 m depth in Turtle Grass, off Panacea, Florida; **E=** *Melongena (Rexmela) corona johnstonei* (Clench and Turner, 1956), length 116 mm, low tide off Steinhatchee, Florida; **F=** *Americoliva nivosa choctaw* Petuch and Myers, 2014, length 37 mm, low tide off Panama City, Florida; **G=** *Aurinia kieneri* (Clench, 1946), length 198 mm, 100 m depth off Dauphin Island, Mobile County, Alabama; **H=** *Aurinia kieneri ethelae* Pilsbry and Olsson, 1953, length 162 mm, 220 m depth off South Pass, Mississippi River Delta, Louisiana (note the characteristic large riblike longitudinal folds that are unique to this subspecies); **I=** *Scaphella junonia johnstoneae* Clench, 1953, length 119 mm, 30 m depth off Apalachicola, Florida; **J=** *Cinguloterebra lindae* (Petuch, 1987), length 68 mm, trawled from 200 m depth south of Apalachicola, Florida (note the large reddish-brown spots, which are characteristic of this species); **K=** *Caricellopsis matchetti* (Petuch and Sargent, 2011), holotype, length 35.9 mm, trawled from 200 m depth west of Naples, Collier County, Florida. Also trawled from south of Apalachicola, Florida; **L** = *Dauciconus aureonimbosus* (Petuch, 1987), holotype length 27 mm, trawled at 150 m depth, 50 km south of Apalachicola, Florida (also found off the Louisiana coast).

Figure 2.15 Index Gastropods of the Texan Subprovince, Carolinian Province.
A= *Cerithideopsis hegewischii* (Philippi, 1848), length 15 mm, on mud flats at low tide, Aransas Pass, Port Aransas, Texas; **B=** *Simnialena uniplicata marferula* Cate, 1973, length 11 mm, on gorgonians washed onto the beach, San Jose Island, Port Aransas, Texas; **C=** *Cinctura lilium* (Fischer von Waldheim, 1807), length 100 mm, 20 m depth off San Jose Island, Port Aransas, Texas; **D=** *Fulguropsis plagosus galvestonense* (Hollister, 1958), length 114 mm, 3 m depth off Anahuac, Bolivar Peninsula, Texas; **E=** *Fulguropsis texanus* (Hollister, 1958), length 90 mm, trawled from 20 m depth off Matagorda Island, Texas; **F=** *Sinistrofulgur pulleyi* (Hollister, 1958), length 154 mm, low tide, West Bay, Galveston Island, Texas; **G=** *Americoliva sayana texana* (Petuch and Sargent, 1986), length 52 mm, low tide, Aransas Pass, Port Aransas, Texas; **H=** *Dalliconus sauros* (Garcia, 2006), holotype, length 29.5 mm, trawled from 60 m depth, 58 km southeast of Port Aransas, Texas; **I, J=** *Stramonita alderi* Petuch and Berschauer, 2020, length 35 mm, Fish Pass Jetty, Mustang Island, Corpus Christi, Nueces County, Texas; **K=** *Scaphella junonia curryi* Petuch and Berschauer, 2019, length 98 mm, trawled by shrimpers from 100 m depth in the Bay of Campeche, northeast of Veracruz, Mexico (also found off Padre Island, Cameron County, Texas).

Figure 2.16 Index Gastropods of the Yucatanean Subprovince, Carolinian Province.
A= *Vokesimurex sallasi* (Rehder and Abbott, 1951), length 44 mm, 50 m depth on the Campeche Bank, south of Alacranes Reef, Yucatan, Mexico; **B**= *Cinctura (Hollisteria) branhamae* (Rehder and Abbott, 1951), length 131 mm, 60 m depth off north of Progreso, Yucatan, Mexico; **C**= *Busycoarctum coarctatum* (Sowerby I, 1825), length 138 mm, 100 m depth off Isla Contoy, Quintana Roo, Mexico; **D**= *Lindafulgur candelabrum* (Lamarck, 1816), length 177 mm, 50 m depth off Progreso, Campeche, Mexico; **E**= *Sinistrofulgur perversum* (Linnaeus, 1758), length 108 mm, 50 m depth south of Cayo Arenas, Yucatan, Mexico; **F**= *Melongena (Rexmela) bispinosa* (Philippi, 1844), length 52 mm, low tide off Isla Holbox, Quintana Roo, Mexico; **G**= *Aristofusus couei* (Petit de la Saussaye, 1853), length 110 mm, 50 m depth south of Alacranes Reef, Yucatan, Mexico; **H**= *Scaphella junonia butleri* Clench, 1953, length 114 mm, 50 m depth off Cabo Catoche, Yucatan, Mexico; **I**= *Gradiconus maya* (Petuch and Sargent, 2011), length 28 mm, Laguna Yalahau, Isla Holbox, Quintana Roo, Mexico; **J**= *Gradiconus sennottorum* (Rehder and Abbott, 1951), length 35 mm, 50 m depth south of Alacranes Reef, Yucatan, Mexico; **K**= *Turbinella wheeleri* Petuch, 1994, length 280 mm, trawled by a shrimper from 25 m depth off Isla Holbox, Quintana Roo State, Yucatan Peninsula, Mexico (originally described as a late Pleistocene fossil from the Everglades region of Florida).

Figure 2.17 Index Gastropods of the Yucatanean Subprovince, Carolinian Province.
A= *Caricellopsis contoyensis* (Emerson, 1979), length 164 mm, trawled from 200 m depth off Cabo Catoche, Quintana Roo, Mexico; **B**= *Americoliva nivosa maya* (Petuch and Sargent, 1986), length 49 mm, 50 m depth off Isla Contoy, Quintana Roo, Mexico; **C**= *Jaspidiconus ixchel* Petuch, Berschauer, and Poremski, 2017, length 15 mm, 2 m depth off Isla Mujeres, Quintana Roo, Mexico; **D**= *Jaspidiconus chaac* Petuch, Berschauer, and Poremski, 2017, length 15 mm, 2.5 m depth off Playa del Carmen, Quintana Roo, Mexico; **E**= *Phyllonotus mexicanus* (Petit de la Saussaye, 1852), length 70 mm, Laguna Yalahua, Isla Holbox, Quintana Roo, Mexico; **F**= *Engina dicksoni* Petuch, 2013, length 20 mm, 100 m depth off Cabo Catoche, Quintana Roo, Mexico; **G**= *Fulguropsis spiratum* (Lamarck, 1816), length 64 mm, 20 m depths on the Campeche Banks, off Progreso, Campeche, Mexico; **H**= *Prunum labiatum* (Kiener, 1841), length 28 mm, 20 m depth off Progreso, Campeche, Mexico; **I**= *Cariboconus kirkandersi* (Petuch, 1987), holotype, length 15 mm, in tide pools, northern tip of Cozumel Island, Quintana Roo, Mexico; **J**= *Scaphella junonia stimpsonorum* Cossignani and Allary, 2019, length 72 mm, trawled from 90 m depth north off Alacranes Reef, Campeche Bank, Yucatan, Mexico; **K**= *Aurinia macginnorum* (Garcia and Emerson, 1987), length 213 mm, 300 m depth off Isla Contoy, Quintana Roo State, Mexico; **L**= *Lindaconus therriaulti* Petuch, 2013, length 62 mm, 50 m depth south of Cayo Arenas, Yucatan, Mexico.

Figure 2.18 Widespread Caribbean Province Gastropods.
A= *Luria cinerea* (Gmelin, 1791), length 24 mm, 10 m depth off Pickles Reef, Plantation Key, Florida Keys, Florida; **B=** *Macrocypraea zebra* (Linnaeus, 1758), length 78 mm, 1 m depth off Cudjoe Key, Florida Keys, Florida; **C=** *Naria acicularis* (Gmelin, 1791), length 24 mm, 2 m depth off Boca de Cangrejos, Puerto Rico; **D=** *Aliger gigas* (Linnaeus, 1758), length 240 mm, 2 m depth off Cape Eleuthera, Eleuthera Island, Bahamas; **E=** *Americoliva reticularis* (Lamarck, 1811), length 45 mm, low tide, Nixes Harbour, South Bimini Island, Bahamas; **F=** *Chicoreus mergus* E. Vokes, 1974, length 34 mm, 50 m depth off Punta Patuca, Honduras; **G=** *Cariboliva scripta* (Lamarck, 1811), length 45 mm, 50 m depth off Punta Patuca, Honduras; **H=** *Atlanticonus granulatus* (Linnaeus, 1758), length 44 mm, 20 m depth off Coxon's Hole, Roatan Island, Honduras; **I=** *Dauciconus daucus* (Hwass, 1792), length 39 mm, 2 m depth off Carriacou Island, Grenadines, Grenada; **J=** *Gladioconus mus* (Hwass, 1792), length 29 mm, 1 m depth off Cape Eleuthera, Eleuthera Island, Bahamas; **K=** *Lindaconus spurius* (Gmelin, 1792), length 46 mm, off Cockburn Harbour, South Caicos Island, Turks and Caicos Islands; **L=** *Americoliva bifasciata* (Kuster, 1878), length 50 mm, 2 m depth off Gonave Island, Haiti; **M=** *Modulus modulus* (Linnaeus, 1758), diameter 13 mm, in Turtle Grass bed, Fajardo, Puerto Rico.

Figure 2.19 Index Gastropods of the Bermudan Subprovince, Caribbean Province.
A= *Entemnotrochus bermudensis* Okutani and Goto, 1983, height 84 mm, 400 m depth off St. David's Island, Bermuda; **B=** *Perotrochus insularis* Okutani and Goto, 1985, height 68 mm, 400 m depth off St. David's Island, Bermuda; **C=** *Timbellus lightbourni* (Harasewych and Jensen, 1979), length 39 mm, holotype, 400 m depth off St. David's Island, Bermuda; **D, E=** *Americoliva nivosa* (Marrat, 1871), length 48 mm, 2 m depth off Blue Horizons Beach, Paget Parish, Bermuda (A Bermudan endemic species; *nivosa*, for over 100 years, was considered to be a "lost" species and was thought to refer to a southern Caribbean shell. The supposedly-undescribed Bermudan shells were named *A. jenseni* by Petuch and Sargent, 1986. However, the recent discovery and illustration of the holotype of *nivosa* shows that *jenseni* and *nivosa* are actually the same shell and that *jenseni* is to be considered a synonym of *nivosa*) **F, G=** *Bermudaconus lightbourni* (Petuch, 1986), length 35 mm, holotype, 400 m depth off St. David's Island, Bermuda; **H, I=** *Jaspidiconus bermudensis* (Clench, 1942), length 43 mm, 2 m depth off Castle Island, Bermuda; **J, K=** *Lightbournus russjenseni* (Lyons and Snyder, 2008), length 24 mm, 400 m depth off St. David's Island, Bermuda.

Figure 2.20 Index Gastropods of the Bahamian Subprovince, Caribbean Province.
A= *Americoliva bahamasensis* (Petuch & Sargent, 1986), length 45 mm, 200 m depth off Grand Bahama Island, Bahamas, found in deep water throughout the eastern Bahamas; **B=** *Laevimodulus honkerorum* (Petuch, 2013), diameter 10 mm, 1 m depth off Tarpum Bay, Eleuthera Island, Bahamas; **C=** *Murexiella deynzerorum* Petuch, 2013, length 21 mm, Green Turtle Cay, Abaco Islands, Bahamas; **D=** *Murexsul zylmanae* (Petuch, 1993), length 38 mm, 10 m depth on reef off Whale Cay, Berry Islands, Bahamas; **E=** *Enaeta cylleniformis* (Sowerby I, 1844), length 21 mm, on sand flats at low tide, Nixes Harbour, South Bimini Island, Bahamas; **F=** *Dauciconus bahamensis* (Vink and Rockel, 1995), length 32 mm, 2 m depth off Cat Cay, Bimini Chain, Bahamas; **G=** *Magelliconus jacarusoi* (Petuch, 1998), length 19.7 mm, 3 m depth off Pork Fish Rocks, Paradise Island, Bahamas; **H=** *Purpuriconus ortneri* (Petuch, 1998), length 27 mm, 2 m depth in coral rubble, off Water Cay, Joulter Cays, Andros Island, Bahamas; **I=** *Purpuriconus stanfieldi* (Petuch, 1998), length 21.3 mm, 2 m depth off Beacon Cay, Exuma Cays, Bahamas; **J=** *Purpuriconus theodorei* (Petuch, 2000), length 25.2 mm, 3 m depth off Pork Fish Rocks, Paradise Island, Bahamas; **K=** *Tuckericonus flavescens* (Sowerby I, 1834), length 24 mm, 1 m depth off Cape Eleuthera, Eleuthera Island, Bahamas.

Figure 2.21 Index Gastropods of the Biminian Infraprovince, Caribbean Province.
A= *Cerithium biminiense* Pilsbry and McGinty, 1949, length 8 mm, on coral rock cliffs near Nixes Harbour, South Bimini Bahamas; **B**= *Enaeta lindae* Petuch, 2013, length 13 mm, by diver at 5 m depth off "Tuna Alley", Cat Cay, Bimini Chain, Bahamas; **C**= *Cariboconus sahlbergi* (da Motta and Harland, 1986), length 19.7 mm, 2 m depth off Turtle Rocks, Bimini Chain, Bahamas; **D**= *Jaspidiconus herndli* Petuch and Myers, 2014, length 13 mm, 3 m depth off Victory Cay, Bimini Chain, Bahamas; **E**= *Jaspidiconus oleiniki* Petuch, 2013, length 15 mm, 3 m depth off Nixes Harbour, South Bimini Island, Bahamas; **F**= *Lindaconus lindae* (Petuch, 1987), length 44 mm, dredged at 400 m depth along the Bimini Wall, off Victory Cay, Bimini Chain, Bahamas; **G**= *Magelliconus zylmanae* (Petuch, 1998), length 17.3 mm, by diver at 3 m depth off Nixes Harbour, South Bimini Island, Bahamas; **H**= *Purpuriconus donnae* (Petuch, 1998), length 26.3 mm, 2 m depth under coral rubble off Sunshine Inn South Bimini Island, Bahamas; **I**= *Purpuriconus richardbinghami* (Petuch, 1993), length 45 mm, 10 m depth off Victory Cay, Bimini Chain, Bahamas; **J**= *Strioterebrum biminiensis* Petuch, 1987, length 23 mm, low tide in Nixes Harbour, South Bimini Island, Bahamas; **K**= *Polystira starretti* Petuch, 2002, length 44 mm, 400 m depth along the Bimini Wall, off Victory Cay, Chain, Bahamas; **L**= *Murexsul sunderlandi* (Petuch, 1987), holotype length 13 mm, 10 m depth, off Cay Sal Bank, Bahamas.

Figure 2.22 Index Gastropods of the Eleutheran and Abacoan Infraprovinces, Caribbean Province.
(Eleutheran Infraprovince) A= *Chicoreus dunni* Petuch, 1987, length 33 mm, 1 m depth in Palmetto Point salt pond, Eleuthera Island, Bahamas; **B=** *Jaspidiconus exumaensis* Petuch, 2013, length 16 mm, 1 m depth off Cape Eleuthera, Eleuthera Island, Bahamas; **C=** *Jaspidiconus marcusi* Petuch, Berschauer, and Poremski, 2016, length 12 mm, 2 m depth off Tarpum Bay, Eleuthera Island, Bahamas; **D, E=** *Purpuriconus abbotti* (Clench, 1942), length 24 mm, 2 m depth off Windermere, Eleuthera Island, Bahamas; **(Abacoan Infraprovince) F=** *Murexsul honkeri* Petuch, 2013, length 21 mm, 2 m depth off Green Turtle Cay, Abaco Islands, Bahamas; **G=** *Polygona paulae* Petuch, 2013, length 31 mm, 2 m depth off Green Turtle Cay, Abaco Islands, Bahamas; **H, I=** *Gradiconus honkerorum* (Petuch and Myers, 2014), holotype, length 14 mm, 2 m depth off Green Turtle Cay, Abaco Islands, Bahamas; **J=** *Jaspidiconus branhamae* (Clench, 1953), length 17 mm, 2 m depth off Green Turtle Cay, Abaco Islands, Bahamas; **K=** *Purpuriconus jucundus* (Sowerby III, 1857), length 34 mm, 2 m depth off Green Turtle Cay, Abaco Islands, Bahamas.

Figure 2.23 Index Gastropods of the Antillean Subprovince and Belizean and Cuban Infraprovinces, Caribbean Province. (Antillean Subprovince) A= *Americoliva antillensis* (Petuch and Sargent, 1986), length 32 mm, 2 m depth off Gonave Island, Haiti; **B=** *Jaspidiconus macintoshi* Petuch, 2013, length 14 mm, 2 m depth off Point Udall, St. Croix Island, U.S. Virgin Islands; **C=** *Magelliconus sphaecelatus* (Sowerby I, 1833), length 13 mm, 10 m depth off Providenciales Island, Turks and Caicos Islands; **D=** *Purpuriconus cardinalis* (Hwass, 1792), length 24 mm, on reef off Gonave Island, Haiti; **E=** *Jaspidiconus tammymyersae* Petuch and Berschauer, 2019, length 15 mm, 8 m depth at night, off Little St. James Island, U.S. Virgin Islands; **F=** *Chicoreus cosmani* Abbott and Finlay, 1979, length 45 mm, on coral rubble, 20 m depth off Ocho Rios, Jamaica; **(Belizean Infraprovince) G=** *Dermomurex coonsorum* Petuch, 2013, length 10 mm, low tide on barrier reef, Glover's Atoll, Belize; **H=** *Jaspidiconus chinchorroensis* Petuch, Berschauer, and Poremski, 2017, length 15 mm, 2 m depth, Banco Chinchorro Atoll, Quintana Roo, Mexico; **I=** *Purpuriconus belizeanus* Petuch and Sargent, 2011, length 16 mm, 10 m depth, Glover's Atoll, Belize; **(Cuban Infraprovince) J, K=** *Macrocypraea (Lorenzicypraea) cervus lindseyi* Petuch, 2013, length 50 mm, 1 m depth off Manzanillo, Golfo de Guacanayabo, Cuba; **L=** *Jaspidiconus prugnaudorum* Petuch and Berschauer, 2018, length 24 mm, 2 m depth off Cayo Anclitas, Jardines de la Reina Archipelago, Cuba; **M =** *Americoliva mooreana* Petuch, 2013, holotype length 20 mm, 310 m depth off Turneffe Islands, Belize.

Figure 2.24 Index Gastropods of the Jamaican, Hispaniolan, Caymanian, and Puerto Rican Infraprovinces, Caribbean Province.

(Jamaican Infraprovince) A= *Cumia sunderlandi* (Petuch, 1995), length 23 mm, 5 m depth off Ocho Rios, Jamaica; **B**= *Americoliva broderipii jamaicensis* (Marrat, 1867), length 26 mm, 2 m depth off Ocho Rios, Jamaica; **(Hispaniolan Infraprovince) C**= *Americoliva broderipii* (Duros de St. Germaine, 1857), length 26 mm, washed ashore at Punta Salinas, Ocoa Bay, Dominican Republic; **D**= *Americoliva broderipii zombia* (Petuch and Sargent, 1986), length 25 mm, on sand flats, 1 m depth off Gonave Island, Haiti; **E**= *Jaspidiconus lindapowersae* Petuch and Berschauer, 2017, length 18 mm, low tide, Punta Salinas, Ocoa Bay, Dominican Republic; **(Caymanian Infraprovince) F**= *Jaspidiconus janapatriceae* Petuch and Berschauer, 2016, length 10 mm, 3 m depth off West Bay, Grand Cayman Island, Cayman Islands; **G**= *Magelliconus explorator* (Vink, 1990), length 9 mm, 10 m depth off West Bay, Grand Cayman Island, Cayman Islands (also found in Jamaica); **(Puerto Rican Infraprovince) H**= *Americoliva broderipii* unnamed subspecies, length 28 mm, 1 m depth off Luquillo Beach, Puerto Rico; **I**= *Jaspidiconus boriqua* Petuch, Berschauer, and Poremski, 2016, length 20 mm, 2 m depth off Cabo Rojo, Puerto Rico; **J**= *Jaspidiconus culebranus* Petuch, Berschauer, and Poremski, 2016, length 19 mm, 2 m depth off Punta Soldado, Culebra Island; **K**= *Jaspidiconus anaglypticus* (Crosse, 1865), length 17 mm, 2 m depth off Cabo Rojo, Puerto Rico.

Figure 2.25 Index Gastropods of the Nicaraguan Subprovince, Caribbean Province.
A= *Ficus villai* Petuch, 1998, length 67 mm, in sand at night, 10 m depth, off Trujillo, Honduras; **B=** *Chicoreus bullisi* E. Vokes, 1974, length 65 mm, 35 m depth off Puerto Cabezas, Nicaragua; **C=** *Vokesimurex garciai* (Petuch, 1987), length 71 mm, 35 m depth off Puerto Cabezas, Nicaragua; **D=** *Vokesimurex samui* (Petuch, 1987), length 78 mm, 2 m depth, off Trujillo, Honduras; **E=** *Pleioptygma helenae* (Radwin and Bibbey, 1972), length 98 mm, 35 m depth off Punta Patuca, Honduras; **F=** *Agaronia hilli* Petuch, 1987, length 28 mm, 1 m depth off Bragman's Bluff, Puerto Cabezas, Nicaragua; **G=** *Gradiconus paschalli* (Petuch, 1998), length 67 mm, 3 m depth off Puerto Cortes, Honduras; **H=** *Purpuriconus kukulcan* (Petuch, 1980), length 26 mm, 10 m depth off Coxon's Hole, Roatan Island, Honduras (red color form); **I=** *Tenorioconus harlandi* (Petuch, 1987), length 33 mm, 10 m depth off Coxon's Hole, Roatan Island, Honduras; **J=** *Hindsiclava tippetti* Petuch, 1987, length 69 mm, 35 m depth off Puerto Cabezas, Nicaragua; **K=** *Dauciconus sunderlandi* (Petuch, 1987), length 27 mm, 20 m depth off Utila Island, Honduras; **L=** *Conomitra lindae* Petuch, 1987, length 9 mm, in coral rubble, 10 m depth off Utila Island, Honduras; **M=** *Enaeta bessei* Petuch, 2013, holotype length 10 mm, 10 m depth on Rosalind Bank, Honduras.

Figure 2.26 Index Gastropods of the Nicaraguan Subprovince, Caribbean Province.
A= *Agaronia leonardhilli* Petuch, 1987, length 44 mm, trawled from 10 m depth off Bluefields, Nicaragua; **B**= *Lamellilatirus sunderlandorum* Lyons and Snyder, 2013, length 33 mm, trawled from 100 m depth off Bluefields, Nicaragua; **C**= *Lindaconus lorenzianus* (Dillwyn, 1817), length 59 mm, in Turtle Grass, 2 m depth off the Caratasca Keys, Honduras; **D**= *Enaeta reevei* Dall, 1907, length 15 mm, in coral rubble, 3 m depth off Coxon's Hole, Roatan Island, Honduras; **E**= *Jaspidiconus allamandi* Petuch, 2013, length 17 mm, in Turtle Grass, 2 m depth off Coxon's Hole, Roatan Island, Honduras; **F**= *Jaspidiconus roatanensis* Petuch and Sargent, 2011, length 14 mm, 10 m depth off Coxon's Hole, Roatan Island, Honduras; **G**= *Dibaphimitra janetae* Petuch, 1987, length 48 mm, 10 m depth, north coast of Roatan Island, Honduras (a distinct species that is often incorrectly synonymized with the Florida Keys endemic *D. florida*, but differs from its Keys congener in having a slender shell with a white aperture and protoconch; see Figure 2.11A); **H**= *Jaspidiconus masinoi* Petuch, Berschauer, and Poremski, 2016, length 13 mm, 7 m depth off Sandy Cay, Utila Cays, Honduras; **I**= *Jaspidiconus sargenti* Petuch, 2013, length 19.2 mm, off Coxon's Hole, Roatan Island, Honduras; **J**= *Cariboconus magnottei* (Petuch, 1987), length 14 mm, in algae, 1 m depth off Coxon's Hole, Roatan Island, Honduras; **K**= *Gradiconus garciai* da Motta, 1982, length 38 mm, trawled from 30 m depth off Puerto Cabezas, Nicaragua; **L**= *Atlanticonus ritae* (Petuch, 1995), holotype, length 27.5 mm, in lobster trap set in 20 m depth, Gorda Bank, Honduras; **M**= *Prunum sunderlandorum* Petuch and Berschauer, 2020, length 12.6 mm, 5 m depth on muddy sand, Trujillo Bay, Trujillo, Honduras.

Figure 2.27 The *Voluta* Species Radiation of the Nicaraguan Subprovince, Caribbean Province.
A= *Voluta demarcoi* Olsson, 1965, holotype, length 84.6 mm, 20 m depth off Punta Patuca, Honduras (a valid taxon; incorrectly called "*polypleura*" by recent workers); **B**= *Voluta ernesti* (Petuch, 1990), holotype, length 54 mm, 20 m depth off Puerto Limon, Costa Rica; **C**= *Voluta hilli* (Petuch, 1987), holotype, length 88 mm, trawled from 30 m depth, off the Miskito Cays, Nicaragua; **D**= *Voluta kotorai* (Petuch, 1981), holotype, length 49 mm, from lobster divers, 20 m depth on Gorda Bank, Honduras; **E**= *Voluta morrisoni* (Petuch, 1980), holotype, length 73.4 mm, from lobster divers, 15 m depth on Rosalind Bank, Honduras; **F**= *Voluta harasewychi* (Petuch, 1987), holotype, length 45 mm, 16 m depth off Corn Island, Nicaragua; **G**= *Voluta polypleura* Crosse, 1876, holotype (Liverpool Museum), length 56 mm, 10 m depth off Caratasca Cays, Honduras; **H**= *Voluta sunderlandi* (Petuch, 1987), pale blue color form, length 52 mm, 10 m depth off Trujillo, Honduras (also found on Utila Island); **I**= *Voluta retemirabila* (Petuch, 1981), holotype, length 75 mm, 10 m depth on Misteriosa Bank, Honduras; **J**= *Voluta garciai* (Petuch, 1981), holotype, length 71 mm, taken by lobster divers, 10 m depth on Gorda Bank, Honduras; **K**= *Voluta virescens* Lightfoot, 1786, length 75 mm, 20 m depth off Cartagena, Colombia; found in both the Nicaraguan and Venezuelan Subprovinces and shown here for comparison with the Nicaraguan endemic volutes.

Figure 2.28 Index Gastropods of the Blasian Infraprovince, Caribbean Province.
A= *Chicoreus hilli* (Petuch, 1990), length 40 mm, 35 m depth, north of El Porvenir, San Blas Archipelago, Panama (often incorrectly synonymized with the widespread Caribbean *Chicoreus mergus,* but differs in having much larger shoulder spines and a more slender body whorl); **B**= *Hemipolygona cuna* (Petuch, 1990), length 45 mm, 10 m depth off Moro Tupo, San Blas Archipelago, Panama; **C**= *Voluta lindae* Petuch, 1987, length 33 mm, 20 m depth off East Holandes Cay, San Blas Archipelago, Panama; **D**= *Prunum leonardhilli* (Petuch, 1990), length 21 mm, 20 m depth off Portobelo, Panama; **E**= *Americoliva ernesti* (Petuch, 1990), length 35 mm, 20 m depth off Portobelo, Panama; **F**= *Atlanticonus glenni* (Petuch, 1993), length 24 mm, 5 m depth off Moro Tupo, San Blas Archipelago, Panama; **G**= *Gradiconus ernesti* (Petuch, 1990), length 29 mm, 35 m depth, north of El Porvenir, San Blas Archipelago, Panama; **H**= *Jaspidiconus kellyae* Petuch, Berschauer, and Poremski, 2017, length 23 mm, 35 m depth off El Porvenir, San Blas Archipelago, Panama; **I**= *Tenorioconus panamicus* (Petuch, 1990), length 27 mm, 35 m depth off El Porvenir, San Blas Archipelago, Panama; **J**= *Knefastia hilli* Petuch, 1990, length 41 mm, 20 m depth off Portobelo, Panama; **K**= *Voluta virescens lacertina* Petuch, 1990, length 34 mm, 20 m depth off Portobelo, Panama (for comparison with the Blasian endemic *Voluta lindae*); **L**= *Torcula howardpetersi* (Petuch and Berschauer, 2020), length 40 mm, 35 m depth off El Porvenir, San Blas Archipelago, Panama; **M**= *Murexiella edwardpauli* Petuch 1990, length 15 mm, trawled at 50 m depth off Portobello, Panama.

Figure 2.29 Index Gastropods of the Venezuelan Subprovince, Caribbean Province.
A= *Muracypraea bicornis* (Sowerby II, 1870), length 58 mm, 10 m depth off Los Monges Islands, Gulf of Venezuela, Venezuela; **B=** *Muracypraea mus* (Linnaeus, 1758), length 42 mm, in Turtle Grass beds at low tide, Amuay Bay, Paraguana Peninsula, Venezuela; **C=** *Muracypraea tristensis* Petuch, 1987, length 61 mm, 100 m depth off Adicora, Paraguana Peninsula, Venezuela; **D=** *Panamurex petuchi* E. Vokes, 1992, length 28 mm, 200 m depth off the Monges Islands, Gulf of Venezuela, Venezuela; **E=** *Amalda tankervillii* (Swainson, 1825), length 72 mm, 2 m depth off Los Taques, Paraguana Peninsula, Venezuela; **F=** *Eburna glabrata* (Linnaeus, 1758), length 75 mm, 20 m depth off Punto Fijo, Paraguana Peninsula, Venezuela; **G=** *Americoliva porcea* (Marrat, 1870), length 57 mm, 1 m depth in Amuay Bay, Paraguana Peninsula, Venezuela; **H=** *Conasprelloides tristensis* (Petuch, 1987), length 45 mm, 100 m depth off Adicora, Paraguana Peninsula, Venezuela; **I=** *Gradiconus gibsonsmithorum* (Petuch, 1986), length 18 mm, 30 m depth off Punto Fijo, Paraguana Peninsula, Venezuela; **J=** *Tenorioconus granarius* (Kiener, 1845), length 45 mm, 30 m depth off Punto Fijo, Paraguana Peninsula, Venezuela; **K=** *Gradiconus parascalaris* (Petuch, 1987), length 41.8 mm, 50 m depth off Los Taques, Paraguana Peninsula, Venezuela.

Figure 2.30 Index Gastropods of the Colombian Infraprovince, Caribbean Province.
A= *Muracypraea bicornis donmoorei* (Petuch, 1979), length 51 mm, 25 m depth off Barranquilla, Colombia; **B=** *Phenacovolva lindae* Petuch, 1987, length 23 mm, 50 m depth off Riohacha, Colombia; **C=** *Ficus lindae* Petuch, 1987, length 85 mm, 10 m depth off Bahia Concha, Santa Marta, Colombia; **D=** *Sconsia lindae* Petuch, 1987, length 70 mm, 40 m depth off Tolu, Gulf of Morrosquillo, Colombia; **E=** *Vokesimurex bayeri* Petuch, 2001, length 100 mm, 40 m depth off Tolu, Gulf of Morrosquillo, Colombia; **F=** *Oniscidia lindae* Petuch, 1987, length 35 mm, 50 m depth off Riohacha, Colombia; **G=** *Amalda williamsoni* (Petuch, 1987), length 27 mm, 500 m depth in the Gulf of Morrosquillo, Colombia; **H=** *Americoliva goajira* (Petuch and Sargent, 1986), length 43 mm, 25 m depth on a sand sea floor, off Barranquilla, Colombia; **I=** *Dauciconus vikingorum* (Petuch, 1993), length 37 mm, trawled by shrimpers from 50 m depth off Barranquilla, Colombia; **J=** *Lindaconus phlogopus* (Tomlin, 1937), length 65 mm, 40 m depth off Tolu, Gulf of Morrosquillo, Colombia; **K=** *Prunum poulosi* Lipe, 1996, length 32 mm, 40 m depth off Tolu, Gulf of Morrosquillo, Colombia; **L=** *Jaspidiconus tayrona* Petuch, Berschauer, and Poremski, 2017, holotype, length 16.9 mm, 1.5 m depth off Tayrona National Park, Magdalena Province, Colombia.

Figure 2.31 Index Gastropods of the Grenadian Subprovince, Caribbean Province.
A= *Murexsul huberti* (Radwin and D'Attilio, 1976), length 20 mm, 2 m depth, Hawksbill Beach, Antigua Island, Antigua and Barbuda; **B=** *Vokesimurex bellus* (Reeve, 1845), length 51 mm, Buccaneer Bay, St. Vincent Island, Lesser Antilles; **C=** *Globivasum globulus* (Lamarck, 1816), length 26 mm, 2 m depth in coral rubble, Jolly Harbour, Antigua Island, Antigua and Barbuda; **D=** *Globivasum whicheri* Petuch, 2013, length 29 mm, 3 m depth off Prickley Pear Cays, Anguilla, Lesser Antilles; **E=** *Enaeta guildingi* (Sowerby I, 1844), length 14 mm, Buccaneer Bay, St. Vincent Island, Lesser Antilles; **F=** *Jaspidella carminiae* Petuch, 1992, length 25 mm, 2 m depth off Esparqui, Los Roques Atoll, Venezuela; **G=** *Voluta musica* Linnaeus, 1758, length 44 mm, 1 m depth off Malmok, Aruba, Dutch Caribbean Islands; **H, I=** *Mitraelyria beauii archeri* (Angas, 1865), length 40 mm, 10 m depth at night, off Pointe Noir, Guadeloupe Island; **J=** *Jaspidiconus jaspideus* (Gmelin, 1791), length 22 mm, 2 m depth in Rockly Bay, Tobago Island, Trinidad and Tobago; **K=** *Vasum capitellum* form *mitis* (Lamarck, 1822), length 51 mm, 2 m depth on rock platform, Jolly Harbour, Antigua Island, Antigua and Barbuda.

Figure 2.32 Index Cone Shells of the Grenadian Subprovince, Caribbean Province.
A= *Tenorioconus caracanus* (Hwass, 1792), length 38.6 mm, Isla Tesoro, Islas Aves de Barlovento, Islas Aves Atoll, Venezuela; **B=** *Tenorioconus cedonulli* (Linnaeus, 1767), length 43 mm, 8 m depth in Buccaneer Bay, St. Vincent Island, Lesser Antilles; **C=** *Tenorioconus dominicanus* (Hwass, 1792), length 41 mm, on a Turtle Grass bed at 10 m depth in Dubuc Bay, Dominica Island, Lesser Antilles; **D=** *Tenorioconus dominicanus grenadensis* (Hwass, 1792), length 40 mm, 20 m depth off Saline Island, Grenadian Grenadines; **E=** *Tenorioconus mappa* (Lightfoot, 1786), length 45 mm, 20 m depth in Rockley Bay, Tobago Island, Trinidad and Tobago; **F=** *Tenorioconus martinicanus* (Hwass, 1792), length 31 mm, 3 m depth, Petit Martinique Island, Vincentian Grenadines; **G=** *Tenorioconus sanguineus* (Kiener, 1850), length 39 mm, 20 m depth off Playa el Chivo, Cumana, Venezuela; **H, I=** *Tenorioconus trinitarius* (Hwass, 1792), length 31 mm, 15 m depth off Punta de Piedras, Margarita Island, Venezuela; **J=** *Poremskiconus beddomei* (Sowerby III, 1901), length 20 mm, 2 m depth off White Island, Carriacou Island, Grenadian Grenadines; **K=** *Jaspidiconus arawak* Petuch and Myers, 2014, length 15 mm, 2 m depth in clean coral sand, White Island, Carriacou Island, Grenadian Grenadines; **L=** *Jaspidiconus berschaueri* Petuch and Myers, 2014, length 14 mm, St. Maarten, Lesser Antilles.

Figure 2.33 Grenadian Subprovince Panbathyal Taxa, Caribbean Province.
A= *Perotrochus sunderlandorum* Petuch and Berschauer, 2017, diameter 58.5 mm, 250 m depth off North Point, Klein Curacao Island, Curacao, Dutch Caribbean Islands; **B**= *Cotonopsis lindae* Petuch, 1988, length 21 mm, 200 m depth off St. James, Barbados (*C. allaryi* Bozetti, 2012, is a synonym); **C**= *Siratus perelegans* (E. Vokes, 1965), length 62 mm, 200 m depth off St. James, Barbados; **D**= *Attenuiconus marileeae* Harasewych, 2014, length 24 mm, 250 m depth off Klein Curacao Island, Curacao, Dutch Caribbean Islands; **E**= *Dalliconus coletteae* Petuch, 2013, holotype, length 20 mm, 300 m depth off St. James, Barbados; **F**= *Sandericonus hunti* (Wils and Moolenbeek, 1979), length 22 mm, 200 m depth off St. James, Barbados; **G**= *Sandericonus perprotractus* (Petuch, 1987), length 56 mm, 200 m depth off St. James, Barbados; **H**= *Sandericonus sanderi* (Wils and Moolenbeek, 1979), length 21 mm, 200 m depth off St. James, Barbados; **I**= *Sandericonus sorenseni* (Sander, 1982), length 23 mm, 200 m depth off St. James, Barbados; **J**= *Sassia (Cymatiella) lewisi* Harasewych and Petuch, 1980, length 14 mm, 300 m depth off St. James, Barbados; **K**= *Americoliva olivacea* (Marrat, 1870), length 33 mm, 50 m depth off Saline Island, Grenadian Grenadines; **L**= *Lindapterys sanderi* Petuch, 1987, length 18 mm, 300 m depth off St. James, Barbados.

Figure 2.34 Index Gastropods of the Martiniquean and Roquesian Infraprovinces, Caribbean Province.
(Martiniquean Infraprovince) A, B= *Timbellus phyllopterus* (Lamarck, 1822), length 73 mm, 20 m depth off Pointe Faula, Le Vauclin, Martinique Island; **C=** *Dauciconus boui* (da Motta, 1988), length 30 mm, 10 m depth off Le Marigot, Martinique Island; **D=** *Dauciconus norai* (da Motta and Raybaudi Massilia, 1992), length 32 mm, 20 m depth off Le Lorrain, Martinique Island; **E=** *Poremskiconus colombi* (Monnier and Limpalaër, 2012), length 18 mm, 10 m depth off La Vauclin, Martinique Island; **F=** *Poremskiconus hennequini* (Petuch, 1993), length 17 mm, 3 m depth off Anse Meunier, Martinique Island; **(Roquesian Infraprovince) G=** *Murexsul duffyi* (Petuch, 1992), length 16 mm, 1 m depth off Crasqui, Los Roques Atoll, Venezuela; **H=** *Americoliva fulgurator bullata* (Marrat, 1871), length 32 mm, 3 m depth off Esparqui, Los Roques Atoll, Venezuela; **I, J=** *Tenorioconus duffyi* (Petuch, 1992), length 41 mm, 1 m depth off Esparqui, Los Roques Atoll, Venezuela; **K=** *Strombina francesae* Gibson-Smith and Gibson-Smith, 1974, length 20 mm, 1 m depth off Crasqui, Los Roques Atoll, Venezuela; **L=** *Roquesia lindae* Petuch, 2013, holotype, length 9 mm, 2 m depth off Crasqui, Los Roques Atoll, Venezuela.

Figure 2.35 Index Gastropods of the Aruban Infraprovince, Caribbean Province.
A= *Murexiella hilli* Petuch, 1987, length 37 mm, 1 m depth, off Malmok, Aruba; **B**= *Eburna balteata* (Swainson, 1825), length 38 mm, 3 m depth off Malmok, Aruba; **C**= *Americoliva fulgurator* (Röding, 1798), length 49 mm, 3 m depth off Malmok, Aruba; **D**= *Arubaconus hieroglyphus* (Duclos, 1833), length 12 mm, 2 m depth off Malmok, Aruba; **E**= *Jaspidiconus booti* Petuch, Berschauer, and Poremski, 2015, length 14 mm, 3 m depth off Malmok, Aruba; **F**= *Jaspidiconus vantwouldti* Petuch, Berschauer, and Poremski, 2015, length 10 mm, 1 m depth off Arashi, Aruba; **G**= *Perplexiconus wendrosi* Tenorio and Afonso, 2013, length 11 mm, 1 m depth off Barcadera, Aruba; **H**= *Tenorioconus aurantius* (Hwass, 1792), length 49 mm, 10 m depth off Klein Bonaire Island, Bonaire, Dutch Caribbean Islands; **I**= *Tenorioconus curassaviensis* (Hwass, 1792), length 32 mm, 3 m depth off Malmok, Aruba; **J**= *Tenorioconus monicae* Petuch and Berschauer, 2015, length 43 mm, 3 m depth off Malmok, Aruba; **K**= *Lindaconus baylei arubaensis* (Nowell-Usticke, 1968), length 45 mm, 2 m depth off Barcadera, Aruba (an endemic Aruban subspecies of the Venezuelan *L. baylei*).

Figure 2.36 Index Gastropods of the Barbadan Infraprovince and the Surinamian Subprovince, Caribbean Province. (Barbadan Infraprovince) A= *Mitra lenhilli* Petuch, 1988, length 19 mm, trawled from 200 m depth off St. James, Barbados; **B=** *Americoliva barbadensis* (Petuch and Sargent, 1986), length 43 mm, trawled from 200 m depth off St. James, Barbados; **C=** *Clathrodrillia petuchi* (Tippett, 1995), length 53 mm, trawled from 200 m depth off St. James, Barbados; **(Surinamian Subprovince) D=** *Poirieria oregonia* (Bullis, 1964), length 45 mm, trawled from 150 m depth off Georgetown, Guyana; **E=** *Siratus springeri* (Bullis, 1964), length 43 mm, trawled from 150 m depth off Georgetown, Guyana; **F, G=** *Phyllonotus guyanensis* Garrigues and Lamy, 2016, length 82 mm, trawled by shrimpers from 35 m depth off Cayenne, French Guiana; **H, I=** *Siratus thompsoni* (Bullis, 1964), length 29 mm, trawled from 100 m depth off Paramaribo, Suriname; **J=** *Conasprelloides brunneobandatus* (Petuch, 1992), length 28 mm, trawled by shrimpers from 50 m depth off Georgetown, Guyana; **K=** *Fasciolaria guyanensis* Lyons and Snyder, 2016, length 112 mm, 35 m depth off Cayenne, French Guiana; **L=** *Kellyconus rachelae* (Petuch, 1988), holotype length 24 mm, trawled by shrimpers from 100 m depth off Boca Araguao, Orinoco River Delta, Venezuela (type locality corrected in Petuch, 2013: 140).

Figure 2.37 Widespread Brazilian Province Gastropods. **A**= *Titanostrombus goliath* (Schroter, 1805), length 310 mm, on *Padina* algae bed, 2 m depth off Iheus, Bahia State, Brazil; **B**= *Luria cinerea brasiliana* Lorenz, 2002, length 23 mm, 3 m depth off Porto Seguro, Bahia State, Brazil; **C**= *Naria acicularis marcuscoltroi* Petuch and Myers, 2015, length mm, 5 m depth off Santos, São Paulo State, Brazil; **D**= *Macrocypraea dissimilis* Schilder, 1924, length 86 mm, 3 m depth off Porto Seguro, Bahia State, Brazil; **E**= *Cyphoma macumba* Petuch, 1979, length 24 mm, 2 m depth on the Timbebas Reefs, Abrolhos Platform, Bahia State, Brazil; **F**= *Siratus tenuivaricosus* (Dautzenberg, 1927), length 61 mm, 50 m depth off Vitoria, Espirito Santo State, Brazil; **G**= *Aurantilaria aurantiaca* (Lamarck, 1816), length 86 mm, low tide, Parcel das Paredes Reefs, Abrolhos Platform, Bahia State, Brazil; **H**= *Turbinella laevigata* Anton, 1839, length 92 mm, low tide, Parcel das Paredes Reefs, Abrolhos Platform, Bahia State, Brazil; **I**= *Vasum cassiforme* (Kiener, 1841), length 89 mm, 20 m depth off Natal, Rio Grande do Norte State, Brazil; **J**= *Morum bayeri* Petuch, 2001, length 23 mm, 3 m depth off Rio do Fogo, Rio Grande do Norte State, Brazil; **K**= *Dauciconus worki* (Petuch, 1998), length 41 mm, 20 m depth off Itaparica Island, Bahia State, Brazil; **L**= *Haliotis aurantia* Simone, 1998, length 14 mm, trawled from 200 m depth off Vitoria, Espirito Santo State, Brazil; **M**= *Murexiella leonardhilli* Petuch, 1987, length 31 mm, 10 m depth in Todos os Santos Bay, Bahia State, Brazil.

Figure 2.38 Index Gastropods of the Cearaian Subprovince, Brazilian Province.
A= *Bayericerithium bayeri* Petuch, 2001, length mm, 1 m depth, off Touros, Rio Grande do Norte State, Brazil; **B**= *Voluta ebraea* Linnaeus, 1758, length 124 mm, 3 m depth off Icapui, Ceará State, Brazil; **C**= *Bullata lilacina* (Sowerby II, 1846), length 27 mm, on coralline algal rubble, 20 m depth off Rio do Fogo, Rio Grande do Norte State, Brazil; **D**= *Oniscidia matthewsi* (Emerson, 1967), length 25 mm, 50 m depth off Fortaleza, Ceará State, Brazil; **E**= *Jaspidiconus damasoi* (Cossignani, 2007), length 11 mm, in coralline algae, 20 m depth off Camocim, Ceará State, Brazil; **F**= *Americoliva circinata jorioi* Petuch, 2013, length 23 mm, 3 m depth off Rio do Fogo, Rio Grande do Norte State, Brazil; **G**= *Conasprelloides hazinorum* Petuch and Myers, 2014, length 60 mm, 50 m depth off Aracaju, Sergipe State, Brazil; **H**= *Jaspidiconus damasomonteiroi* Petuch and Myers, 2014, length 18 mm, 20 m depth off Camocim, Ceará State, Brazil; **I**= *Poremskiconus mariaodeteae* Petuch and Myers, 2014, length 22 mm, 20 m depth off Camocim, Ceará State, Brazil; **J**= *Poremskiconus smoesi* Petuch and Berschauer, 2016, length 20 mm, 50 m depth off Camocim, Ceará State, Brazil; **K**= *Poremskiconus tourosensis* Petuch and Berschauer, 2018, length 21 mm, 5 m depth off Touros, Rio Grande do Norte State, Brazil; **L**= *Jaspidiconus serafimi* Petuch and Berschauer, 2019, length 16 mm, 12 m depth off Rio do Fogo, Rio Grande do Norte State, Brazil; **M**= *Morum mariaodeteae* Petuch and Berschauer, 2020, length 18 mm, 35 m depth on rocky sea floor, off Camocim, Ceará State, Brazil.

Figure 2.39 Index Gastropods of the Cearaian Subprovince and the Noronhan and Trindadean Infraprovinces, Brazilian Province.

(Cearaian Subprovince) A= *Jaspidiconus toincabrali* Petuch and Berschauer, 2019, length 19 mm, 35 m depth off Tibau Norte, Rio Grande do Norte State, Brazil; **B=** *Jaspidiconus icapui* Petuch and Berschauer, 2018, length 15 mm, 35 m depth off Icapui, Ceará State, Brazil; **C=** *Poremskiconus fonsecai* Petuch and Berschauer, 2016, length 15 mm, in coralline algae, 10 m depth off Rio do Fogo, Rio Grande do Norte State, Brazil; **D=** *Poremskiconus mauricioi* (Coltro, 2004), length 21 mm, 5 m depth off Camocim, Ceará State, Brazil; **E=** *Jaspidiconus joanae* Petuch and Berschauer, 2018, length 16 mm, 5 m depth off Rio do Fogo, Rio Grande do Norte State, Brazil; **(Noronhan Infraprovince) F=** *Nerita (Ritena) chlorostoma deturpensis* Vermeij, 1970, length 34 mm, low tide, Baia do Sancho, Fernando de Noronha Island; **G=** *Echinolittorina vermeiji* (Bandel and Kadolsky, 1982), height 9 mm, on rocks along the high tide line, Baia dos Porcos, Fernando de Noronha Island; **H=** *Malea noronhensis* Kempf and Matthews, 1969, length 65 mm, 5 m depth, in Baia do Sueste, Fernando de Noronha Island; **I=** *Leucozonia ponderosa* Vermeij and Snyder, 1998, length 38 mm, 1 m depth, in Baia do Sueste, Fernando de Noronha Island; **J=** *Polygona vermeiji* Petuch, 1988, length 29 mm, 1 m depth, in Baia do Sancho, Fernando de Noronha Island; **K=** *Enaeta leonardhilli* Petuch, 1988, length 12 mm, low tide, Praia de Atalaia, Fernando de Noronha Island; **(Trindadean Infraprovince, Bahian Subprovince) L=** *Macrocypraea mammoth* Simone and Cavallari, 2019, holotype, length 111 mm, Trindade Island, off Espirito Santo State, Brazil.

Figure 2.40 Index Gastropods of the Bahian Subprovince, Brazilian Province.
A= *Murexiella iemanja* Petuch, 1979, length 7 mm, 10 m depth, off the Timbebas Reefs, Abrolhos Platform, Bahia State, Brazil (in some of the recent literature this Brazilian endemic has been incorrectly referred to as *M. glypta*, a related species that is found only in the Carolinian Province); **B**= *Pustulatirus ogum* (Petuch, 1979), length 38 mm, Parcel das Paredes Reefs, Abrolhos Platform, Bahia State, Brazil; **C**= *Plicoliva zelindae* (Petuch, 1979), length 29 mm, 1 m depth on Pedra Lixa Reefs, Abrolhos Platform, Bahia State, Brazil; **D**= *Morum berschaueri* Petuch and Myers, 2015, length 36 mm, 20 m depth off Guarapari, Espirito Santo State, Brazil; **E**= *Bullata bullata* (Born, 1778), length 62 mm, 20 m depth off Ilheus, Bahia State, Brazil; **F**= *Americoliva circinata* (Marrat, 1870), length 58 mm, low tide, Coroa Vermelha Reefs, Abrolhos Platform, Bahia State, Brazil; **G**= *Conasprelloides coltrorum* Petuch and Myers, 2014, length 73 mm, 50 m depth off Guarapari, Espirito Santo State, Brazil; **H**= *Dauciconus riosi* (Petuch, 1986), length 49 mm, 50 m depth off Guarapari, Espirito Santo State, Brazil; **I**= *Pygmaepterys oxossi* (Petuch, 1979), length 12 mm, 5 m depth in coral rubble, off Alcobaca, Bahia State, Brazil; **J**= *Jaspidiconus ericmonnieri* Petuch and Myers, 2014, length 47 mm, 50 m depth off Guarapari, Espirito Santo State, Brazil; **K**= *Polygona bayeri* (Petuch, 2001), length 35 mm, 5 m depth off Itaparica Island, Bahia State, Brazil; **L**= *Cancellaria petuchi* Harasewych, Petit, and Verhecken, 1992, length 38 mm, 10 m depth in the Guarapari Channel, Guatrapari, Espirito Santo State, Brazil.

Figure 2.41 Index Cone Shells of the Bahian Subprovince and Itaparican Infraprovince, Brazilian Province. **A=** *Jaspidiconus crabosi* Petuch and Berschauer, 2018, length 16 mm, low tide, Morere, Boipeba Island, Bahia State, Brazil; **B=** *Jaspidiconus itapua* Petuch and Berschauer, 2018, length 12 mm, 1 m depth off Farol de Itapua, Itapua, Bahia State, Brazil; **C=** *Jaspidiconus josei* Petuch and Berschauer, 2016, length 20 mm, 2 m depth off Praia de Guarajuba, Bahia State, Brazil; **D=** *Jaspidiconus keppensi* Petuch and Berschauer, 2018, length 10 mm, 3 m depth off Alcobaca, Abrolhos Platform, Bahia State, Brazil; **E=** *Jaspidiconus ramosorum* Petuch and Berschauer, 2019, length 22 mm, 20 m depth of sand and coral rubble, Vicosa Reefs off Nova Vicosa, Bahia State, Brazil; **F=** *Jaspidiconus poremskii* Petuch and Myers, 2014, length 12 mm, 1 m depth off Corumbau, Bahia State, Brazil; **G=** *Poremskiconus brasiliensis* (Clench, 1942), length 20 mm, 3 m depth off Guarapari, Espirito Santo State, Brazil; **(Itaparican Infraprovince) H=** *Jaspidiconus pomponeti* Petuch and Myers, 2014, length 10 mm, 1 m depth in Baia Itapajipe, off Ribeira, Todos os Santos Bay, Bahia State, Brazil; **I=** *Jaspidiconus henckesi* (Coltro, 2004), length 15 mm, low tide, off Cacha Pregos, Itaparica Island, Bahia State, Brazil; **J=** *Jaspidiconus marinae* Petuch and Myers, 2014, length 12 mm, 1 m depth in sand, off Mar Grande, Itaparica Island, Bahia State, Brazil; **K=** *Jaspidiconus ogum* Petuch and Myers, 2014, length 15 mm, low tide off Vera Cruz, Itaparica Island, Bahia State, Brazil.

Figure 2.42 Index Gastropods of the Abrolhosian Infraprovince, Brazilian Province.
A, B= *Coltroconus bianchii* Petuch and Berschauer, 2018, length 8 mm, 150 m depth on the Minerva Seamount, southern Bahia State, Brazil; **C=** *Coltroconus bodarti* (Coltro, 2004), length 13 mm, 20 m depth off Ilha Santa Barbara, Abrolhos Archipelago; **D=** *Coltroconus delucai* (Coltro, 2004), length 11 mm, 20 m depth off the Parcel das Paredes Reefs, Abrolhos Platform, Bahia State, Brazil; **E=** *Coltroconus henriquei* Petuch and Myers, 2014), length 10 mm, 35 m depth on the Royal Charlotte Bank, Abrolhos Platform, Bahia State, Brazil; **F=** *Coltroconus iansa* (Petuch, 1979), length 14 mm, 5 m depth, off Pedra Lixa Reefs, Abrolhos Platform, Bahia State, Brazil; **G=** *Coltroconus schirrmeisteri* (Coltro, 2004), length 8 mm, 50 m depth on Sulfur Bank, Abrolhos Platform, Bahia State, Brazil; **H=** *Poremskiconus bertarolleae* (Costa and Simone, 1997), length 23 mm, 3 m depth off Ponta Cumuruxuatiba, Bahia State, Brazil; **I=** *Poremskiconus abrolhosensis* (Petuch, 1978), length 22 mm, 20 m depth, off the Parcel das Paredes, Abrolhos Platform, Bahia State, Brazil (the holotype of *abrolhosensis* is a juvenile specimen of what was later named *baiano* by Coltro, 2004); **J=** *Poremskiconus cargilei* (Coltro, 2004), length 28 mm, 10 m depth off Coroa Vermelha Reefs, Abrolhos Platform, Bahia State, Brazil; **K=** *Poremskiconus uhlei* Petuch. Coltro and Berschauer, 2020, length 19 mm, 10 m depth off Guaratibas Reefs, Abrolhos Platform, Bahia State, Brazil; **L=** *Morum damasoi* Petuch and Berschauer, 2020, length 18 mm, 25 m depth on coral rubble, off Parcel das Paredes Reefs, Abrolhos Platform, Bahia State, Brazil.

Figure 2.43 Widespread Paulinian Province Gastropods.
A= *Xenophalium iheringi* (Carcelles, 1953), length 53 mm, 50 m depth off Tramandai, Rio Grande do Sul State, Brazil; **B=** *Charonia marylenae* Petuch and Berschauer, 2020, length 114 mm, 50 m depth off Cabo Frio, Rio de Janeiro State, Brazil; **C=** *Coronium elegans* Simone, 1996, length 55 mm, 50 m depth off Itajai, Santa Catarina State, Brazil; **D=** *Apertifusus frenguellii* (Carcelles, 1953), length 96 mm, 100 m depth off Guaratuba, Parana State, Brazil; **E=** *Buccinanops cochlidium* (Dillwyn, 1817), length 52 mm, in sand, low tide, Tramandai, Rio Grande do Sul State, Brazil; **F=** *Odontocymbiola americana* (Reeve, 1856), length 35 mm, 50 m depth off Cabo Frio, Rio de Janeiro State, Brazil; **G=** *Zidona dufresnei* (Donovan, 1823), length 158 mm, 50 m depth off Guaruja, São Paulo State, Brazil; **H=** *Olivancillaria urceus* (Röding, 1798). length 58 mm, low tide, Itajai, São Paulo State, Brazil; **I=** *Lamniconus clerii* (Reeve, 1844), length 51 mm, 50 m depth off Arraial do Cabo, Rio de Janeiro State, Brazil; **J=** *Lamniconus lemniscatus* (Reeve, 1849), length 52 mm, 50 m depth off Arraial do Cabo, Rio de Janeiro State, Brazil; **K=** *Pachycymbiola brasiliana* (Lamarck, 1811), length 96 mm, 2 m depth off Tramandai, Rio Grande do Sul State, Brazil.

Figure 2.44 Index Gastropods of the Janeiran Subprovince, Paulinian Province.
A= *Odontocymbiola cleryana* (Petit de la Saussaye, 1856), length 47 mm, 50 m depth off Arraial do Cabo, Rio de Janeiro State, Brazil; **B**= *Lamniconus tostesi* (Petuch, 1986), length 34 mm, trawled by fishermen from 100 m depth off Cabo Frio, Rio de Janeiro State, Brazil; **C**= *Lamniconus xanthocinctus* (Petuch, 1980), length 42 mm, 100 m depth off Ilha Grande, Rio de Janeiro State, Brazil; **D**= *Sandericonus carioca* (Petuch, 1986), length 38 mm, 50 m depth off Arraial do Cabo, Rio de Janeiro State, Brazil; **E**= *Lamniconus clenchi* (Martins, 1943), length 43 mm, 50 m depth off Farol de São Thome, Rio de Janeiro State, Brazil; **F**= *Lamniconus patriceae* Petuch and Myers, 2014, length 45 mm, 150 m depth off Ubatuba, São Paulo State, Brazil; **(Macaean Infraprovince) G**= *Odontocymbiola macaensis* Calvo and Coltro, 1997, length 46 mm, 50 m depth off Macae, Rio de Janeiro State, Brazil; **H**= *Odontocymbiola saotomensis* Calvo and Coltro, 1997, length 51 mm, 100 m depth off Farol de São Thome, Rio de Janeiro State, Brazil; **I**= *Jaspidiconus simonei* Petuch and Myers, 2014, length 20 mm, low tide, Arraial do Cabo, Rio de Janeiro State, Brazil; **J**= *Lamniconus petestimpsoni* Petuch and Berschauer, 2016, length 52 mm, 100 m depth off Farol de São Thome, Rio de Janeiro State, Brazil; **K**= *Goniofusus damasoi* Petuch and Berschauer, 2016, length 81 mm, 50 m depth off Arraial do Cabo, Rio de Janeiro State, Brazil.

Figure 2.45 Index Gastropods of the Janeiran Subprovince, Paulinian Province.
A= *Cyphoma versicolor* Fehse, 2003, length 19 mm, 3 m depth off Ilhabela, São Paulo State, Brazil; **B**= *Ranella barcellosi* Matthews, Rios, and Coelho, 1973, length 136 mm, 100 m depth, off Itajai, Santa Catarina State, Brazil; **C**= *Strombus worki* Petuch, 1993, length 81 mm, 2 m depth off Ilhabela, São Paulo State, Brazil; **D**= *Odontocymbiola simulatrix* Leal and Bouchet, 1989, length 103 mm, 250 m depth off Itajai, Santa Catarina State, Brazil; **E**= *Apertifusus mariaodeteae* Petuch and Berschauer, 2016, length 51 mm, 200 m depth off Itajai, Santa Catarina State, Brazil; **F**= *Leucozonia granulilabris* (Vermeij and Snyder, 2003), length 57 mm, 2 m depth off Ilhabela, São Paulo State, Brazil; **G**= *Americoliva circinata tostesi* (Petuch, 1987), length 51 mm, 1 m depth off Ilhabela, São Paulo State, Brazil; **H**= *Olivancillaria auricularia* (Lamarck, 1810), length 40 mm, low tide on sand, Imbituba, Santa Catarina State, Brazil; **I**= *Odontocymbiola simulatrix nana* Allary and Cossignani, 2015, length 76 mm, trawled from 200 m depth off Tramandai, Rio Grande do Sul State, Brazil; **J**= *Dalliconus edpetuchi* Monnier, Limpalaër, Roux, and Berschauer, 2015, length 63 mm, 250 m depth off Itajai, Santa Catarina State, Brazil; **K**= *Conasprelloides capricorni* (van Mol, Tursch, and Kempf, 1967), length 45 mm, 250 m depth off Itajai, Santa Catarina State, Brazil; **L**= *Trophon pelseneeri* (E.A. Smith, 1915), length 24 mm, trawled from 200 m depth off Tramandai, Rio Grande do Sul State, Brazil; **M**= *Cyphoma guerrinii* Fehse, 2001, length 19 mm, on gorgonians, 10 m depth off Caraguatatuba, São Paulo State, Brazil.

Figure 2.46 Index Gastropods of the Uruguayan Subprovince and Platensian Provinciatone, Paulinian Province. **(Uruguayan Subprovince) A=** *Coronium acanthodes* (Watson, 1882), length 72 mm, 100 m depth off Punta del Este, Uruguay; **B=** *Buccinanops duartei* Klappenbach, 1961, length 26 mm, low tide on sand bars, La Paloma, Uruguay; **C=** *Adelomelon martensi* (Strebel, 1906), length 161 mm, 50 m depth off Isla de Lobos, Maldonado, Uruguay; **D=** *Lamniconus carcellesi* (Martins, 1945), length 47 mm, 25 m depth off La Paloma, Uruguay; **E=** *Olivancillaria contortuplicata* (Reeve, 1850), length 32 mm, in the beach surf zone, Punta del Este, Uruguay; **F=** *Olivancillaria teaguei* Klappenbach, 1961, length 18 mm, in the beach surf zone, Punta del Este, Uruguay; **G=** *Hanetia rushii* (Pilsbry, 1897), length 29 mm, under rocks at low tide, La Paloma, Uruguay; **(Platensian Provinciatone) H=** *Adelomelon ancilla* (Lightfoot, 1786), length 160 mm, trawled from 30 m depth off La Paloma, Uruguay; **I=** *Odontocymbiola subnodosa* (Leach, 1814), length 80 mm, 80 m depth off Mar del Plata, Buenas Aires Province, Argentina; **J=** *Provocator corderoi* (Carcelles, 1947), length 42 mm, 100 m depth south of Mar del Plata, Argentina; **K=** *Volvarina warreni* (Marrat, 1876), length 20 mm, 100 m depth off Mar del Plata, Argentina; **L=** *Duplicaria gemmulata* (Kiener, 1839), length 46 mm, low tide, Arenas Verdes, Necochea, Argentina.

CHAPTER 3. Eastern Atlantic Region

The Eastern Atlantic Region spans the western coasts of Europe (Eurasia) and Africa, from Portugal south to southern Namibia. This large region encompasses three separate subregions that are relevant to this book: the **Northeastern Atlantic Paratropical Subregion**, the **Eastern Atlantic Tropical Subregion**, and the **Southeastern Atlantic Paratropical Subregion**. These three subregions encompass two warm temperate provinces, two tropical provinces, eight subprovinces, and one provinciatone. The provinces include (from north to south); the paratropical **Mediterranean Province**, the tropical **Verdesian Province**, the tropical **Guinean Province**, and the paratropical **Namibian Province**. The spatial arrangement of these provincial units is shown here on Figure 3.1. This core of warm temperate and tropical areas is bounded in the north by a provinciatonal zone, transitional with the subarctic Norwegian Province, and another transitional zone in the south with the South African Region (the Namaquan Provinciatone). A number of wide-ranging gastropods demarcate the boundaries of the Eastern Atlantic Region, some of which include the ovulids *Neosimnia spelta* and *Pseudosimnia carnea*, the charoniid *Charonia lampas*, the ranellid *Monoplex corrugatum*, the muricid *Pagodula fraseri*, the nassariid *Himia wolffi*, and the turrid *Fusiturris similis* (shown here, with other index species, on Figure 3.6). These gastropods occur in the Mediterranean, Verdesian, and Guinean Provinces and a few, such as *Himia wolffi*, range all the way to the Namibian Province.

Northeastern Atlantic Paratropical Subregion

Extending from Cabo Espichel, Portugal to Cape Bojador (Cap Boujdour), Western Sahara, and encompassing the entire Mediterranean Sea and also the northern Macaronesian Islands (the Canary, Madeira, Selvagens, and Azores Islands), the Northeastern Atlantic Paratropical Subregion contains only a single major biogeographical unit, the Mediterranean Province. The oceanographic history of the Mediterranean Province is extremely complicated and still is poorly-known. The most dramatic period during the Cenozoic history of the Mediterranean Province occurred in the late Miocene (Messianian Age), when sea levels dropped so low that all of the Mediterranean Sea actually dried up, becoming a series of salt flats and hypersaline lakes. At this time, the shallow Gibraltar Sill was emergent and formed a wall-like isthmus that separated the Mediterranean from the Atlantic. Later, during the higher sea level stands of the early Pliocene, the Gibraltar Sill was overtopped and the Mediterranean Sea re-filled during the Zanclean Flood event, again reconnecting to the Atlantic Ocean. At this time, warm water conditions had also returned to the newly-flooded sea and this new subtropical marine climate allowed migrants from tropical West Africa to move northward and invade the Mediterranean area. Later, during the early Pleistocene Epoch with its series of glacial stages, cold sea temperatures and drops in sea level again returned, causing the extinction of most of the Pliocene tropical West African migrants. The cold-tolerant descendants of some of these Pliocene West African genera are still extant in the Mediterranean Province and produce the paratropical characteristic of the present molluscan fauna.

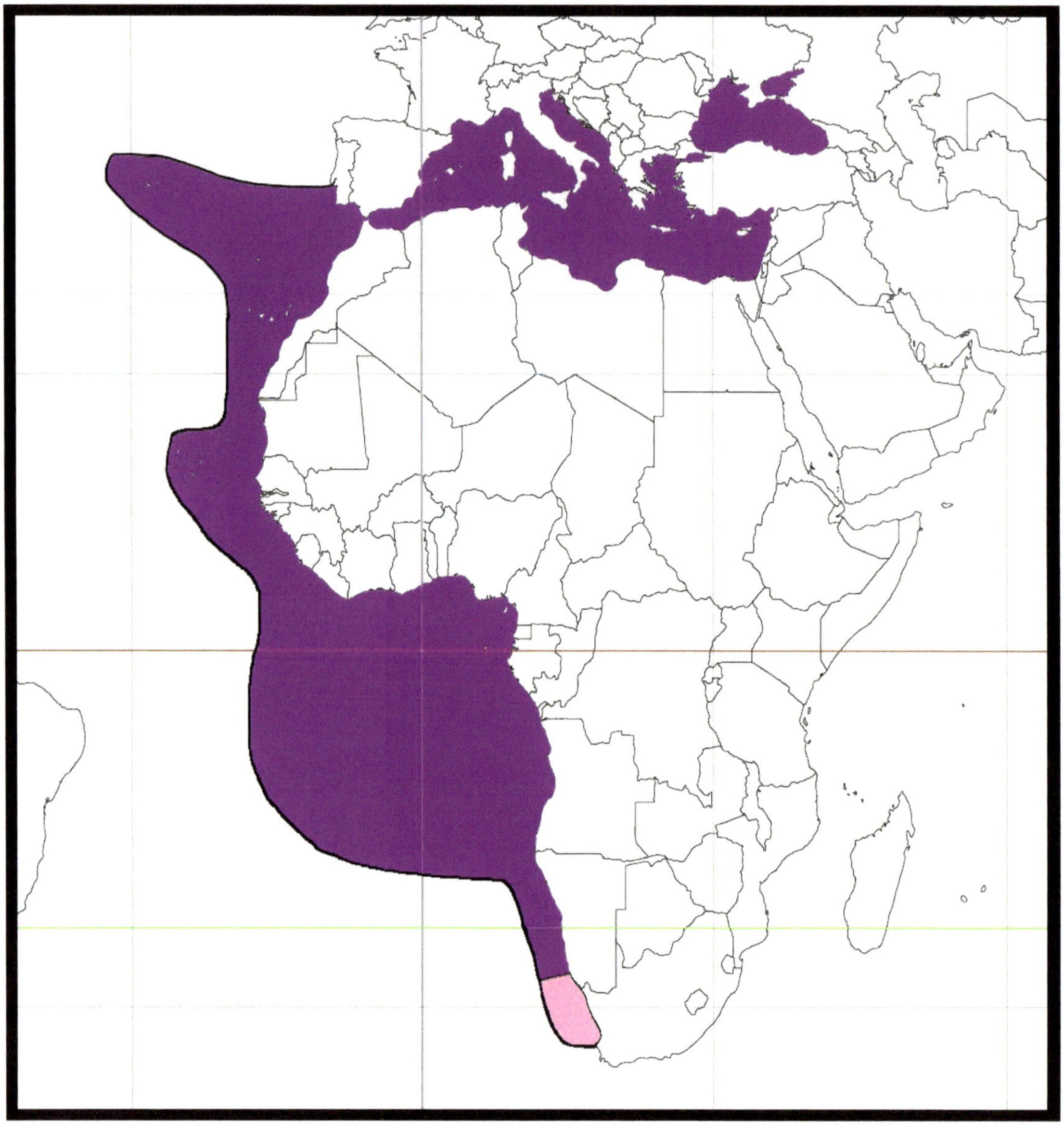

Figure 3.1 Map of the **Eastern Atlantic Region**, showing the areal extents of the combined tropical and paratropical subregions (purple). The faunal transition zone between the Namibian and South African Provinces, the **Namaquan Provinciatone**, is shown in the far south (light rose).

During the mid-Pleistocene, sea levels again dropped precipitously due to glacial ice build-up, and this caused the Mediterranean basins to become separated from the Atlantic Ocean due to the emergence of the narrow Gibraltar Sill. Although essentially having become an immense salt water lake, the Pleistocene Mediterranean Sea never formed hypersaline water conditions due to the large amount of fresh water input coming from rivers such as the Nile, Po, Ebro, and Rhone. Even with normal oceanic salinities having been preserved during the Pleistocene, the Mediterranean shrank to almost half its current size and was actually divided into two separate salt water lakes; an eastern inland sea composed of the interconnected Levantine and Ionian Basins, and a western inland sea composed of the interconnected Alboran, Algerian, Ligurian, and Tyrrhenian Basins. The shallow banks in the Strait of Sicily would have been exposed during this time, forming a connecting land bridge between Tunisia, Sicily, and southern

Italy. The two isolated inland seas evolved their own separate molluscan faunas and the remnants of this faunal separation are still seen in the modern Mediterranean molluscan assemblages (as two distinct subprovinces). During maximum sea level lows (100-200 m below present sea level), most of the Adriatic Sea was completely dry and supported forests and grasslands. Only the southernmost end retained a basin that was deep enough to create a refuge for marine faunas. The Aegean Sea area was also partially emergent, with only a few small shallow marine refuges along the northern and central areas.

Based on their shared fauna with the Mediterranean basins, the northern Macaronesian Archipelagoes (the Canary, Selvagens, Madeira, and Azores Islands) are here considered to be part of the Mediterranean Province and are given subprovincial status. Being outside of the enclosed Mediterranean Sea, these Macaronesian islands were exposed to different environmental conditions during the Pleistocene and evolved different molluscan faunas, often with high levels of endemism. During the Late Pleistocene, the sea temperatures surrounding these offshore islands fell below warm temperate limits, causing the regional extinction (extirpation) of prominent tropical molluscan residents such as *Thetystrombus latus* (Lecointre, Tinkler, and Richards, 1967). This classic Guinean Province index species (discussed later in this chapter) was once abundant on the Canary Islands but is now absent from the local ecosystems. The southernmost part of Macaronesia, the Cape Verde Islands, constitutes a molluscan faunal province all to itself and will be discussed separately later in this chapter.

Mediterranean Molluscan Province

Named for the Mediterranean Sea, the Mediterranean Molluscan Province contains one of the most distinctive and unusual molluscan faunas found any where on Earth. The province spans all eight of the Mediterranean Sea basins (the Aegean, Levantine, Ionian, Adriatic, Tyrrhenian, Ligurian, Algerian, and Alboran) and also the coast of Portugal, the coast of Morocco to Cape Agadir, and the Canary, Madeira, Selvagens, and Azores Islands. The Azores Islands, far out in the North Atlantic, house only a very few Mediterranean species and are here considered to be only a highly impoverished outlier of the main Mediterranean fauna. As a result of this geographical isolation during the Pleistocene, coupled with extreme environmental changes, the Mediterranean Province now contains four subprovinces and seven infraprovinces. Two of these subprovinces, the **Ionian** (with its Aegean, Levantine, Adriatic, and Libyan Infraprovinces) and the **Algerian** (with its Sicilian and Alboranian Infraprovinces) are confined to the Mediterranean Sea area and are influenced by the continental environments of southern Europe and northern Africa. The other two subprovinces, the **Canarian** and **Madeiran**, are confined to isolated Macaronesian archipelagoes in the open Atlantic Ocean. The Canarian fauna, in particular, contains a large number of West African taxa and acts as a transition zone between the Mediterranean and Guinean Provinces. Numerous Red Sea taxa have also invaded the Mediterranean Sea through the Suez Canal and have become established as part of the Ionian Subprovince ecosystems. Referred to as “Lessepsian

Migrants" (named for Ferdinand de Lesseps, who oversaw the building of the Suez Canal), some of these invaders include the cowries *Naria turdus*, *Palmadusta lentiginosa dancalica*, and *Purpuradusta notata*, the muricid *Murex forskali*, the abalone *Haliotis pustulata*, and the strombid *Conomurex persicus*. Most have been given varietal names (forms) by several European workers. Widespread Mediterranean gastropod index taxa are shown here on Figure 3.7.

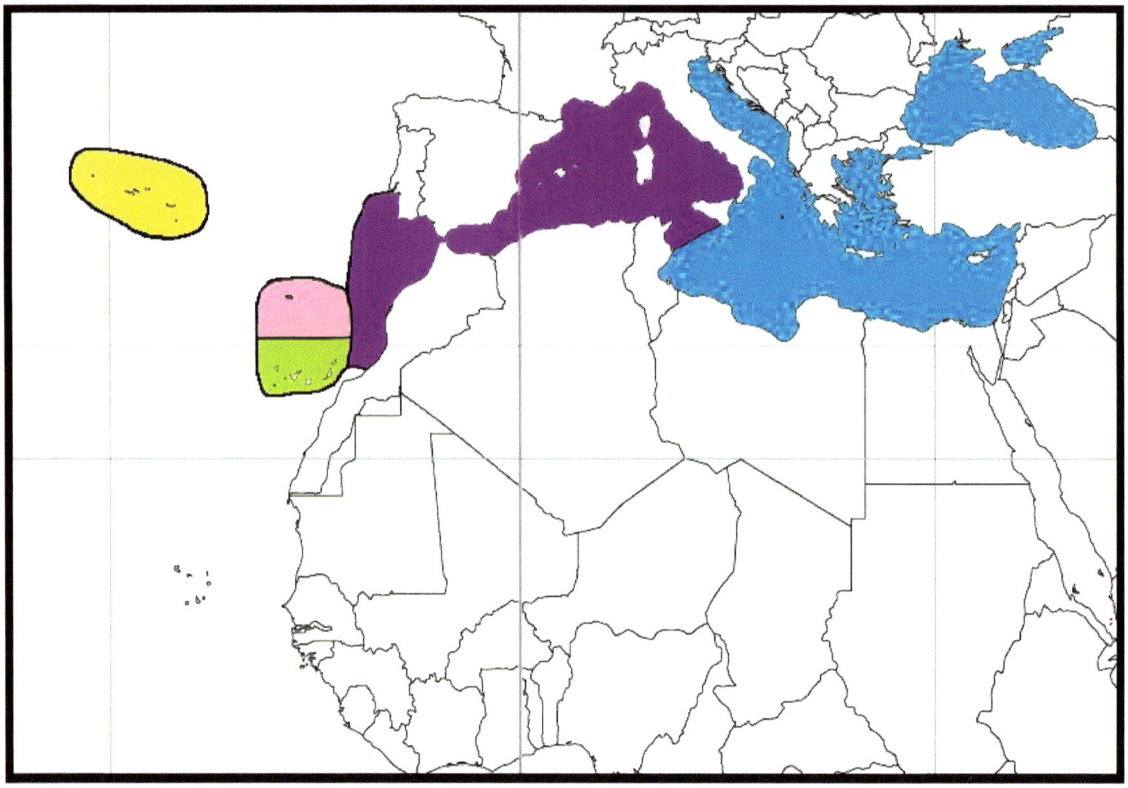

Figure 3.2 Map of the Mediterranean Molluscan Province, showing the areal extents of its subprovinces: the **Ionian Subprovince** (blue), the **Algerian Subprovince** (purple), the **Canarian Subprovince** (lime green), the **Madeiran Subprovince** (pink), and the Azores Islands (yellow).

Of primary interest within the Mediterranean molluscan fauna is a large species radiation of the conid genus *Lautoconus*. More than 90 species-level taxa have been assigned to this complex by previous workers, most of which have been considered, traditionally, to be synonyms of *L. ventricosus* (Gmelin, 1791). This single-taxon viewpoint, however, has been overturned by more recent detailed studies, some of which incorporate both biochemical and anatomical analyses. This research by modern workers has indicated that at least 20 of the synonymized taxa may actually represent valid species that belong to a close-knit complex of siblings. Each basin within the Mediterranean Sea seems to have evolved its own cluster of *Lautoconus* species, primarily due to geographical isolation caused by the sea level low stands during the late Pleistocene. We show 13 members of this species radiation here on Figures 3.7-3.10, most of which we are listing as "subspecies" of *L. ventricosus*. Future research may demonstrate that most, if not all, of these taxa are valid full species.

Ionian Subprovince

The Ionian Subprovince, named for the Ionian Sea of the eastern Mediterranean, encompasses four separate oceanographic basins (classically referred to as "Seas"): the Aegean, the Levantine, the Adriatic, and the Ionian (see Figure 3.2). As a result of lowered sea levels during the late Pleistocene, the Ionian area was much reduced in size, with the Adriatic Sea and much of the Aegean Sea being dry land. During this time, the Levantine Basin and the main Ionian Basin remained the deepest areas of the eastern Mediterranean and acted as refugia for many early Pleistocene and Pliocene gastropods. Some of these, such as the muricid *Trunculariopsis pecchiolianus*, were originally described as Plio-Pleistocene fossils but are now known to be extant along the Turkish coast. Each of the four basins of the Ionian Subprovince area contains its own separate infraprovince, each with its own endemic fauna of *Lautoconus* cone shells, *Ocinebrina* muricids, calliostomatids, and fasciolariids. Characteristic widespread index gastropods of the Ionian Subprovince include the abalone *Haliotis mykonosensis*, the muricid *Murexsul cevikeri*, and the conid *Lautoconus vayssierei*. These are illustrated here on Figure 3.8 of the Iconography at the end of this chapter. As shown by Demir (2003), the molluscan fauna of the adjacent Black Sea is simply a highly impoverished version of the Ionian malacofauna, containing only 47 gastropod species in total. Of these, the large Japonic Province and Chinese muricid *Rapana venosa*, which was accidentally introduced by ship bilge water, has dominated the intertidal areas and has decimated the resident mussel populations. The extreme faunal impoverishment of the Black Sea and Sea of Azov is due to it having been a large fresh water lake up to around 12,000 years ago, when it flooded with sea water through the newly-opened Bosporus Strait. Only the hardiest Mediterranean species can survive within the Black Sea, with its cold winter sea temperatures, fresh water input, and hydrogen sulfide-saturated water below 20 m depth.

Of the four Ionian infraprovinces, the **Aegean Infraprovince** occurs in the warmest area of the entire Mediterranean Sea and contains the largest number of endemic taxa. Some of these include the fasciolariid *Aegeofusinus rolani* (one of six Aegean-restricted species of the endemic genus *Aegeofusinus*), the giant fissurellid limpet *Diodora giannispadai*, the muricid *Ocinebrina aegeensis*, and the large conid *Lautoconus ventricosus gaudiosus* (shown here on Figure 3.8). Of special interest within the Aegean Infraprovince is the presence of an endemic Triton's Trumpet (Charoniidae), *Charonia seguenzae*, which is most often collected along the northern coast of Crete and the Dodecanese Islands. This large predatory gastropod is closely related to *Charonia variegata* from the tropical western Atlantic and West African offshore islands (Cape Verdes, Canaries, Madeira, Annobon, and St. Helena), but differs in being a smaller and more slender shell with distinctly angled varices, producing a "crooked" appearance to the spire whorls. Recent studies of the DNA of *C. seguenzae* also show that it is a full valid species, distinct from *C. variegata*. This endemic triton represents a relict species from the Pleistocene interglacial times that has managed to survive by finding a refugium within the warm, shallow, enclosed Aegean Sea (see Doxa and Divanach, 2013 for details on the ecology and systematics of *C. seguenzae*).

The largest infraprovincial area of the Ionian Subprovince, the **Levantine Infraprovince**, extends along the coast of Asia Minor and North Africa, from Turkey to Cyrenaica, Libya, and also includes the island of Cyprus. Of primary interest here is a species radiation of the conid genus *Lautoconus*, including *L. ventricosus trunculus* (endemic to Cyprus), *L. ventricosus pretunculus* and *L. ventricosus subviridis* (Turkey, Lebanon, and Israel) (shown here on Figure 3.8). The northernmost Ionian evolutionary hot spot, the **Adriatic Infraprovince**, includes the entire Adriatic Sea basin and Dalmatian Islands and houses a small fauna of distinctive endemic gastropods. During the late Pleistocene sea level lows, the Adriatic Sea dried up and only the southernmost deeper water area, off the Albanian coast, maintained oceanic conditions. This small area acted as a refugium during that time, with the surviving endemic taxa later re-colonizing the Adriatic as sea levels began to rise and the basin was re-flooded. Some of these Adriatic endemics include the muricid *Ocinebrina ingloria*, the calliostomatid *Calliostoma virescens*, and the conid *Lautoconus ventricosus adriaticus* (shown on Figure 3.9). The fourth Ionian evolutionary hot spot, the **Libyan Infraprovince**, extends from Tunisia eastward to the Cyrenaica coast of Libya and encompasses the Gulfs of Gabes and Sidra and the Maltese and Pelagie Archipelagoes. As in the other two Ionian infraprovinces, the Libyan hot spot houses a small radiation of endemic cone shells, including species such as *Lautoconus ventricosus arenarius*, *L. ventricosus exilis* (both illustrated on Figure 3.9), and *L. ventricosus rubens*, and an unusual fauna of spiny ocenebrine muricids, including *Ocinebrina hispidula* (Figure 3.9). The Maltese Archipelago, containing Gozo, Comino, Cominotto, and Malta Islands, also houses a number of endemic gastropods, including the small abalone *Haliotis stomatiaeformis* and the trochid *Gibbula nivosa* (both illustrated on Figure 3.9).

Algerian Subprovince

The Algerian Subprovince, named for the Algerian Basin, encompasses the entire western half of the Mediterranean Sea, from the Strait of Sicily, northern Sicily, and western Italy in the east to Portugal and northern Morocco in the west, and including Corsica, Sardinia, the Balearic Islands, and the Tyrrhenian, Ligurian, Balearic, and Alboran Seas. The subprovince actually encompasses three separate deep basins, the Tyrrhenian, Alboran, and Algerian, and each of these contains a distinctive set of endemic taxa; several of these are shown here on Figures 3.9 and 3.10. Within this large, complex oceanographic system, two evolutionary hot spots have developed; the **Sicilian Infraprovince** (including Sicily and Sardinia) and the **Alboranian Infraprovince** (the Alboran Sea between Spain and Morocco and the Atlantic coasts of Portugal and northern Morocco). Both of these infraprovinces house large complexes of endemic cone shells, containing species such as *Lautoconus ventricosus galloprovincialis* (Ligurian Sea coastline), *L. emisus* and *L. emisus marmoratus* (endemic to Sicily and Sardinia), *L. desidiosus* (Portugal and Lampedusa Island), and *L. ventricosus persistens* (Portugal and Alboran Sea area). The Alboran Sea area also contains the only volute shells found in the Mediterranean Sea, *Cymbium olla* and *Ampulla priamis*, a distinctive complex of endemic muricids including *Ocinebrina* *brevirobusta* and *Ocinebrina nicolai*, an endemic cowrie, *Zonaria pyrum insularum* (and the color form *nigromarginata*), and a complex of fasciolariids that includes *Fusinus clarae* (endemic to Sicily) and *Fusinus albacarinoides* (endemic to southern Spain). The distinctive Mediterranean cowrie, *Schilderia achatidea*, is also found primarily within the Alboran Sea basin, but does extend northward along the Balearic Sea coast as far as the Gulf of Marseilles (Lorenz, 2017).

Canarian Subprovince

The **Canarian Subprovince**, named for the Canary Islands off northwestern Africa, comprises the largest insular outlier of the Mediterranean Province. This elongated archipelago is composed of seven high volcanic islands (Lanzarote, Tenerife, Gran Canaria, Hierro, Gomera, La Palma, and Fuerteventura) and numerous smaller islets, and extends from just offshore of the southern Moroccan coast westward into the open Atlantic. Because of its proximity to the West African coast, the malacofauna of the Canarian Subprovince contains a small component of Guinean Province species and acts as a weak provinciatonal area between the Mediterranean and Guinean Provinces. Some of the classic African species that are Canarian residents include the conid *Genuanoconus genuanus*, the muricids *Bolinus cornutus* and *Murexiella bojadorensis*, the cassid *Cypraecassis senegalica*, the marginellid *Marginella glabella*, and the volutes *Cymbium cymbium* and *Cymbium tritonis* (illustrated in Hernandez, Rolan, Swinnen, *et al*., 2011). Besides these extraprovincial residents, the subprovince is characterized by a high level of local endemism, almost reaching provincial status. Some of the more important Canarian endemics include the muricid *Ocinebrina leukos*, the abalone *Haliotis coccinea canariensis*, the cowrie *Naria spurca cascabullorum*, the trochid *Gibbula aurantia* (also found in the Selvagens Islands), and the cone shells *Kalloconus siamensis* (also found in the Selvagens and Madeira Islands) and *Varioconus guanche*. These are illustrated here on Figure 3.11. The classic Canarian endemic cone, *Varioconus guanche*, is found only on the central and easternmost islands of the archipelago and has developed a number of distinct populations. One of these, *Varioconus guanche nitens* (Figure 3.11 I), is restricted to Lanzarote Island and appears to represent incipient speciation on the eastern islands, possibly the vanguard of a future species radiation. All of these West African and Canarian endemic species co-occur with classic Mediterranean species such as the calliostomatid *Ampullotrochus granulatum,* the turbinid *Bolma rugosa*, the cerithiid *Cerithium vulgatum*, and the volute *Ampulla priamis*, as well as the widespread Eastern Atlantic species illustrated on Figure 3.6.

Madeiran Subprovince

The open oceanic area north of the Canary Islands, encompassing Madeira Island, the Desertas Islands, and Porto Seguro Island of the Madeiran Archipelago, and also the Selvagens Islands constitutes a separate biogeographical unit that we refer to here as the **Madeiran Subprovince**. Although housing a very impoverished molluscan fauna (see Segers, Swinnen, and De Prins, 2009 for a survey of the entire fauna), the Madeiran Subprovince has evolved a number of interesting endemic taxa, some of which include the patellid *Patella tenuis*, the fissurellid *Emarginula paivana*, the abalone *Haliotis coccinea* (Figure 3.11 J), the aclidid *Aclis vitrea*, the alvaniid *Alvania spreta*, the triviid *Niveria grohorum*, and the muricid *Ocinebrina inordinata*.

Eastern Atlantic Tropical Subregion

Extending from Cape Bojador (Cap Boujdour), Western Sahara, south to the mouth of the Kunene River, on the Angola-Namibia border, and including the Cape Verde Archipelago and St. Helena and Ascension Islands, the Eastern Atlantic Tropical Subregion encompasses the largest area of tropical oceanographic conditions in the Atlantic Ocean. Although immense in size, the subregion encompasses only two molluscan faunal provinces, the **Verdesian Province** (the Cape Verde Archipelago) and the **Guinean Province** (the entire coastline from Western Sahara to Namibia and the offshore islands). The former is spatially small and compact and does not have well-defined subprovinces or infraprovinces. The latter is expansive and faunally complex, with five distinct subprovinces (the **West Saharan Subprovince**, **Senegalian Subprovince**, **Biafran Subprovince**, **Angolan Subprovince**, and **Helenean Subprovince**) and at least four infraprovinces. The spatial arrangement of these provincial units is shown here on Figure 3.1. A number of wide-ranging tropical gastropods can be used to define the absolute limits of the Eastern Atlantic Tropical Subregion and some of these include the strombid *Thetystrombus latus*, the muricid *Hexaplex rosarium*, and the harpid *Harpa doris* (illustrated on Figure 3.17).

Verdesian Molluscan Province

At the southern end of the Macaronesian Ecoregion, and extending westward off the Cape Verde Peninsula of Senegal, the isolated Cape Verde Archipelago has formed sets of habitats and ecosystems that are unique within the Atlantic Ocean. The islands themselves are geologically old, dating from the Cretaceous, and have remained geologically stable for most of the Cenozoic Era. The Cape Verde Archipelago, in its present form, comprises ten large islands arranged in two separate chains; a northern and eastern chain referred to as the Ilhas do Barlavento ("Windward Islands", with Santo Antão, São Vicente, Santa Luzia, São Nicolau, Sal, and Boa Vista Islands) and a southern chain referred to as the Ilhas do Sotavento ("Leeward Islands", with Brava, Fogo, Santiago, and Maio Islands). Rising and falling sea levels during glacial and interglacial stages (from the Miocene to the Pleistocene) appear to be the main cause of speciation on the Cape Verdes, primarily due to the resultant genetic isolation on widely-scattered islands and within sheltered bays on individual islands. Rapid eustatic fluctuations during the Pleistocene, in particular, created an entirely new and distinctive molluscan fauna, filled with endemic taxa at both the specific and generic levels. Preliminary analyses of percentages of endemism in the key tropical index families have shown that the molluscan faunas of the Cape Verdes have levels of endemism exceeding 70% in most groups, demonstrating that the entire archipelago represents a distinct molluscan faunal province. This species-rich biogeographical unit was named the **Verdesian Province** by the senior author (Petuch, 1975a).

The variety of environments on these islands, coupled with their geographical isolation from each other, has allowed for the evolution of entire species complexes in several families. Of primary interest on these island chains is an exceptionally large species radiation of the Cape Verde endemic conid genus *Africonus* (see Petuch, 1975a) with at least 85 known taxa being distributed on different islands across the archipelago. This complex of closely-related cones represents the largest evolutionary explosion in the family Conidae known from anywhere on Earth, and is particularly amazing considering the geographically-small area of the Verdesian Province. In our present treatment of the genus *Africonus*, we follow the taxonomy established by Abalde, Tenorio, Afonso, *et al*. (2017) and Tenorio, Abalde, *et al.* (2020) but consider all of the taxa to be full species at this time, and not species and subspecies as they propose. Of the 73 species and subspecies illustrated by Abalde, *et al*., we show 42 endemic *Africonus* taxa here on Figures 3.12 through 3.16. Other important Verdesian gastropod species radiations are seen in the buccinid genus *Euthria*, with at least 24 endemic species having evolved on the various islands of the archipelago (two species shown here on Figure 3.12 and 3.13) and in the endemic genus *Viridifusus* (with at least three species). The island chains also house numerous other endemics, including the cowries *Zonaria picta* and *Schilderia achatidea verdensis*, the large cassid *Cassis norai*, three olivid species including *Strephona flammulata verdensis* (shallow intertidal areas), *Strephona dolicha* (deep water off the Leeward Cape Verdes), and *Anazola boavistensis*, the endemic abalone *Haliotis fernandesi*, the patellid *Patella lugubris*, the muricids *Hexaplex rosarium pomiformis* ("*insularum*"), *Stramonita haemostoma oceanica*, and *Favartia burnayi*, and a small radiation of the endemic cone shell subgenus *Kalloconus (Trovaoconus)* including species such as *K. ateralbus*, *K. venulatus*, *K. pseudonivifer*, *K. trochulus*, and *K. marimaris*. Many of these are illustrated on Figures 3.12, 3.13, and 3.14.

Four of the Cape Verde Islands contain large and highly localized species radiations of the genera *Africonus*, *Kalloconus (Trovaoconus)*, and *Euthria* and appear to be separate evolutionary hot spots. These include São Vicente, Sal, and Boa Vista in the Barlavento Islands and Maio in the Sotavento Islands. Future faunal analyses may show that these islands should be recognized as separate infraprovinces or even subprovinces of the Verdesian Province. A review of the Verdesian *Euthria* species radiation was given by Fraussen and Swinnen (2016).

Guinean Molluscan Province

Named for the Guinea Coast of West Africa (traditionally the area extending from Senegal to Benin), the Guinean Molluscan Province is the single largest tropical biogeographic unit in the Atlantic Ocean. Extending from Cape Bojador (Cap Boujdour), Western Sahara southward to the Kunene River mouth on the Namibia-Angola border, this immense province encompasses numerous tropical environments, including open desert coasts dominated by sand beaches, huge

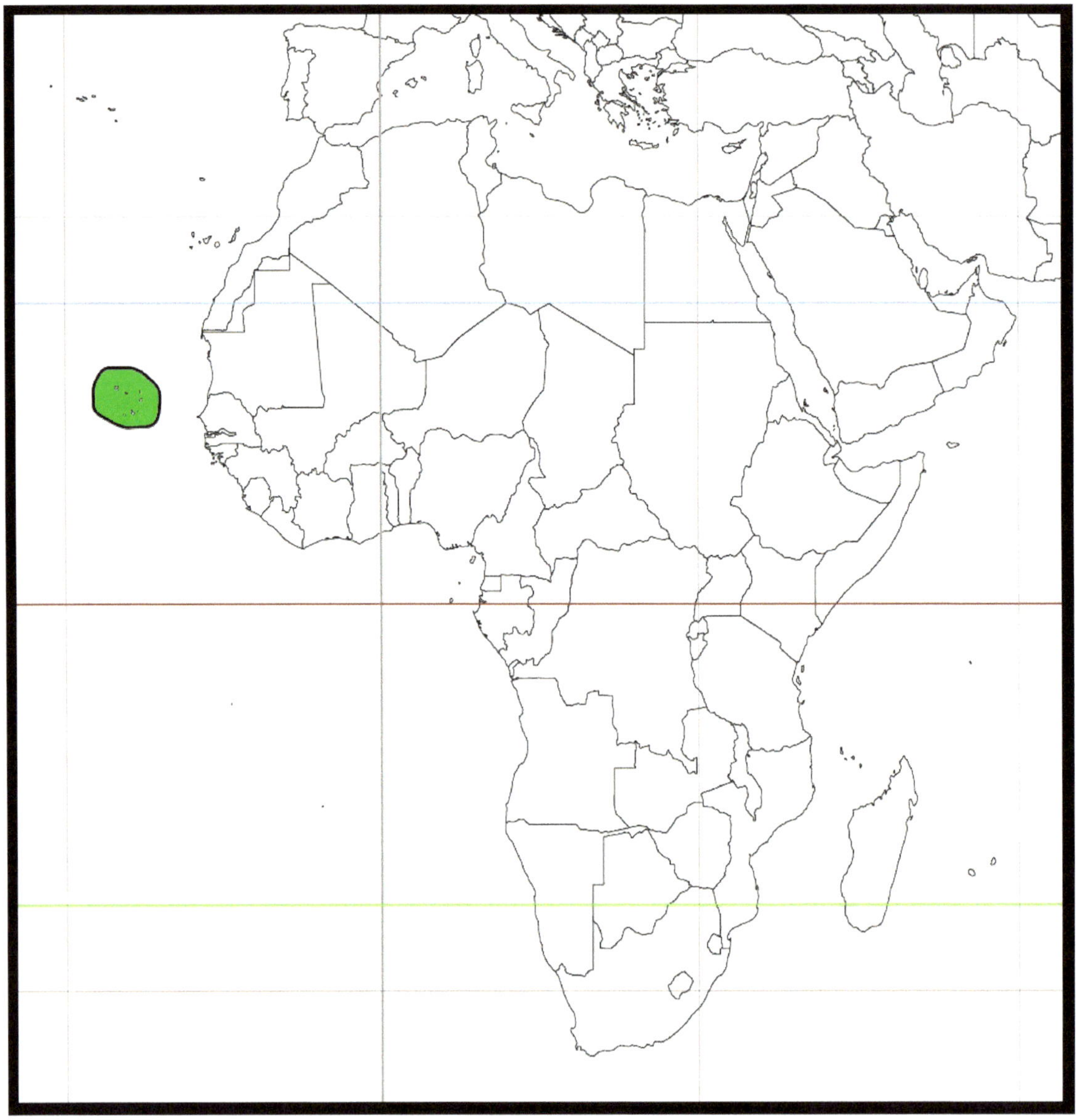

Figure 3.3 Map of the **Verdesian Molluscan Province** (green), showing the spatial configuration of the Cape Verde Archipelago.

brackish water estuaries and coastal lagoons, rocky headlands and basaltic rock platforms, and isolated offshore islands fringed with coral reefs. Because of separate, geographically-isolated faunas and numerous evolutionary hot spots, the Guinean Province can be divided into five distinct subprovinces, the **West Saharan**, **Senegalian**, **Biafran**, **Angolan**, and **Helenean**. Two main areas of evolution and speciation occur within the Guinean Province; a northern area centered on the Cape Verde Peninsula of Senegal and a southern area centered on the coast of Angola (see Lima *et al.*, 2014 for a discussion of the cone shell faunas). These unusually-rich and mutually-exclusive faunas are here incorporated into the Senegalian Subprovince and the Angolan Subprovince, respectively. The northern West Saharan Subprovince and the intervening Biafran Subprovince are both much less species-rich and are seen to contain impoverished malacofaunas when compared with those of Senegal and Angola. Some of the classic widespread Guinean index species are shown here on Figure 3.17.

The Guinean malacofauna closely resembles the faunas of the late Miocene and Pliocene Paratethys Sea, a shallow tropical ocean that covered much of southern and central Europe and the Mediterranean Sea area. Paratethys fossil shell beds contain many of the classic index genera found in the Guinean Province, including the clavatulid *Clavatula*, the turrid *Fusiturris*, the cancellariid *Bivetiella*, the cypraeids *Trona* and *Zonaria*, and the conid *Kalloconus*. During the cold ocean temperatures of the Pleistocene, these tropical groups, and many others, became extinct within the Mediterranean Sea area but survived farther south in the warmer water areas along the West African coast. The Pleistocene Guinean Province, during the coldest times, served as a refugium for many of the European Paratethyan gastropods. Some of these migrated back into the Mediterranean Sea in the late Pleistocene, when warmer sea temperatures returned, and these are now established as part of the Mediterranean Province malacofauna. Many other Paratethyan genera, however, were unable to re-invade the Mediterranean area and have remained along the tropical African coast, where they have undergone new species radiations. The Recent Guinean Province retains an archaic Mio-Pliocene appearance because of these resident relictual taxa.

West Saharan Subprovince

Named for the country of Western Sahara (a disputed territory presently claimed by Morocco), this northernmost subprovince of the Guinean Province extends from Cabo Bojador, Western Sahara south to near St. Louis, Senegal. Because of the cooler water conditions and limited habitats along the Sahara Desert coast, the West Saharan malacofauna is impoverished and lacks several of the widespread Guinean index taxa such as the cypraeids *Zonaria zonaria* and *Trona stercoraria,* the strombid *Thetystrombus latus*, the olivid *Strephona flammulata*, and the conid *Kalloconus pulcher*. Instead, the West Saharan area has evolved a highly distinctive molluscan fauna all of its own. The rocky shorelines along many areas of the West Saharan Subprovince, particularly near and within the large, fjord-like Dahkla Bay, shelter an interesting intertidal fauna that is unique in West Africa. Some of the more important endemic rocky intertidal gastropods include the muricids *Ocinebrina miscowichae*, *Jaton hemitripterus*, and *Hexaplex hoplites* and the conid *Varioconus saharicus*. The adjacent intertidal sand flats, which dominate coastal areas such as the interior of Dakhla Bay, also house a rich endemic fauna, supporting large aggregations of the cone shell *Kalloconus byssinus*, the cysticid *Persicula pseudocingulata*, the large volute *Cymbium marocanus*, and the marginellid *Marginella irrorata* (some of these are shown here on Figure 3.18).

Senegalian Subprovince

Named for the country of Senegal, this subprovince extends from St. Louis, northern Senegal south to near Cape Three Points, adjacent to Takoradi, Ghana. Of the five subprovinces of the Guinean Province, the Senegalian contains the richest molluscan fauna found anywhere in West Africa. Besides having the full complement of widespread Eastern Atlantic species and all of the classic Guinean Province index species, the Senegalian Subprovince also contains two species-rich infraprovinces that

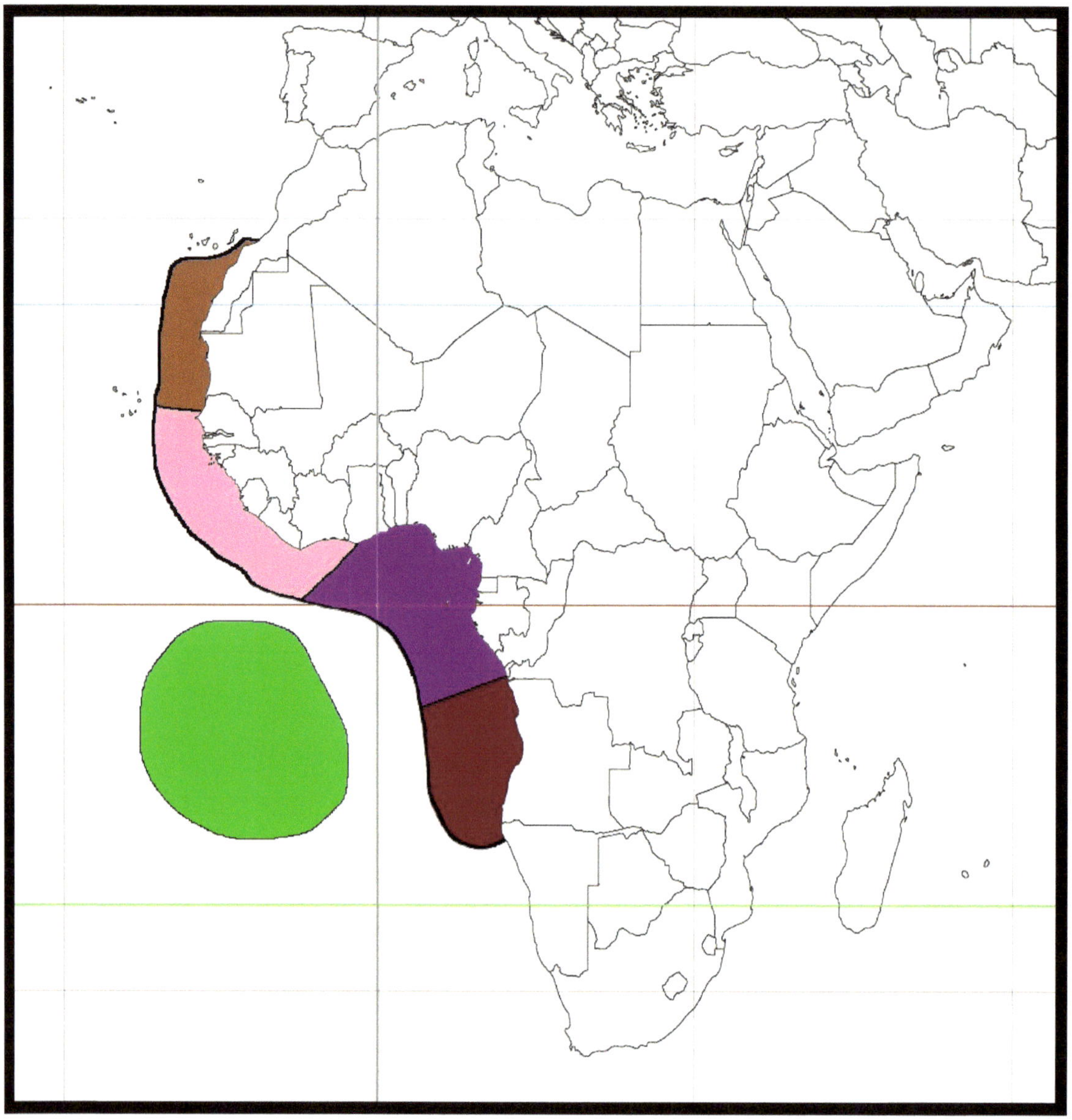

Figure 3.4 Map of The Guinean Molluscan Province, showing the areal extents of its subprovinces: the **West Saharan Subprovince** (brown), the **Senegalian Subprovince** (light rose), the **Biafran Subprovince** (purple), the **Angolan Subprovince** (dark rose), and the **Helenean Subprovince** (green).

are filled with highly localized endemic taxa. Of all the Guinean subprovinces, the Senegalian houses the richest fauna of cowrie shells, with eight species and subspecies being found within the subprovincial boundaries. Of these, five members of the genus *Zonaria* are endemic, including *Z. sanguinolenta*, *Z. petitiana*, *Z. petitiana pseudopyrum*, *Z. pyrum senegalensis*, and *Z. zonaria gambiensis*. All of these, along with the widespread *Z. zonaria*, form the single largest species radiation of their genus seen anywhere in the Eastern Atlantic (see Figures 3.19, 3.20, and 3.22). A large and distinctive fauna of *Cymbium* volutes has also evolved in the Senegalian Subprovince, including species such as *C. glans*, *C. cymbium, C. senegalensis*, *C. cucumis*, *C. marmoratum*, *C. gracile*, *C. fragile*, and several undescribed species (some are shown here on Figure 3.19).

Within the Senegalian Subprovince, the area around Cap-Vert (the Cape Verde Peninsula) of Senegal is of special interest and represents a highly localized evolutionary hot spot. Referred to here as the **Gorean Infraprovince** (named for Goree Island in the Bay of Dakar), this center of high biodiversity extends from Yoff Tongor on the northern side of Cap-Vert, around the peninsula and the Bay of Dakar, to Somone on the Petit Côte. This area, with its extensive basaltic rock platforms and reefs, isolated rocky islands, and interspersed sandy bays, supports a highly endemic and localized gastropod fauna, some of which include the muricid *Jaton flavidus*, the cowrie *Zonaria sanguinolenta*, and the marginellid *Marginella aurantia*. The most prominent species radiation of the Gorean Infraprovince is seen in the cone shell genus *Varioconus*, where over 18 species are present, often abundantly, on the basaltic reefs and sandy bays around Cap-Vert. Some of these geographically-restricted endemics include *Varioconus mercator*, *V. echinophilus* (which lives under sea urchins on exposed rock reefs; see Petuch, 1975b), *V. tacomae*, *V. hybridus*, *V. bruguieresi*, *V. cloveri*, *V. dorotheae*, *V. pineaui*, *V. unifasciatus*, *V. belairensis*, and many others. Seventeen of these endemic cones are shown here on Figures 3.20 and 3.21. Along West Africa, only Angola has a similar large cone shell radiation, but the component species are spread over the entire Angolan coastline. The Gorean *Varioconus* complex, on the other hand, is unique among cone shell faunas in that it is composed of so many species that are packed into a single geographically very small area.

Farther south, beyond the limits of the Gorean Infraprovince, a second evolutionary hot spot has recently been discovered near the mouth of the Gambia River in the country of Gambia (Petuch and Berschauer, 2017). Referred to here as the **Gambian Infraprovince** (named for Gambia), this small hot spot was found to occur on a series of algae-covered rock platforms and mud flats along the open sea coast just south of the Gambia River mouth. The infraprovincial gastropod fauna of the Gambian coast is ecologically and oceanographically isolated from the Gorean Infraprovince by the intervening Saloum River and its extensive system of deltas and brackish lagoons. The large amount of freshwater effluent from this river system, coupled with the effluent of the Gambia River, produces a wide low-salinity (oligohaline) physiological barrier for marine mollusks, effectively separating the faunas of Cap-Vert and Gambia. As a result of this genetic isolation, the gastropod fauna of the geographically-small Gambian Infraprovince has a very high percentage of endemism and is typified by species such as the clavatulid *Pusionella milleri subgranulatus,* the naticids *Natica royi* and *N. collaria gambiae*, and the terebriform nassariid *Bullia* (*Naytia*) *turrita* unnamed subspecies (see Figure 3.22). Of particular interest within the Gambian Infraprovince is a large fauna of highly-restricted cone shells (Conidae) of the genus *Varioconus* that, in many ways, is an analog of the larger *Varioconus* radiation of the Gorean Infraprovince to the north. This Gambian radiation is found only on isolated low rock platforms scattered along the Gambian coast at subtidal depths and at least ten endemic species are now known to inhabit these disjunct algae-covered reefs. Some of these include *Varioconus rikae*, *V. gambiensis*, *V. fernandi*, *V. orri*, and *V. wolof* (all shown here on Figure 3.22).

Biafran Subprovince

The central subprovince of the Guinean Province is here named for the Bight of Biafra, the wide embayment at the eastern end of the Gulf of Guinea and which is bounded by Nigeria, Cameroon, Equatorial Guinea, and Gabon and the island nation of São Tome e Principe. The Biafran Subprovince extends from Cape Three Points, Ghana, southward to the Congo River Delta on the border of the Democratic Republic of the Congo and Angola, and encompasses the offshore islands of Bioko, Principe, São Tome, and Annobon. Being located in the highest-rainfall area of Africa, the habitats and environments of this subprovince are heavily impacted by the immense amount of freshwater effluent that is discharged from the numerous rivers that flow down from the mountains to the coastline. Some of the larger rivers, which have extensive delta systems and act as ecological barriers to marine organisms, include the Niger, Cross, Akpa Yafe, Wouri, Sanage, Nyong, Ogooue, and the Congo. The Niger-Cross River delta system, alone, extends for hundreds of kilometers and this major biotic barrier manifests itself in an abrupt and obvious faunal change. To the west of this delta complex, species-rich communities of Senegalese Subprovincial endemics abound, but east of the delta complex, these species disappear and their communities are replaced by impoverished Biafran molluscan assemblages. Although less species-rich than their Senegalese counterparts, the muddy lagoonal molluscan communities of the Biafran Subprovince do contain interesting endemic species such as the olivids *Agaronia razetoi* and *Anazola biraghii*, the volute *Cymbium pachyus*, the dwarf cowrie *Trona stercoraria cameroonica*, and several species of the Biafran endemic muricid genus *Africanella*, including *A. coseli* and *A. isaacsi* (all shown here on Figure 3.23).

The offshore, deeper water molluscan communities of the Biafran Subprovince also contain a number of distinctive and characteristic endemic species and subspecies, such as the cowries *Zonaria angelicae* and *Zonaria petitiana petiformis*, the extremely spiny muricid *Homalocantha melanomathos chinii* (all shown here on Figure 3.23), and the volutes *Cymbium congoensis* and *C. coenyei*. The offshore islands of Bioko and Annobon, and the island nation of São Tome e Principe, altogether, house an evolutionary hot spot that we refer to here as the **Tomean Infraprovince** (named for São Tome Island). These volcanic islands, being far enough offshore to be out of the influence of fresh water river effluent, have clear, open-oceanic water conditions and are the only areas in all of the Guinean Province that have extensive coral reef growth. These corals are all western Atlantic species whose larvae have ridden the eastward-flowing North Equatorial Countercurrent across the Atlantic to the offshore islands. Only a few coral species have been able to colonize the Tomean reefs, with the most abundant reef-building taxa being the Antler Coral *Acropora palmata*, the Star Coral *Montastrea cavernosa*, and the Starlet Coral *Siderastrea siderea*. Although impoverished, these unique coral reef systems have produced extensive carbonate environments that shelter a small endemic molluscan fauna, including the marginellid *Insulamarginella spinacea*, the cysticid *Gibberula thomensis* (both shown here on Figure 3.23), and the cone shell *Varioconus tamsianus*. Many other small endemic gastropods are known from these Gulf of

Guinea coral reefs, but most are still undescribed. Future research on the Tomean malacofauna, and the virtually unexplored coralline environments, will undoubtedly uncover many more new and interesting endemic mollusks.

Angolan Subprovince

Named for the country of Angola, this subprovince is as species-rich, and has as high a level of endemism, as does its northern counterpart in Senegal. Being south of the Congo River and in a semi-arid region, the Angolan Subprovince does not have the massive freshwater input seen in the Biafran Subprovince and has normal oceanic salinities throughout its coastal areas. Much of the shoreline of Angola is composed of large high cliffs with intervening sandy beaches, and small isolated bays, both of which will produce genetic isolation in low vagility gastropod families. Because of these multiple dispersal barriers and the resultant allopatric speciation, the Angolan Subprovince has evolved a very distinctive molluscan fauna with a high level of endemicity. Some of the typical Angolan index species include the cowrie *Zonaria angolensis*, the muricids *Orania angolensis*, *Jaton sinespina*, and *Muricopsis punctata*, the naticid *Natica rocquignyi*, the marginellid *Marginella lucani*, a large fauna of turroidean gastropods, including clavatulids such as *Clavatula conica* and *Pusionella compacta* and pseudomelatomids such as *Crassispira oliva*, and the endemic cone shell genus and species *Pseudonoduloconus carnalis* (all shown here on Figure 3.24). In the deep water areas off northern Angola (continental slope-bathyal zone), the rarely-seen volute, *Athleta emmanuelae*, has recently been rediscovered and is now known to be part of the Angolan malacofauna (see Figure 3.24K). Of particular interest in the Angolan Subprovince is a large species radiation of the conid genus *Varioconus*, with at least 32 known species, making this the second-largest Eastern Atlantic cone shell fauna after that of the Cape Verde Islands (25 *Varioconus* species are shown here on Figures 3.24, 3.25, and 3.26).

Oceanographically, the Angolan Subprovince is divided into two distinct water temperature regimes: a warm northern area, extending from the Congo River south to approximately Baia Farta, influenced by a southward-flowing gyre off the warm Guinea Current (the Gulf of Guinea Gyre); and a cooler water southern area, extending from Baia Farta south to the Cunene River on the Namibian border, that is influenced by the cold, subantarctic northward-flowing Benguela Current. These two oceanographic areas have their own distinctive molluscan faunas, with the highest levels of endemism being seen in the genus *Varioconus*, enough for each to qualify as a separate infraprovincial evolutionary hot spot. The northern, warm water evolutionary center is here referred to as the **Luandan Infraprovince** (named for Luanda, in northern Angola) and it contains some of the classic *Varioconus* species that were named over 150 years ago, such as *V. variegatus*, *V. bulbus*, *V. obtusus*, *V. africanus*, and *V. zebroides*, as well as many recently-discovered species such as *V. lineopunctatus*, *V. nunesi*, *V. equiminaensis*, *V. naranjus*, and *V. annagretae* (most shown here on Figure 3.25). The southern, cooler water evolutionary hot spot is here referred to as the **Namibe Infraprovince** (named for Namibe Province of southern Angola) and it contains a remarkable conid fauna that has only recently been investigated. This exploratory work has led to the description of many new *Varioconus* species, such as *V. trovaoi*, *V. eusebioi*, *V. petuchi*, *V. flavusalbus*, and *V. babaensis*, along with the re-discovery of several well-known older taxa such as *V. chytreus* and *V. fuscolineatus* (all shown here on Figure 3.26). Like the Cape Verde Islands *Africonus* species, most of the Angolan *Varioconus* species are restricted to small isolated bays or small stretches of coastline, demonstrating that they have direct development.

Helenean Subprovince

Located in the central South Atlantic, between Angola and Brazil, the Helenean Subprovince (named for St. Helena Island) includes both St. Helena and the more northerly Ascension Island. These two remote islands, which are actually emergent peaks rising off the Mid-Atlantic Ridge, represent the only insular subprovincial component of the Guinean Province. Essentially an outlier of the mainland Guinean fauna, the Helenean Subprovince contains only a few of the widespread Guinean taxa, such as the muricid *Thais nodosa*, the cassid *Cypraecassis senegalica*, the littorinid *Nodilittorina miliaris*, and the naticid *Naticarius dillwyni* (see Rosewater, 1975). As would be expected on remote, isolated islands, the overall malacofauna of the subprovince is relatively impoverished. Ascension Island also houses a few more African taxa than does its sister island, especially the strombid *Thetystrombus latus*, which is absent on St. Helena. Besides the Guinean faunal component, the islands of the Helenean Subprovince have also evolved a distinctive endemic gastropod fauna that includes species such as the cowries *Naria sanctaehelenae* (Ascension Island only), *Naria sanctaehelenae bonapartei* (St. Helena only), and *Luria lurida oceanica*, the muricid *Stramonita rustica bicarinata*, the nassariid *Nassarius sanctaehelenae*, the buccinid *Tritonidea consanguineus*, the harpid *Harpa doris robusta*, and the cone shell *Varioconus jourdani* (most shown here on Figure 3.27). The remote Tristan da Cunha Island to the south of St. Helena, which is under the influence of the cold Subantarctic West Wind Drift water mass, houses a very impoverished malacofauna with only a few local endemics. Primary among these endemics are the ranellids *Argobuccinum tristanensis* and *Cymatona philomelae* (Figure 3.27H).

Southeastern Atlantic Paratropical Subregion

Extending from the Cunene River Mouth on the Angola-Namibia border southward to the West Coast Peninsula of the Republic of South Africa, the Southeastern Atlantic Paratropical Subregion contains only two faunal units, the Namibian Province and the broad transitional Namaquan Provinciatone. Unlike the Guinean Province to the north, which is bathed in the warm waters of the Guinea Current, the oceanography of the Namibian Province is under the influence of the cold Benguela Current. Originating in the very cold subantarctic West Wind Drift area, the Benguela Current deflects northward at the southern tip of Africa and flows along the coast of Namibia. Near the Cunene River area of northernmost Namibia, the current abruptly deflects westward, where it eventually warms and blends into the Equatorial Current. The Benguela Current comes closest to land along the stretch of coastline extending from St. Helena Bay, South Africa, to north of Luderitz, Namibia, and this area has the coldest ocean temperature found anywhere on the African continent. The water temperatures of the Guinea and Benguela Currents cause Namibia to have two separate oceanographic regimes; a warm-temperate area extending from the Cunene River southward to Luderitz, and a colder water area running along the entire Namaqualand coast, from Luderitz south to the West Coast Peninsula.

Namibian Molluscan Province

Named for the country of Namibia, this province extends from the Cunene River south to near Luderitz, Namibia, and contains the famous "Skeleton Coast" north of Swakopmund. Although only marginally paratropical, with water temperatures often below 20 degrees Celsius, the Namibian Province contains an impoverished but highly-endemic molluscan fauna. These low water temperatures prevent many classic tropical families, such as the Potamididae, Strombidae, Cassidae, Melongenidae, Olividae, and Harpidae, from becoming established within the provincial boundaries. The paratropical nature of the province, however, is underscored by the presence of endemic tropical-type species such as the cowrie *Cypraeovula namibiensis*, the aporrhaid *Aporrhais pesgallinae*, the cone shell *Sciteconus patens*, the marginellids *Kaokomarginella stuarti*, *Roseamarginella nimbosa*, and *Prunum walvisianum*, and the pseudomelatomid *Comitas saldanhae* (all shown on Figures 3.28, 3.29, 3.30). These paratropical genera and species occur along with endemic cold water taxa such as the naticid *Euspira massieri*, the limpet *Patella swakopmundensis*, the muricid *Nucella ovalis*, the nassariid *Nassarius scopalarcus*, and the trochid *Callumbonella namibiensis*, demonstrating the mixed nature of the Namibian fauna. Of particular interest within the Namibian Province is a newly-discovered deep water endemic radiation of the volutid genus *Athleta*, including species such as *Athleta massieri*, *A. easoni*, and *A. disparilis* (all shown here on Figure 3.29).

South of Luderitz, the Benguela Current is at its closest to the African mainland and the water temperatures fall to near 10 degrees Celsius for most of the year. This colder water area extends along the entire Namaqualand coast and acts as a broad faunal transition zone between the typical Namibian fauna and that of the South African Province (Cape Province, Transkei, and Natal areas of South Africa) and is referred to as the **Namaquan Provinciatone**. Here, many of the Namibian provincial index taxa, such as *Cypraeorbis namibiensis*, *Roseamarginella nimbosa*, and *Nucella ovalis* occur together with classic South African taxa such as the limpet *Cymbula compressa* (found on *Ecklonia* kelp), the fasciolariid *Lugubrularia lugubris*, the muricids *Vaughtia purpuroides* and *Trochia cingulata*, the ranellid *Fusitriton murrayi*, the buccinid *Burnupena cincta limbosa*, and the clavatulid *Clionella*. As is characteristic of these types of faunal transition zones, several provinciatonal endemics occur along with the Namibian and South African faunal components, including the volute *Athleta lutosa* and the cone shell *Sciteconus mozambicus macei*. The very cold, stormy, and high surf environments of the Namaqualand coast, together, act as an ecological barrier that separates the faunas of the Namibian Province from those of the South African Province (see Chapter 10).

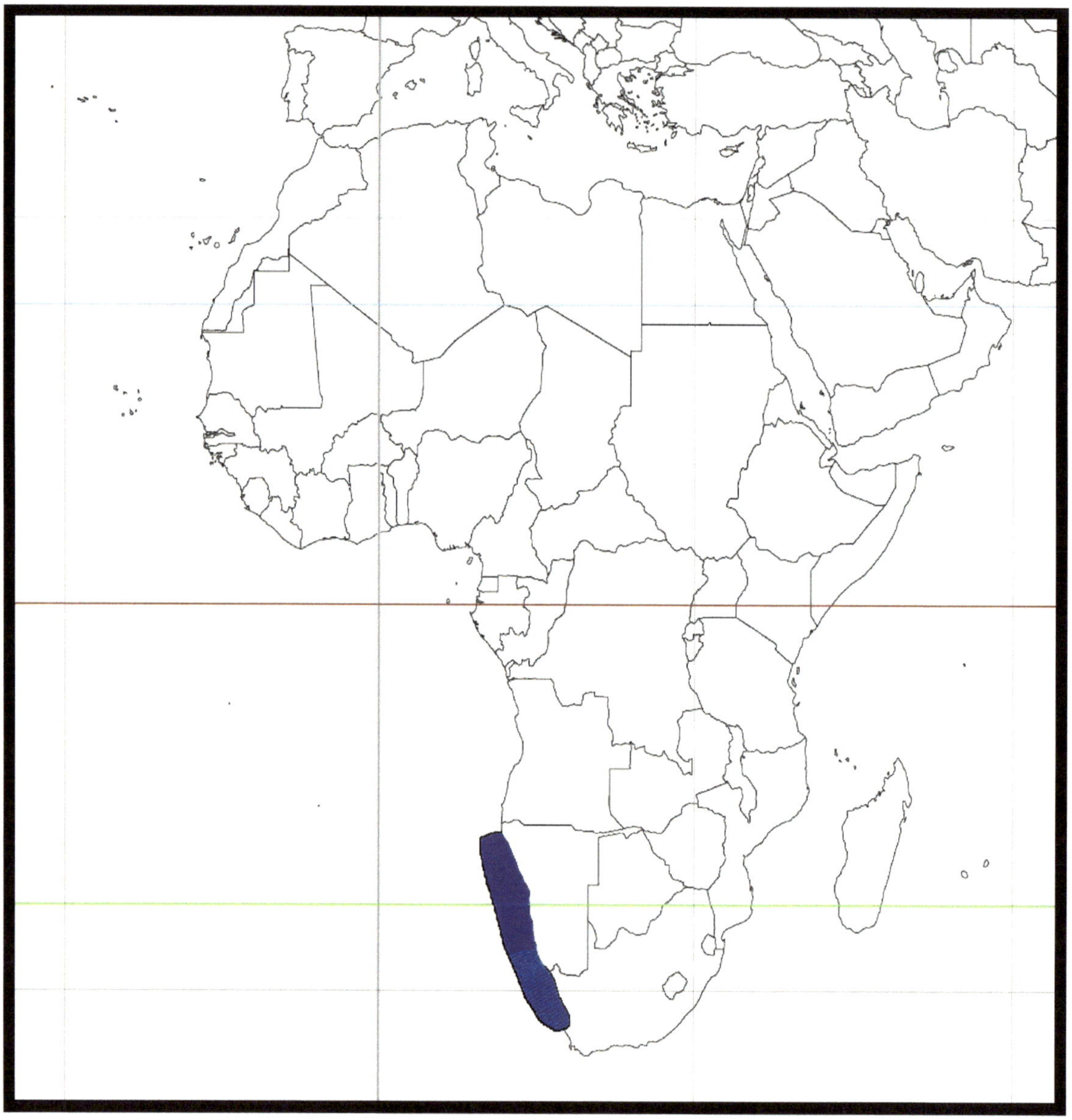

Figure 3.5 Map of the **Namibian Molluscan Province** (blue), showing the areal extent of the **Namaquan Provinciatone** (cross-hatched blue).

ICONOGRAPHY OF GASTROPODS OF THE EASTERN ATLANTIC REGION

(Principal Index Gastropods are shown on Figures 3.6 to 3.29)

Figure 3.6 Widespread Index Gastropods of the Eastern Atlantic Region, occurring in the Mediterranean, Verdesian, and Guinean Provinces.
A= *Neosimnia spelta* (Linnaeus, 1758), length 16 mm, 10 m depth off Palermo, Sicily, Italy; **B**= *Pseudosimnia carnea* (Poiet, 1789), length 15 mm, found on octocorals at 20 m depth off Malaga, Spain; **C**= *Charonia lampas* (Linnaeus, 1758), length 209 mm, 20 m depth off Cape Agadir, Morocco; **D**= *Monoplex corrugatum* (Lamarck, 1816), length 75 mm, 5 m depth off Syracusa, Sicily, Italy; **E**= *Semicassis saburon* (Bruguiere, 1792), length 56 mm, 50 m depth off Malaga, Spain; **F**= *Pagodula fraseri* (Knudsen, 1956), length 44 mm, 200 m depth off Madaleine Island, Cape Verde Peninsula, Senegal; **G**= *Hima wolffi* (Knudsen, 1956), length 30 mm, 50 m depth off Swakopmund, Namibia; **H**= *Episcomitra zonata* (Marryat, 1819), length 68 mm, 5 m depth off Messina, Strait of Messina, Sicily, Italy; **I**= *Bivetiella cancellata* (Linnaeus, 1758), length 34 mm, 20 m depth off Palermo, Sicily, Italy; **J**= *Fusiturris similis* (Bivona, 1838), length 49 mm, 20 m depth off Malaga, Spain; **K**= *Cumia reticulata* (Blainville, 1829), length 18 mm, under rocks, 2 m depth off Mondello, Sicily.

Figure 3.7 Widespread Mediterranean Province Gastropods.
A= *Luria lurida* (Linnaeus, 1758), length 51 mm, 3 m depth off Khania, Crete, Greece; **B**= *Naria spurca* (Linnaeus, 1758), length 31 mm, under rocks in 2 m depth off Haifa, Israel; **C**= *Zonaria pyrum* (Gmelin, 1791), length 39 mm, on sponges, 5 m depth off Izmir, Turkey; **D**= *Aporrhais pespelicani* (Linnaeus, 1758), length 45 mm, trawled from 10 m depth off Palermo, Sicily; **E**= *Aporrhais serresianus* (Michaud, 1828), length 55 mm, trawled from 25 m depth off Malaga, Spain; **F**= *Bolinus brandaris* (Linnaeus, 1758), length 70 mm, 10 m depth, Gulf of Taranto, off Taranto, Italy; **G**= *Truncularioposis trunculus* (Linnaeus, 1758), length 80 mm, 2 m depth, Algeciras, Spain; **H**= *Tarantinaea lignaria* (Linnaeus, 1758), length 44 mm, 2 m depth off Siracusa, Sicily; **I**= *Euthria cornea* (Linnaeus, 1758), length 66 mm, 2 m depth on rocks, Gulf of Taranto off Taranto, Italy; **J**= *Lautoconus ventricosus* (Gmelin, 1791), length 24 mm, on rocks and algae, 1 m depth off Almeria, Spain; **K**= *Fusinus pulchellus* (Philippi, 1844), length 39 mm, 20 m depth off Taranto, Gulf of Taranto, Italy.

Figure 3.8 Index Gastropods of the Ionian Subprovince and the Aegean and Levantine Infraprovinces, Mediterranean Province.

(Widespread Ionian Subprovince) A= *Haliotis mykonosensis* Owen, Hanavan, and Hall, 2001, length 38 mm, 3 m depth off Evia Island, Karystos, Greece; **B=** *Murexsul cevikeri* (Houart, 2000), length 23 mm, under rocks, 3 m depth off Khania, Crete, Greece; **C=** *Lautoconus vayssierei* (Pallary, 1906), length 27 mm, in sponges, 5 m depth off Cala Francese, Lampedusa Island, Pelagie Islands, Italy; **(Aegean Infraprovince) D=** *Charonia seguenzae* (Aradas and Benoit, 1870), length 210 mm, 14 m depth off Evia Island, Karystos, Greece; **E=** *Ocinebrina aegeensis* Aissaoui, Barco, and Oliverio, 2017, length 12 mm, 40 m depth off Nea Peramos, Greece; **F=** *Aegeofusinus rolani* (Buzzurro and Ovalis, 2005), length 13 mm, 3 m depth off Evia Island, Karystos, Greece; **G=** *Diodora giannispadai* (Aissaoui, Puillandre, and Bouchet, 2017), length 46 mm, under rocks in a *Posidonia* sea grass bed, 6 m depth off Halkidiki, Kasandra, Marmaras, Greece; **H=** *Lautoconus ventricosus gaudiosus* (Nicolay, 1978), length 36 mm, 3 m depth off Andros Island, Cyclades Islands, Greece, Aegean Sea; **(Levantine Infraprovince) I=** *Truncylariopsis pecchiolianus* (Ancona, 1871), length 50 mm, trawled from 15 m depth off Bodrum, Turkey; **J=** *Lautoconus ventricosus pretunculus* (Monterosato, 1917), length 24 mm, 2 m in rock rubble and sponges, Haifa, Israel; **K=** *Lautoconus ventricosus trunculus* (Monterosato, 1900), length 23 mm, on rocks and sponges, 5 m depth off Ormidia, Cyprus (endemic to Cyprus); **L=** *Lautoconus ventricosus subviridis* (de Gregorio, 1885), length 25 mm, 2 m depth off El Gharam Beach, Marsa Matruh, Egypt.

Figure 3.9 Index Gastropods of the Adriatic and Libyan Infraprovinces, Ionian Subprovince, and the Algerian Subprovince, Mediterranean Province.
(Adriatic Infraprovince) A= *Calliostoma virescens* Coen, 1933, height 12 mm, 2 m depth off Osjak Island, Bay of Gradina, Dalmatia, Croatia; **B=** *Ocinebrina ingloria* Crosse, 1865), length 18 mm, 3 m depth off Chioggia, Gulf of Venice, Italy; **C=** *Lautoconus ventricosus adriaticus* (Nardo, 1847), length 29 mm, 2 m depth off Osjak Island, Bay of Gradina, Dalmatia, Croatia; **(Libyan Infraprovince) D=** *Ocinebrina hispidula* (Pallary, 1904), length 13 mm, 5 m depth off Al Uqaylah, Gulf of Sirte, Libya; **E=** *Lautoconus ventricosus exilis* (Gaglini in Settepassi, 1985), length 25 mm, 3 m depth off Al Qaylah, Gulf of Sirte, Libya; **F=** *Lautoconus ventricosus arenarius* (Monterosato, 1917), length 25 mm, low tide, Cheikh Yahya, Djerba, Tunisia; **G=** *Gibbula nivosa* (A.Adams, 1853), width 7 mm, 10 m depth on *Posidonia* sea grass, St. Thomas Bay, Malta (endemic to Malta); **H=** *Haliotis stomatiaeformis* Reeve, 1846, length 23 mm, on rocks, 4 m depth in St. Paul's Bay, Malta (endemic to Malta); **(Algerian Subprovince) I=** *Schilderia achatidea* (Gray, 1837), length 41 mm, 40 m depth off Malaga, Spain; **J=** *Aptyxis syracusanus* (Linnaeus, 1758), length 42 mm, 100 m depth in the Gulf of Castellum Mare, Sicily; **K=** *Lautoconus ventricosus galloprovincialis* (Locard, 1886), length 36 mm, 2 m depth off Plage Tahiti, St. Tropez, France (endemic to the Ligurian Basin).

Figure 3.10 Index Gastropods of the Sicilian and Alboranian Infraprovinces, Algerian Subprovince, Mediterranean Province.
(Sicilian Infraprovince) A= *Bolinus brandaris cagliaritanus* Settepassi, 1977, length 58 mm, 5 m depth in Laguna de Marceddi, Sardinia (endemic to Sardinia); **B=** *Fusinus clarae* Russo and Renda, 2013, length 42 mm, trawled from 50 m depth off Bagheria, Gulf of Palermo, Sicily; **C=** *Lautoconus emisus* (de Gregorio, 1885), length 31 mm, 2 m depth off Mondello, Sicily; **(Alboranian Infraprovince) D=** *Zonaria pyrum insularum* Schilder, 1928, length 36 mm, on rocks and sponges, 2 m depth off Cadiz, Spain; **E=** *Ocenebra brevirobusta* Houart, 2000, length 27 mm, 80 m depth off Essaouira, Morocco; **F=** *Ocinebrina nicolai* Monterosato, 1884, length 13 mm, trawled from 120 m depth off Tangier, Morocco; **G=** *Fusinus albacarinoides* Hadorn, Afonso, and Rolan, 2009, length 21 mm, 50 m depth, south of Almeria, Spain (restricted to the Alboran Sea); **H=** *Ampulla priamis* (Gmelin, 1791), length 53 mm, trawled from 200 m depth off Farol, Algarve, Portugal; **I=** *Cymbium olla* (Linnaeus, 1758), length 132 mm, in sand, 20 m depth off Olhao, Portugal; **J=** *Lautoconus desidiosus* (Adams, 1854), length 21 mm, in algae at extreme low tide, Olhao, Portugal; **K=** *Lautoconus ventricosus persistens* (Kobelt, 1906), length 24 mm, under rocks at extreme low tide, Olhao, Portugal; **L=** *Ocinebrina purpuroidea* Pallary, 1920, length 11 mm, low tide under rocks, Essaouira, Morocco (confined to the outer coast of Morocco).

Figure 3.11 Index Gastropods of the Canarian Subprovince and Madeiran Subprovince, Mediterranean Province. **(Canarian Subprovince) A=** *Haliotis coccinea canariensis* Nordsieck, 1975, length 51 mm, 2 m depth off Los Silos, Tenerife Island, Canary Islands; **B=** *Gibbula aurantia* Nordsieck and Garcia-Talavera, 1982, diameter 8 mm, low tide off Playa del Socorro, Puerto de la Cruz, Tenerife Island, Canary Islands; **C=** *Patella crenata* d'Orbigny, 1840, length 39 mm, Playa del Socorro, Puerto de la Cruz, Tenerife Island, Canary Islands; **D=** *Naria spurca cascabullorum* Garcia-Talavera, Dionis, and Gomez, 1986, length 18 mm, 100 m depth off Las Galletas, Tenerife Island, Canary Islands; **E=** *Truncularìopsis hoplites canariensis* Nordsieck, 1975, length 77 mm, 20 m depth off Puerto del Carmen, Lanzarote Island, Canary Islands; **F=** *Ocinebrina leukos* Houart, 2000, length 14 mm, 1 m depth off Playa del Socorro, Puerto de la Cruz, Tenerife Island, Canary Islands; **G=** *Kalloconus siamensis* (Hwass, 1792), length 99 mm, 5 m depth in sand off Maspalomas, Gran Canaria Island, Canary Islands, (= *K. canariensis*); **H=** *Varioconus guanche* (Lauer, 1993), length 40 mm, low tide off Candelaria, Tenerife Island, Canary Islands; **I=** *Varioconus guanche nitens* (Lauer, 1993), length 26 mm, low tide off Puerto del Carmen, Lanzarote Island, Canary Islands; **(Madeiran Subprovince) J=** *Haliotis coccinea* Reeve, 1846, length 41 mm, low tide off Camera de Lobos, Madeira Island, Madeira Archipelago.

Figure 3.12 Index Gastropods of the Verdesian Province.
A= *Haliotis fernandesi* Owen and Afonso, 2012, length 26 mm, low tide off Curral Velho, Boa Vista Island, Cape Verde Islands; **B=** *Kalloconus (Trovaoconus) venulatus* (Hwass, 1792), length 35 mm, 2 m depth off Sal Rei, Boa Vista Island, Cape Verde Islands; **C=** *Zonaria picta* (Gray, 1824), length 28 mm, 2 m depth off Sal Rei, Boa Vista Island, Cape Verde Islands; **D=** *Cassis norai* Prati Musetti, 1995, length 146 mm, 20 m depth off Sal Rei, Boa Vista Island, Cape Verde Islands; **E=** *Favartia burnayi* Houart, 1981, length 24 mm, 2 m depth off Pedra Lume, Sal Island, Cape Verde Islands; **F=** *Leucozonia triserialis* (Lamarck, 1822), length 18 mm, low tide, Ilheu de Sal Rei, Boa Vista Island, Cape Verde Islands; **G=** *Viridifusus buxeus* (Reeve, 1847), length 37 mm, 2 m depth off Tarrafal, Santiago Island, Cape Verde Islands; **H=** *Euthria boavistensis* von Cosel, 1982, length 36 mm, 2 m depth off Curral Velho, Boa Vista Island, Cape Verde Islands; **I=** *Euthria rolani* von Cosel, 1982, length 29 mm, 2 m depth off Calhau, São Vicente Island, Cape Verde Islands; **J=** *Anazola boavistensis* Burnay and Da Conceicao, 1986, length 25 mm, 2 m depth, off Sal Rei, Boa Vista Island, Cape Verde Islands; **K=** *Strephona flammulata verdensis* (Petuch and Sargent, 1986), length 28 mm, 2 m depth off Sal Rei, Boa Vista Island, Cape Verde Islands (incorrectly synonymized with *Strephona dolicha*, which is a much smaller, narrower, and more elongated species from deep water off the southern Cape Verdes); **L=** *Africonus vulcanus* Tenorio and Afonso, 2004, length 21 mm, 3 m depth off Derrubado Bay, Boa Vista Island, Cape Verde Islands.

Figure 3.13 Index Gastropods of the Verdesian Province.
A= *Euthria bernardi* Fraussen and Rolan, 2003, length 44 mm, 3 m depth, Derrubado, Boa Vista Island, Cape Verde Islands; **B=** *Trunculariopsis rosarium pomiformis* (Locard, 1897), length 55 mm, Barril, São Nicolau Island, Cape Verde Islands; **C=** *Africonus josephinae* (Rolan, 1980), length 23 mm, 1 m depth off Sal Rei, Boa Vista Island, Cape Verde Islands; **D=** *Africonus kersteni* Tenorio, Afonso, and Rolan, 2008, length 21 mm, 2 m depth, Juncalinho, São Nicolau Island, Cape Verde Islands; **E=** *Africonus luquei* (Rolan and Trovao, 1990), length 24 mm, low tide, Derrubado, Boa Vista Island, Cape Verde Islands; **F=** *Africonus fuscoflavus* (Rockel, Rolan, and Monteiro, 1980), length 23 mm, 2 m depth off Curral Velho, Boa Vista Island, Cape Verde Islands; **G=** *Africonus messiasi* (Rolan and Fernandes, 1990), length 22 mm, 1 m depth off Sal Rei, Boa Vista Island, Cape Verde Islands; **H=** *Africonus irregularis* (Sowerby II, 1858), length 30 mm, 3 m depth, Navio Quebrado Bay, Maio Island, Cape Verde Islands; **I=** *Africonus iberogermanicus* (Rockel, Rolan, and Monteiro, 1980), length 27 mm, 2 m depth off Derrubado, Boa Vista Island, Cape Verde Islands; **J=** *Kalloconus (Trovaoconus) pseudonivifer* (Monteiro and Tenorio, 2004), length 43 mm, 3 m depth, Curral Velho, Boa Vista Island, Cape Verde Islands; **K=** *Patella lugubris* Gmelin, 1791, length 36 mm, low tide, Sal Rei, Boa Vista Island, Cape Verde Islands; **L=** *Africonus derrubado* (Rolan and Fernandes, 1990), length 21 mm, 2 m depth off Derrubado, Boa Vista Island, Cape Verde Islands.

Figure 3.14 Index Cone Shells of the Verdesian Province.
A= *Africonus boavistensis* (Rolan and Fernandes, 1990), length 23 mm, 1 m depth in rock crevices, off Derrubado, Boa Vista Island, Cape Verde Islands; **B=** *Africonus crotchii* (Reeve, 1849), length 22 mm, 2 m depth off Sal Rei, Boa Vista Island, Cape Verde Islands; **C=** *Africonus damottai* (Trovao, 1979), length 25 mm, 2 m depth off Derrubado, Boa Vista Island, Cape Verde Islands; **D=** *Africonus delanoyae* (Trovao, 1979), length 27 mm, 2 m depth off Sal Rei, Boa Vista Island, Cape Verde Islands; **E=** *Africonus diminutus* (Trovao and Rolan, 1986), length 11 mm, 1 m depth off Praia Santa Monica, Boa Vista Island, Cape Verde Islands; **F=** *Africonus evorai* Monteiro, Fernandes, and Rolan, 1995, length 23 mm, 1 m depth off Praia Curralinho, Boa Vista Island, Cape Verde Islands; **G=** *Africonus longilineus* (Rockel, Rolan, and Monteiro, 1980), length 24 mm, 3 m depth off Calheta Funda, Sal Island, Cape Verde Islands; **H=** *Africonus roeckeli* (Rolan, 1980), length 25 mm, 3 m depth off Praia Chaves, Boa Vista Island, Cape Verde Islands; **I=** *Africonus salreiensis* (Rolan, 1980), length 21 mm, 2 m depth off Ilheu de Sal Rei, Boa Vista Island, Cape Verde Islands; **J=** *Africonus swinneni* Tenorio, Afonso, Cunha, and Rolan, 2014, length 25 mm, 3 m depth off Derrubado, Boa Vista Island, Cape Verde Islands; **K=** *Africonus teodorae* (Rolan and Fernandes, 1990), length 19 mm, 3 m depth in Teodora Bay, Boa Vista Island, Cape Verde Islands; **L=** *Kalloconus (Trovaoconus) trochulus* (Reeve, 1844), length 43 mm, 2 m depth off Derrubado, Boa Vista Island, Cape Verde Islands.

Figure 3.15 Index Cone Shells of the Verdesian Province.
A= *Africonus antoniomonteiroi* (Rolan, 1990), length 23 mm, 2 m depth off Murdeira, Sal Island, Cape Verde Islands; **B=** *Africonus cuneolus* (Reeve, 1843), length 22 mm, 2 m depth off Pedra Lume, Sal Island, Cape Verde Islands; **C=** *Africonus felitae* (Rolan, 1990), length 11 mm, 3 m depth off Murdeira, Sal Island, Cape Verde Islands; **D=** *Africonus fontonae* (Rolan and Trovao, 1990), length 24 mm, 3 m depth off Santa Maria, Sal Island, Cape Verde Islands; **E=** *Africonus mordeirae* (Rolan and Trovao, 1990), length 25 mm, 3 m depth off Murdeira, Sal Island, Cape Verde Islands; **F=** *Africonus pseudocuneolus* (Rockel, Rolan, and Monteiro, 1980), length 24 mm, 2 m depth off Murdeira, Sal Island, Cape Verde Islands; **G=** *Africonus regonae* Rolan and Trovao, 1990, length 25 mm, 2 m depth off Palmeira, Sal Island, Cape Verde Islands; **H=** *Africonus serranegrae* Rolan, 1990, length 22 mm, 5 m depth off Serranegra, Sal Island, Cape Verde Islands; **I=** *Kalloconus (Trovaoconus) ateralbus* (Kiener, 1845), length 33 mm, in rock rubble, 2 m depth off Pedra Lume, Sal Island, Cape Verde Islands; **J=** *Africonus anthonyi* Petuch, 1975, length 6 mm, low tide off Serra Negra, Sal Island, Cape Verde Islands; **K=** *Africonus nelsonandradoi* (Cossignani and Fiadeiro, 2015), length 22 mm, 1 m depth off Calhetinha, Sal Island, Cape Verde Islands; **L=** *Africonus melissae* (Tenorio, Afonso, and Rolan, 2008), length 21 mm, 2 m depth off Palmeira, Sal Island, Cape Verde Islands.

Figure 3.16 Index Cone Shells of the Verdesian Province.
A= *Africonus fantasmalis* (Rolan, 1990), length 23 mm, 3 m depth off Calhetinha, Maio Island, Cape Verde, Islands; **B=** *Africonus galeao* (Rolan, 1996), length 25 mm, 2 m depth off Calheta, Maio Island, Cape Verde Islands; **C=** *Africonus raulsilvai* (Rolan, Monteiro, and Fernandes, 1998, length 23 mm, 1 m depth off Soca, Maio Island, Cape Verde Islands; **D=** *Africonus infinitus* (Rolan, 1990), length 21 mm, 1 m depth off Riberia Joao, Maio Island, Cape Verde Islands; **E=** *Africonus maioensis* (Trovao, Rolan, Ilidio, and Felix-Alves, 1990), length 28 mm, 2 m depth on rock reef, Calhetinha, Maio Island, Cape Verde Islands; **F=** *Africonus borgesi* (Trovao, 1979), length 25 mm, 2 m depth off Ponta da Cruz, Santa Luzia Island, Cape Verde Islands; **G=** *Africonus curralensis* (Rolan, 1986), length 24 mm, 1 m depth off Ponta da Cruz, Santa Luzia Island, Cape Verde Islands; **H=** *Africonus verdensis* (Trovao, 1979), length 21 mm, 1 m depth off Ponta da Cruz, Santa Luzia Island, Cape Verde Islands; **I=** *Africonus decoratus* (Rockel, Rolan, and Monteiro, 1980), length 28 mm, 2 m depth off Baia das Gatas, São Vicente Island, Cape Verde Islands; **J=** *Africonus freitasi* Tenorio, Rolan, Pires, *et al.*, 2018, length 12 mm, 2 m depth off Baia das Gatas, São Vicente Island, Cape Verde Islands; **K=** *Africonus lugubris* (Reeve, 1849), length 21 mm, 2 m depth off Lazareto, São Vicente Island, Cape Verde Islands.

Figure 3.17 Widespread Guinean Province Gastropods.
A= *Trona stercoraria* (Linnaeus, 1758), length 37 mm, low tide, N'Gor, Senegal; **B**= *Zonaria zonaria* (Gmelin, 1791), length 28 mm, low tide, N'Gor, Senegal; **C**= *Thetystrombus latus* (Gmelin, 1791), length 111 mm, 1 m depth off Port Gentil, Gabon; **D**= *Aporrhais elegantissima* Parenzan, 1970, length 35 mm, 50 m depth off Abidjan, Ivory Coast (often incorrectly referred to as "*Aporrhais pesgallinae*", which is actually a different species from the Namibian Province; see Figure 3.28 G, H); **E**= *Aporrhais senegalensis* Gray, 1838, length 21 mm, 30 m depth off Dakar, Senegal; **F**= *Hexaplex megacerus* (Sowerby II, 1834), length 51 mm, 2 m depth in Cape Lopez Bay, Port-Gentil, Gabon; **G**= *Truncularopsis rosarium* (Röding, 1798), length 57 mm, 2 m depth off M'Bour, Senegal; **H**= *Harpa doris* Röding, 1798, length 62 mm, 50 m depth off Abidjan, Ivory Coast; **I**= *Strephona flammulata* (Lamarck, 1811), length 37 mm, low tide on a sand flat, Joal Fadiout, Senegal; **J**= *Genuanoconus genuanus* (Linnaeus, 1758), length 51 mm, 50 m depth off Abidjan, Ivory Coast; **K**= *Kalloconus pulcher* (Lightfoot, 1786), length 124 mm, low tide, Joal Fadiout, Senegal; **L**= *Jaton decussatus* (Gmelin, 1791), length 42 mm, 1 m depth off N'Gor, Senegal.

Figure 3.18 Index Gastropods of the West Saharan Subprovince, Guinean Province.
A= *Phorcus sauciatus* (Koch, 1845), height 19 mm, on rocks at low tide, Oum Lbouer, Dakhla, Western Sahara (also found in the Canary Islands); **B**= *Trunculariopsis hoplites* (Fischer, 1876), length 54 mm, low tide on rocks, Dakhla Bay, Dakhla, Western Sahara; **C, D**= *Jaton hemitripterus* (Lamarck, 1816), length 38 mm, low tide on rocks, Dakhla Bay, Dakhla, Western Sahara; **E**= *Ocinebrina miscowichae* Pallary, 1920, length 21 mm, under rocks at low tide, Dakhla Bay, Dakhla, Western Sahara (also found in the Canary Islands); **F**= *Schilderia achatidea longinqua* (Schilder and Schilder, 1938), length 43 mm, trawled from 50 m depth off Dakhla Bay, Western Sahara; **G**= *Persicula pseudocingulata* Tournier, 1997, length 19 mm, low tide on sand bars, Dakhla Bay, Dakhla, Western Sahara; **H**= *Kalloconus byssinus* (Röding, 1798), length 48 mm, low tide on sand bars, Dahkla Bay, Dakhla, Western Sahara; **I, J**= *Varioconus saharicus* Petuch and Berschauer, 2016, length 32 mm, under rocks at low tide, Dakhla Bay, Dakhla, Western Sahara (note the deep purple-black interior of the aperture); **K**= *Marginella irrorata* Menke, 1828, length 41 mm, low tide in sand, Nouadhibou, Mauretania.

Figure 3.19 Index Gastropods of the Senegalian Subprovince, Guinean Province.
A= *Afer pseudofusinus* Frauseen and Hadorn, 2000, length 62 mm, 30 m depth off Cayar, Senegal; **B=** *Zonaria petitiana* (Crosse, 1872), length 25 mm, 30 m depth off M'Bour, Senegal; **C=** *Zonaria petitiana pseudopyrum* Bergonzoni, 2013, length 24 mm, trawled from 50 m depth off Kaloum, Conakry; **D=** *Purpurellus gambiensis* (Reeve, 1845), length 60 mm, 50 m depth off Dakar, Senegal; **E=** *Cymbium cymbium* (Linnaeus, 1758), length 80 mm, low tide, Joal Fadiout, Senegal; **F=** *Cymbium glans* (Gmelin, 1791), length 110 mm, 2 m depth, M'Bour, Senegal; **G=** *Cymbium marmoratum* Link, 1807, length 100 mm, 2 m depth, N'Gor, Senegal; **H=** *Glabella adansoni* (Kiener, 1834), length 26 mm, low tide, Joal Fadiout, Senegal; **I=** *Marginella desjardini* Marche-Marchad, 1957, length 47 mm, 50 m depth off Dakar, Senegal; **J=** *Monteiroconus ambiguus gernanti* (Petuch, 1975), length 35 mm, 50 m depth off Dakar, Senegal; **K=** *Zonaria pyrum senegalensis* Schilder, 1928, length 40 mm, 20 m depth off N'Gor, Senegal.

Figure 3.20 Index Gastropods of the Gorean Infraprovince, Guinean Province.
A= *Zonaria sanguinolenta* (Gmelin, 1791), length 24 mm, low tide, Goree Island, Dakar, Senegal; **B=** *Jaton flavidus* (Jousseaume, 1888) (variant with reduced scaly sculpture), length 38 mm, under rocks at low tide, Goree Island, Dakar, Senegal; **C=** *Varioconus bruguieresi* (Kiener, 1846), length 37 mm, 1 m depth off Goree Island, Dakar, Senegal; **D=** *Varioconus cloveri* (Walls, 1978), length 38 mm, 1 m depth off Goree Island, Dakar, Senegal; **E=** *Varioconus echinophilus* (Petuch, 1975), length 19 mm, in an eroded depression under an *Echinometra* sea urchin, off N'Gor, Senegal; **F=** *Varioconus hybridus* (Kiener, 1847), length 38 mm, low tide, N'Gor, Senegal; **G=** *Varioconus mercator* (Linnaeus, 1758), length 37 mm, low tide, Pointe Almadies, Dakar, Senegal; **H=** *Varioconus pineaui* (Pin and Tack, 1995), length 24 mm, in rubble, 2 m depth, off Popenguine, Petit Cote, Senegal; **I=** *Marginella aurantia* Lamarck, 1822, length 23 mm, under rocks at low tide, N'Gor, Cap Vert, Senegal; **J=** *Haliotis marmorata rosacea* Reeve, 1846, length 89 mm, on rocks at low tide, N'Gor, Cap Vert, Senegal; **K=** *Pusionella nifat* (Bruguiere, 1789), length 50 mm, 10 m depth off Goree Island, Senegal.

Figure 3.21 The *Varioconus* Species Radiation of the Gorean Infraprovince, Guinean Province.
A= *Varioconus belairensis* (Pin and Tack, 1989), length 37 mm, 2 m depth off Bel Air, Dakar, Senegal; **B**= *Varioconus cacao* (Ferrario, 1983), length 32 mm, low tide, Saly, Petit Cote, Senegal; **C**= *Varioconus dorotheae* Monnier and Limpalaër, 2010, length 41 mm, on sponges and rocks, 10 m depth off Madeleine Island, Cape Verde Peninsula, Senegal; **D**= *Varioconus franciscanus* (Hwass, 1792), length 59 mm, 2 m depth off Goree Island, Dakar, Senegal; **E**= *Varioconus guinaicus lamarcki* (Kiener, 1845), length 42 mm, under rocks at low tide, N'Gor, Senegal; **F**= *Varioconus reticulatus* (Born, 1778), length 26 mm, on a basaltic platform off N'Gor, Senegal; **G**= *Varioconus senegalensis* Gulden, Moolenbeek, and Goud, 2017, length 26 mm, low tide, between Ndayane and Popenguine, Senegal; **H**= *Varioconus tacomae* (Boyer and Pelorce, 2009), length 36 mm, 2 m depth off Plage des Amoureux, Goree Island, Dakar, Senegal; **I**= *Varioconus taslei* (Kiener, 1845), length 39 mm, 2 m depth off N'Gor, Senegal; **J**= *Varioconus unifasciatus* (Kiener, 1845), length 39 mm, 2 m depth off Plage des Amoureux, Goree Island, Dakar, Senegal; **K**= *Varioconus trencarti* (Nolf and Verstraeten, 2008), length 32 mm, low tide, Somone, Senegal.

Figure 3.22 Index Gastropods of the Gambian Infraprovince, Guinean Province.
A= *Zonaria zonaria gambiensis* (Shaw, 1909), length 26 mm, 3 m depth off Tanji, Gambia; **B**= *Natica collaria gambiae* Recluz, 1844, width 27 mm, low tide off Brufut, Gambia; **C**= *Natica royi* Pin, 1992, width 22 mm, low tide off Brufut, Gambia; **D**= *Varioconus fernandi* Petuch and Berschauer, 2018, length 37 mm, low tide on a rock reef, Tanji, Gambia; **E**= *Varioconus gambiensis* Petuch and Berschauer, 2018, length 33 mm, low tide on a rock reef, Tanji, Gambia; **F**= *Varioconus orri* (Ninomiya and daMotta, 1982), length 21 mm, low tide on a rock reef, Tanji, Gambia; **G**= *Jaton rikae* Petuch and Berschauer, 2019, length 18 mm, low tide on a rock reef off Gunyur, Gambia; **H**= *Varioconus rikae* Petuch and Berschauer, 2018, length 24 mm, low tide on a rock reef, Tanji, Gambia; **I**= *Varioconus wolof* Petuch and Berschauer, 2018, length 22 mm, low tide on a rock reef, Tanji, Gambia; **J**= *Bullia* (*Naytia*) *turrita* Gray, 1839 (possibly an unnamed subspecies), length 27 mm, low tide off Bakau, Gambia; **K**= *Pusionella milleti subgranulatus* (Petit de la Saussaye, 1851), length 30 mm, 5 m depth off Sanyang, Gambia.

Figure 3.23 Index Gastropods of the Biafran Subprovince, Guinean Province.
A= *Trona stercoraria cameroonica* Lorenz, 2017, length 29 mm, in mud and algae, 1 m depth off Londgi, Cameroon; **B**= *Zonaria angelicae* (Clover, 1974), length 22 mm, trawled from 50 m depth off Kribi, Cameroon; **C**= *Zonaria petitiana petiformis* Lorenz and Hubert, 1993, length 28 mm, 50 m depth off Malabo, Bioko Island, Bight of Biafra; **D**= *Africanella coseli* (Houart, 1989), length 27 mm, on rocks in 1 m depth, off Port-Gentil, Gabon; **E**= *Africanella isaacsi* (Houart, 1984), length 27 mm, on oysters, Bata, Equatorial Guinea; **F**= *Homalocantha melanomathos chinii* Biraghi, 1984, length 37 mm, on oysters and shell rubble in 2 m depth off Kribi, Cameroon; **G**= *Cymbium pachyus* (Pallary, 1930), length 75 mm, in sand and mud, 5 m depth off Kribi, Cameroon; **H**= *Agaronia razetoi* Terzer, 1992, length 38 mm, on mud flats at low tide, south of Limbe, Cameroon; **I**= *Anazola biraghii* Bernard and Nicolay, 1984, length 58 mm, 20 m depth off Kribi, Cameroon; **(Tomean Infraprovince) J**= *Insulamarginella spinacea* (Gofas and Fernandes, 1988), length 12 mm, 2 m depth off Porto Alegre, São Tome Island, São Tome e Principe; **K**= *Gibberula thomensis* (Tomlin, 1919), length 7 mm, 2 m depth off Porto Alegre, São Tome Island, São Tome e Principe.

Figure 3.24 Index Gastropods of the Angolan Subprovince, Guinean Province.
A= *Zonaria angolensis* (Odhner, 1923), length 32 mm, found on a Pen Shell bed, 10 m depth off Luanda, Angola; **B=** *Crassispira oliva* Fernandes, Rolan, and Otero-Schmitt, 1995, length 27 mm, 2 m depth in Lobito, Benguela Province, Angola; **C=** *Natica rocquignyi* Fischer-Piette, 1942, diameter 19 mm, 30 m depth off Porto Amboim, Cuanza Sul Province, Angola; **D=** *Jaton sinespina* Vermeij and Houart, 1996, length 35 mm, 2 m depth off Lobito, Benguela Province, Angola; **E=** *Muricopsis punctata* Houart, 1990, length 11 mm, low tide Lobito, Benguela Province, Angola; **F=** *Orania angolensis* Odhner, 1922, length 21 mm, 10 m depth off Luanda, Luanda Province, Angola; **G=** *Marginella lucani* (Jousseaume, 1884), length 15 mm, 10 m depth in sand off Cacuaco, Luanda Province, Angola; **H=** *Pseudonoduloconus carnalis* (Sowerby III, 1879), length 40 mm, 10 m depth off Mocamedes, Namibe Province, Angola; **I=** *Clavatula conica* (Edwards, 1857), length 36 mm, 15 m depth off Cacuaco, Luanda Province, Angola; **J=** *Pusionella compacta* Strebel, 1914, length 23 mm, 30 m depth off Cacuaco, Luanda Province, Angola; **K=** *Athleta emmanuelae* (Rosso, 1976), length 60 mm, trawled by fisherman from 450 m depth off N'zeto, Zaire Province, Angola (a northern Angolan endemic that was mistakenly said to be from Senegal and Mozambique); (**Luandan Infraprovince**) **L=** *Varioconus negroides* (Paes da Franca *in* Kaicher, 1977), length 25 mm, 3 m depth off Chapeu Armado, São Nicolau, Angola.

Figure 3.25 The *Varioconus* Species Radiation of the Luandan Infraprovince, Angolan Subprovince.
A= *Varioconus annagretae* (Schonherr, 2018), length 31 mm, 3 m depth in Limagens Bay, Benguela Province, Angola; **B**= *Varioconus africanus* (Kiener, 1848), length 25 mm, 2 m depth in rock crevices, Lobito, Benguela Province, Angola; **C**= *Varioconus lineopunctatus* (Kaicher, 1977), length 34 mm, in sand, 2 m depth off Lobito, Benguela Province, Angola; **D**= *Varioconus zebroides* (Kiener, 1845), length 27 mm, under rocks in 3 m depth, Baia dos Elefantes, Benguela Province, Angola; **E**= *Varioconus naranjus* (Trovao, 1975), length 24 mm, in sand, 2 m depth in Baia Azul, Benguela Province, Angola; **F**= *Varioconus variegatus* (Kiener, 1845), length 22 mm, 5 m depth off Lobito, Benguela Province, Angola; **G**= *Varioconus nunesi* (Schonherr, 2018), length 33 mm, 4 m depth in Equimina Bay, Benguela Province, Angola; **H**= *Varioconus equiminaensis* (Schonherr, 2018), length 34 mm, in rock crevices, 4 m depth in Equimina Bay, Benguela Province, Angola; **I**= *Varioconus allaryi* (Bozzetti, 2008), length 27 mm, 5 m depth off Santo Antonio, Benguela Province, Angola; **J**= *Varioconus tenuilineatus* (Rolan and Rockel, 2001), length 24 mm, 1 m depth, Baia dos Elefantes, Benguela Province, Angola; **K**= *Varioconus albuquerquei* (Trovao, 1975), length 15 mm, low tide, Cape Santa Maria, Benguela Province, Angola; **L**= *Varioconus micropunctatus* (Rolan and Rockel, 2000), length 23 mm, 2 m depth in Baia dos Elefantes, Benguela Province, Angola.

Figure 3.26 The *Varioconus* Species Radiation of the Namibe Infraprovince, Angolan Subprovince.
A= *Varioconus bocagei* (Trovao, 1978), length 24 mm, 5 m depth, Lucira Bay, Namibe Province, Angola; **B**= *Varioconus babaensis* (Rolan and Rockel, 2001), length 24 mm, 2 m depth in Baia do Baba, Namibe Province, Angola; **C**= *Varioconus trovaoi* (Rolan and Rockel, 2000), length 24 mm, 5 m depth in Baia do Bom Fim, Lucira, Namibe Province, Angola; **D**= *Varioconus cepasi* (Trovao, 1975), length 34 mm, in sand, 5 m depth off Mocamedes, Namibe Province, Angola; **E**= *Varioconus chytreus* (Melvill, 1884), length 33 mm, 3 m depth off Lucira, Namibe Province, Angola; **F**= *Varioconus flavusalbus* (Rolan and Rockel, 2000), length 23 mm, 2 m depth in Baia de Charungo, Namibe Province, Angola; **G**= *Varioconus fuscolineatus* (Sowerby III, 1905), length 30 mm, 3 m depth off Bentiaba, Namibe Province, Angola; **H**= *Varioconus nobrei* (Trovao, 1975), length 17 mm, 3 m depth off Lucira, Namibe Province, Angola; **I**= *Varioconus petuchi* Monteiro, Afonso, Tenorio, *et al.*, 2014, length 28 mm, 2 m depth in Baia do Baba, Namibe Province, Angola; **J**= *Varioconus alexandrinus* (Kaicher, 1977), length 24 mm, 1 m depth in Baia do Baba, Namibe Province, Angola (*tevesi* and *musivus* are synonyms); **K**= *Varioconus eusebioi* (Schonherr, 2018), length 22 mm, low tide, Saco Mar, Namibe Province, Angola; **L**= *Varioconus franciscoi* (Rolan and Rockel, 2000), length 22 mm, 5 m depth in Baia do Cesar, Namibe Province, Angola.

Figure 3.27 Index Gastropods of the Helenean Subprovince, Guinean Province.
A= *Luria lurida oceanica* Schilder, 1930, length 36 mm, 5 m depth, off Sandy Bay, St. Helena Island; **B=** *Nassarius sanctaehelenae* A.Adams, 1852, length 9 mm, in sand pockets, 10 m depth in James Bay, St. Helena Island; **C=** *Naria sanctaehelenae* (Schilder, 1930), length 24 mm, 10 m depth, off English Beach, Ascension Island; **D, E=** *Naria sanctaehelenae bonapartei* Lorenz, 2017, length 31 mm, 5 m depth off Lot's Wife's Ponds, St. Helena Island; **F=** *Naria sanctaehelenae bonapartei* Lorenz, 2017, (large variant) length 34 mm, 5 m depth off Sandy Bay, St. Helena Island; **G=** *Cymatona philomelae* (Watson, 1880), length 25 mm, 50 m depth off Sandy Point, Tristan da Cunha Island, South Atlantic; **H=** *Harpa doris robusta* Röding, 1798, length 40 mm, 10 m depth off English Beach, Ascension Island; **I, J=** *Varioconus jourdani* (da Motta, 1984), length 27 mm, 2 m depth off Lot's Wife's Ponds, St. Helena Island.

Figure 3.28 Index Gastropods of the Namibian Province.
A= *Callumbonella namibiensis* Rolan, Gonzalez-Porto, and de Matos-Pita, 2009, diameter 12 mm, 400 m depth, between Luderitz and Walvis Bay, Namibia; **B, C=** *Patella swakopmundensis* Massier, 2009, width 67 mm, low tide on rocks, Langstrand, Swakopmund, Namibia; **D, E=** *Cypraeovula namibiensis* Massier, 2006, length 26 mm, along the open coast north of Oranjemund, Namibia; **F=** *Euspira massieri* Petuch and Berschauer, 2018, height 36 mm, 400 m depth off Walvis Bay, Namibia; **G, H=** *Aporrhais pesgallinae* Barnard, 1963, length 38 mm, 350 m depth off Walvis Bay, Namibia (often confused with the similar *Aporrhais elegantissima*, which occurs farther north along West Africa; see Figure 3.17 D); **I=** *Nucella ovalis* (Blainville, 1832), length 37 mm, low tide, north of the desalinization plant at Wlotzkasbaken, Namibia; **J=** *Nassarius scopularcus* (Barnard, 1959), length 22 mm, on the beach, 3 km north of Walvis Bay, Namibia; **K=** *Sciteconus patens* (Sowerby III, 1903), length 87 mm, 250 m depth off Luderitz, Namibia.

Figure 3.29 Index Gastropods of the Namibian Province.
A= *Athleta disparilis* (Rehder, 1969), length 36 mm, trawled from 400 m depth south of Luderitz, Namibia; **B, C**= *Athleta easoni* Petuch and Berschauer, 2017, length 70 mm, 400 m depth off Walvis Bay, Namibia; **D, E**= *Athleta massieri* Petuch and Berschauer, 2017, length 47 mm, 300 m depth of Walvis Bay, Namibia; **F, G**= *Kaokomarginella stuarti* (Kilburn, 1977), length 13 mm, low tide, Meobbaai, central Namib Desert coast, Namibia; **H**= *Prunum walvisianum* (Tomlin, 1902), length 14 mm, 100 m depth off Walvis Bay, Namibia; **I, J**= *Roseamarginella nimbosa* (Veldsman, 2013), length 26 mm, along the open coast north of Oranjemund, Namibia; **K**= *Comitas saldanhae* (Barnard, 1958), length 58 mm, 300 m depth off Luderitz, Namibia; **L**= *Athleta lutosa* (Koch, 1948), length 73 mm, 400 m depth off Oranjemund, Namibia.

CHAPTER 4. Eastern Pacific Region

The Eastern Pacific Region spans the western coasts of North and South America, from the Aleutian Islands of Alaska south to the Wollaston Islands of Chile, and also offshore island chains such as the Galapagos, Cocos, Clipperton, Malpelo, and the Revillagigedos. This immense region mostly encompasses areas that have cold Arctic and Subantarctic oceanographic conditions, from Alaska to Oregon in the north and from northern Chile to the Strait of Magellan area in the south. Within this extensive region, only two centrally-located, proportionally-smaller areas are relevant to this book: the **Northeastern Pacific Paratropical Subregion** and the **Eastern Pacific Tropical Subregion**. The combined area of these two subregions only encompasses one paratropical province, one tropical province, 9 subprovinces, and one provinciatone. The provinces include the **Californian Province** in the north and the **Panamic Province** in the south. The spatial arrangement of these provinces is shown here on Figure 4.1. Unlike the provincial arrangement of the western Atlantic, with two central eutropical provinces (the Caribbean and Brazilian) being bounded by a paratropical province to the north and south (the Carolinian and Paulinian), the Eastern Pacific differs in having one central eutropical province (the Panamic) that is bounded by a paratropical province only in the north (the Californian). This asymmetry is caused by the cold, northward-flowing Humboldt Current, which comes very close to shore and bathes the entire coastline south of northernmost Peru in subantarctic water. Classic eutropical/paratropical gastropod families such as the Cypraeidae, Conidae, Strombidae, Melongenidae, and Ficidae all disappear abruptly south of Paita, Peru, and are replaced by the cold water molluscan fauna of the Peruvian Province. Only one tropical family has managed to survive in this subantarctic water mass, the Olividae and its single representative species, *Felicioliva peruviana*.

A number of wide-ranging gastropods demarcate the boundaries of the northern section of the Eastern Pacific Region (the Aleutian, Oregonian, and Californian Provinces), including the haliotids *Haliotis kamtschatkana* and *Haliotis walallensis*, the eucyclidids *Bathybembix bairdii* and *Cidarina cidaris*, the calliostomatid *Calliostoma annulatum*, the astraeine turbinid *Pomaulax gibberosus*, the muricids *Ceratostoma foliatum*, *Nucella emarginata*, and *Paciocinebrina lurida*, the buccinid *Neptunea (Sulcosipho) tabulata*, and the olivellids *Callianax biplicata* and *C. alectona* (some shown here on Figure 4.4). The provinces of the southern section of the region (the Panamic and Peruvian) have virtually no species in common, changing abruptly from a subantarctic fauna to a tropical one in a very narrow boundary area at Paita, Peru. This is the sharpest faunal transition between provinces known from anywhere on Earth.

Northeastern Pacific Paratropical Subregion

Ranging from Point Conception, California, to southern Baja California Sur, Mexico, the Northeastern Pacific Paratropical Subregion contains only a single major biogeographical unit, the Californian Province. Unlike the Panamic Province to the south, the molluscan faunas of the Californian Province grade directly into those of

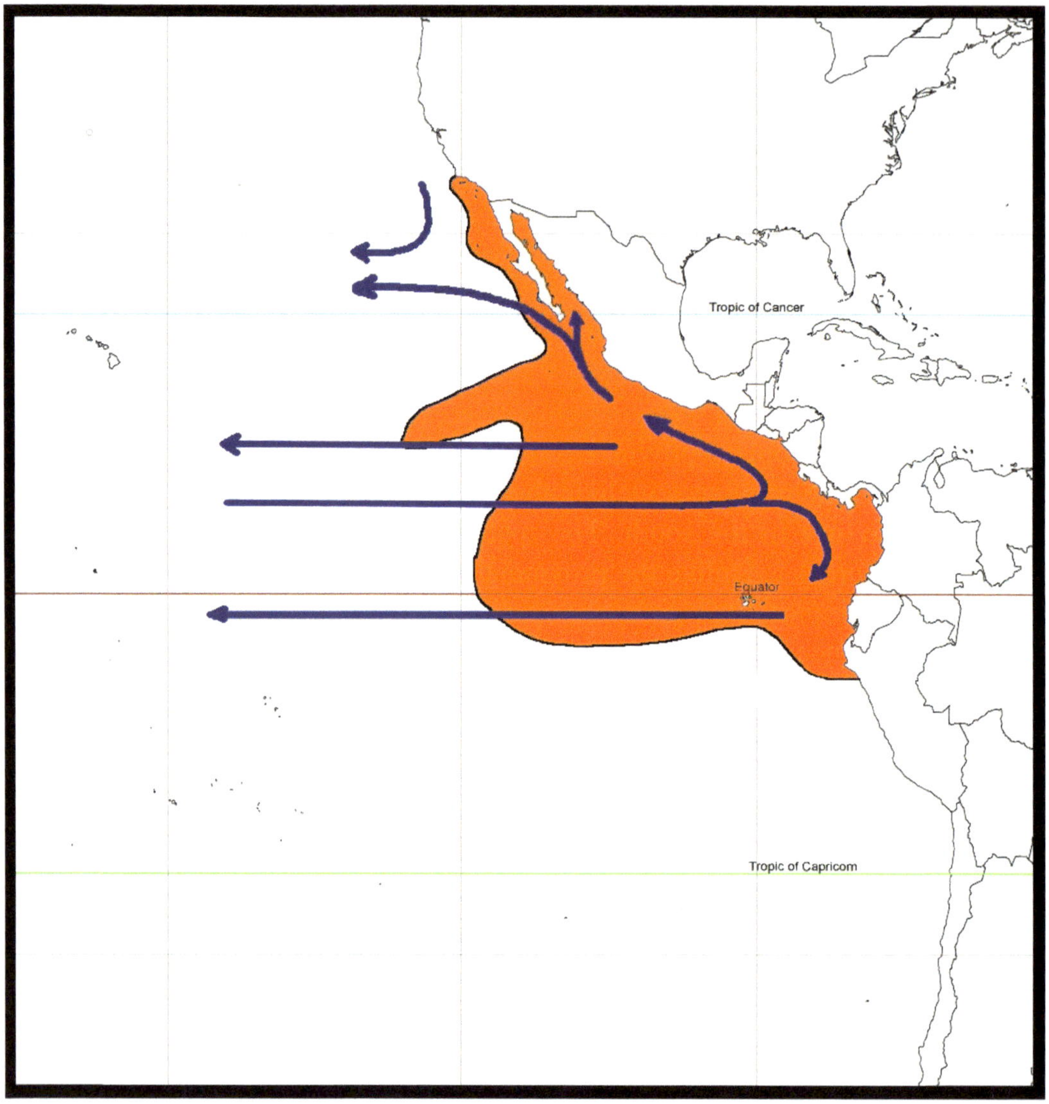

Figure 4.1 Map of the Eastern Pacific Ocean and the Eastern Pacific Region, showing the areal extent of the combined **Northeastern Paratropical Subregion** and the **Tropical Eastern Pacific Subregion** and the primary warm water currents.

the cold water Northeastern Pacific Temperate Subregion (Oregonian Province) to the north, over a broad area that extends from southern Oregon south to Point Conception. This transitional zone, the **Northern Californian Provinciatone**, houses a small number of endemic gastropods (provinciatonal endemics), the most important being the tegulid *Tegula montereyi* (Figure 4.5L). These endemics occur along with wide-ranging cold-temperate Oregonian Province species, such as the muricids *Boreotrophon tripherus*, *Boreotrophon raymondi*, *Scabrotrophon lasius*, and *Paciocinebrina atropurpurea*, the olivellids *Callianax pycna*, the naticid *Callinaticina oldroydii*, and the pseudomelatomid *Antiplanes catalinae*. Within the provinciatone, these cold water species are sympatric with tropically-derived species such as the cypraeid *Neobernaya spadicea* (Monterey southward), the ovulid *Neosimna vidleri*, the mitrid *Atrimitra idae*, and the cone shell *Californiconus californicus* (San Francisco southward). The full complement of Californian Province index taxa first appears at Monterey Bay and the gastropod fauna becomes increasingly more species-rich farther to the south.

Californian Molluscan Province

Extending from Point Conception, California, in the north to Punta Abreojos, Baja California Sur, Mexico, in the south and encompassing offshore islands such as the Channel Islands, Santa Catalina, Guadalupe, Cedros, San Benito, and others, the Californian Province is a classic paratropical province, containing only a partial complement of tropically-derived molluscan families. Although lacking important eutropical families such as the Modulidae, Strombidae, Harpidae, Xenophoridae, Cassidae, Ficidae, and Melongenidae, the province does contain other tropically-derived groups such as the families Cypraeidae (*Neobernaya spadicea*), Potamididae (*Cerithidea californica*), Bursidae (*Crossata californica*), Terebridae (*Terebra pedroana*), and Conidae (*Californiconus californicus*). Based upon the work of Berschauer and Clark (2018), the Californian Province is now known to be a far more complicated biogeographical unit than was originally thought, comprising three separate subprovinces that are here named the **Diegan**, **Cedrosian**, and **Guadalupean**. The combined area of these subprovinces houses an amazingly-rich molluscan fauna, with the most note-worthy component being nearly 50 different muricid species, the second-largest muricid fauna known from paratropical seas anywhere on Earth. Some of these classic Californian species, which all belong to Eastern Pacific endemic genera, include *Paciocinebrina circumtexta*, *Acanthinucella spirata*, *Roperia poulsoni*, *Maxwellia gemma*, *Maxwellia santarosana*, *Calcitrapessa leeana*, and *Forreria belcheri* (all shown here on Figures 4.6, 4.7, and 4.8). Other important Californian endemic taxa include the monotypic cone shell genus *Californiconus* (*C. californicus*), the monotypic cowrie genus *Neobernaya* (*N. spadicea*), the turbinid genus *Megastraea* (with *M. undosa* and *M. turbanica*), and the monotypic fissurellid genus *Megathura* (*M. crenulata*). The Californian Province also houses the single largest abalone (Haliotidae) fauna found in a major biogeographical unit known from anywhere on Earth, containing 11 species and subspecies as well as the largest abalone species, *Haliotis rufescens* (all shown here on Figures 4.4 through 4.8). For a detailed overview and iconography of the Californian Province molluscan fauna, see Berschauer and Clark, 2018.

Diegan Subprovince

Named for San Diego, California, this northern subprovince of the Californian Province extends from Point Conception, California, south to Cabo Colonet, Baja California Norte, Mexico, and encompasses the entire area of the Southern California Bight (Point Conception to Punta Banda, Mexico; Berschauer and Clark, 2018: 3, 122). Oceanographically, this area contains cooler water temperatures than do the areas south of Cabo Colonet and is under the influence of cold water upwellings and gyre-like giant eddies that form on the lee sides of the Channel Islands. This stable paratropical marine environment has allowed for the evolution of a distinctive mixed molluscan fauna, composed of classic widespread Californian Province living sympatrically with a large number of endemic taxa. Some of the more important

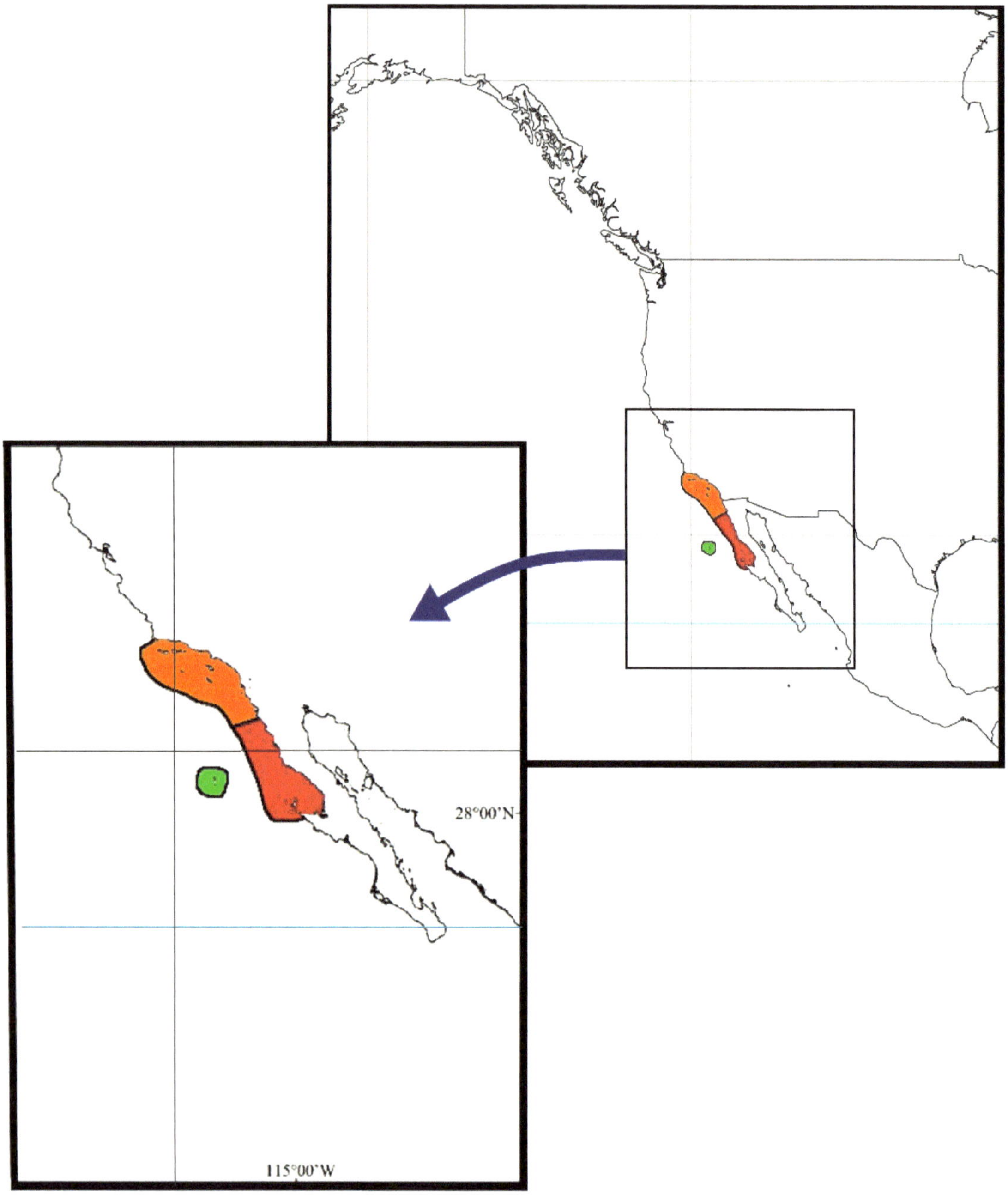

Figure 4.2 Map of the Californian Molluscan Province, showing the areal extents of its subprovinces: the **Diegan Subprovince** (orange), the **Cedrosian Subprovince** (red), and the **Guadalupean Subprovince** (green).

Diegan endemic gastropods include the tegulids *Stearnsia regina* (belonging to the monotypic genus *Stearnsia*) and *Tegula umbilicatum*, the fasciolariid *Harfordia chucksnelli*, a very large endemic muricid fauna containing species such as *Paciocinebrina barbarensis*, *Paciocinebrina grippi*, *Paciocinebrina beta*, *Scabrotrophon cerritensis*, *Scabrotrophon grovesi*, *Boreotrophon peregrinus*, *Boreotrophon triangulatus*, *Boreotrophon bentleyi*, and *Boreotrophon keepi*, the mitrid *Atrimitra catalinae*, the olivellid *Callianax diegensis*, and the pseudomelatomid *Crassispira semiinflata* (some shown here on Figures 4.6 and 4.7).

Cedrosian Subprovince

Named for Cedros Island, off Punta Eugenia, Viscaino Peninsula, Baja California, this southern subprovince of the Californian Province extends from Cabo Colonet, Baja California south to near Cabo Abreojos, Baja California Sur. Being outside of the Diegan cold water upwellings and eddies, the Cedrosian Subprovince contains warmer oceanographic conditions, particularly in the extreme far southern end, where it grades into the Magdalenan Subprovince of the eutropical Panamic Province (discussed later in this chapter). Although containing almost the full complement of widespread Californian species, the Cedrosian Subprovince also contains numerous endemic taxa, some of which are confined to very small areas. A classic example of these geographically-restricted Cedrosian taxa is seen in the fasciolariid *Hesperaptyxis negusi*, which is endemic to San Benito Island near the far southernmost boundary of the subprovince. Other Cedrosian endemics include the abalone *Haliotis fulgens turveri*, the large pseudolivids *Macron orcutti*, *Macron aethiops*, and *Macron aethiops* form *kellettii*, the fasciolariid *Callifusus edjanssi*, and the large muricid *Austrotrophon pinnatus*. Some of these are illustrated here on Figure 4.8. Several prominent Cedrosian taxa also range into the southern part of the Diegan Subprovince, producing a separate component of faunal overlap. Most noteworthy of these overlap species are the turbinid *Megastraea turbanica* and the unusual three-spined muricid *Calcitrapessa leeana* (both shown here on Figure 4.8).

Guadalupean Subprovince

Located 400 km southwest of the city of Ensenada, Baja California, Mexico, the isolated volcanic island of Guadalupe contains the only completely insular subprovince of the Californian Province. Composed of the remnants of two overlapping shield volcanoes that had formed on a sea floor spreading center, this basaltic island, and a few small adjacent rocky islets, rises abruptly out of depths of over 4000 m. The steep cliffs that form much of the coastline continue underwater, resulting in sheer submarine cliffs and a sharply-angled slope. These abrupt drop-offs, and the adjacent high-productivity water conditions and abundant fish populations, afford the perfect habitat for abundant sharks, including a large resident population of Great White Sharks. The neritic habitats available to marine mollusks on Guadalupe Island are very limited, comprising only rocky intertidal, high wave-action exposed cliffs and subtidal basaltic rock platforms. These limited types of biotopes, coupled with the wide geographical isolation from the Baja California mainland, have allowed only a very few types of Californian Province gastropods to become established on the island. Some of these include the littorinid *Littorina planaxis* and the fasciolariid *Barbarofusus kobelti*. Although containing a highly-impoverished malacofauna, a small component of endemic Guadalupean species has managed to evolve on the limited spatial configuration of this insular subprovince. Most noteworthy of these is a unique fauna of endemic abalones, including *Haliotis corrugatus oweni*, *Haliotis cracherodii californiensis*, and *Haliotis fulgens guadalupensis* (shown here on Figure 4.8). These occur together with other island endemics such as the tegulid *Tegula gallina multifilosa*, the vermetid *Dendropoma* cf.

lituella (probably an unnamed species), the turbinid *Pomaulax gibberosus guadalupeanus*, the small fasciolariid *Barbarofusus guadalupensis*, and the muricid *Paciocinebrina seftoni* (some of these are shown here on Figure 4.8).

Eastern Pacific Tropical Subregion

Extending from the southern tip of Baja California, Mexico, to Paita, Peru, and including the entire Gulf of California, Cocos Island, Clipperton Island, and the Galapagos Islands, the Eastern Pacific Tropical Subregion encompasses only a single eutropical molluscan province, the **Panamic Province**. Although adjacent to the Polynesian Province of the Indo-Pacific Super-Region in the west, the resident molluscan faunas of the tropical Eastern Pacific are much more closely related to the malacofaunas of the Caribbean and Brazilian Provinces of the tropical western Atlantic. The gastropod taxa of these three eutropical American provinces all had common ancestors in the Miocene-Pliocene **Gatunian Paleoprovince** (named for the Gatun Formation of Panama) of the ancient Caribbean Sea, which extended across the then-flooded Isthmus of Panama as a single giant faunal province (see Petuch, 2004 for details on American paleobiogeography and paleoceanography). Once Panama closed-off during the late Pliocene, the ancient Caribbean fauna was divided into two large sections; the western side evolved into the Recent Panamic Province and the eastern side evolved into the Recent Caribbean and Brazilian Provinces. This shared common ancestry is seen in the presence of many ubiquitous gastropod genera, found in all three Recent sister provinces, such as the strombid *Strombus*, the cypraeid *Macrocypraea*, the ovulid *Cyphoma*, the muricids *Phyllonotus*, the volute *Enaeta*, the olivid *Americoliva*, and the cone shell *Dauciconus*. For a detailed overview of the Panamic molluscan fauna, see Myra Keen's classic work, *Sea Shells of Tropical West America* (1971).

Panamic Molluscan Province

Named for the country of Panama, this eutropical province extends from Punta Abreojos, Baja California Sur, Mexico, south to Isla Lobos de Tierra, Paita District, Peru, and encompasses the offshore islands of Clipperton, Cocos, Malpelo, and Gorgona and the archipelagos of the Revillagigedos, Tres Marias, and Galapagos. Throughout its entire extent, the Panamic malacofauna is relatively homogenous in composition, with most species having large ranges that cover the entire province. Some of these classic widespread Panamic index species, which range from Baja California, Mexico, to northern Peru, include the eocypraeid *Jenneria pustulata*, the cypraeid *Pseudozonaria arabicula*, the tonnid *Malea ringens*, the muricid *Homalocantha oxyacantha*, the fasciolariid *Opeatostoma pseudodon*, the harpid *Harpa crenata*, and the cone shells *Gradiconus monilifer* and *Ximeniconus ximenes* (all shown here on Figures 4.9 and 4.10). Besides the extensive area defined by these wide-ranging species, several smaller areas of localized endemism also occur along the north-south gradient of the province. These areas here are recognized as six previously-undescribed subprovinces, including the **Magdalenan**, **Cortezian**, **Jaliscoan**, **Cocosian**, **Chiriquian**, and **Ecuadorian**. Four infraprovincial evolutionary hot spots are also known to exist within these subprovinces and these will be described in the following sections. Some of the more prominent widespread Panamic gastropods are illustrated here on Figures 4.9 and 4.10.

Besides the ubiquitous Panamic fauna, the province also houses a large number of distinctive endemic genera and species complexes, such as the cypraeid genus *Pseudozonaria*, the muricid genus *Zacatrophon*, the olivid genera *Felicioliva*, *Vullietoliva*, *Porphyria*, and *Strephonella*, the fasciolariid genus *Hesperaptyxis*, and the conid genera *Ductoconus*, *Ximeniconus*, *Globiconus*, and *Pyruconus*. These endemic Panamic groups co-occur with a small number of important genera and species that derived from the Indo-Pacific Super-Province to the west. Some of these include a cowrie of the *Luria isabella* species complex (*L. isabellamexicana*), the muricid *Quoyula monodonta*, and cone shells of the genera *Cylinder* (*C. dalli*, a "textile cone"), *Harmoniconus* (*H. nux*), *Tesselliconus* (*T. edaphus*), and *Miliariconus* (*M. tiaratus*). These Indo-Pacific-derived taxa became established on the Western American coast in the early Pleistocene after the closing of the Isthmus of Panama, as evidenced by their absence from the Recent faunas of the Caribbean and Brazilian Provinces and from the western Atlantic Pliocene fossil record. Several wide-ranging Indo-Pacific species have also become established on the offshore islands and along the western coast of Costa Rica and Panama, including the cowrie *Talparia talpa*, the cone shells *Virroconus ebraeus* and *Virroconus chaldeus*, the terebrid *Oxymeris maculata* (on Cocos, Malpelo, and Gorgona Islands and the Revillagigedo Islands), and the mitrid *Mitra mitra* (on Cocos, Malpelo, and Gorgona Islands and the Perlas Islands). The long-lived planktotrophic veligers of these Indo-Pacific taxa can cross the immense Eastern Pacific barrier by utilizing the accelerated eastward-flowing warm currents during times of "El Niño"-Southern Oscillation warm water events.

Another characteristic of the Panamic malacofauna is the presence of many species that have close analogues in the tropical Western Atlantic. These species pairs, or "cognate species", had a common ancestor in the Pliocene Gatunian Paleoprovince and were bisected into two mirror-image taxa when the Panamanian land bridge developed. After 3 million years of genetic isolation and genetic drift, these cognate pairs have diverged morphologically but still retain enough similarities to be recognizable as trans-isthmian twins. Some of the more important of these include:

1. (Cypraeidae) *Macrocypraea cervinetta* (Panamic)--*Macrocypraea zebra* (Caribbean)
2. (Ovulidae) *Cyphoma emarginatum* (Panamic)--*Cyphoma signatum* (Caribbean)
3. (Cassidae) *Cypraecassis coarctatum* (Panamic)--*Cypraecassis testiculus* (Caribbean); *Semicassis centiquadrata* (Panamic)--*Semicassis granulatum* (Caribbean)
4. (Strombidae) *Strombus gracilior* (Panamic)--*Strombus pugilis* (Caribbean); *Titanostrombus galeatus* (Panamic)--*Titanostrombus goliath* (Brazilian)
5. (Muricidae) *Murexiella humilis* (Panamic)--*Murexiella hilli* (Caribbean); *Phyllonotus peratus* (Panamic)--*Phyllonotus pomum* (Caribbean); *Plicopurpura pansa*

(Panamic)--*Plicopurpura patula* (Caribbean); *Stramonita biserialis* (Panamic)--*Stramonita floridana* (Caribbean)
6. (Fasciolariidae) *Triplofusus princeps* (Panamic)--*Triplofusus papillosus* (Carolinian)
7. (Melongenidae) *Melongena patula* (Panamic)--*Melongena melongena* (Caribbean)
8. (Buccinidae) *Truncaria filosa* (Panamic)--*Truncaria lindae* (Caribbean)
9. (Strombinidae) *Cotonopsis mendozana* (Panamic)--*Cotonopsis lindae* (Caribbean)
10. (Conidae) *Gradiconus scalaris* (Panamic)--*Gradiconus parascalaris* (Caribbean); *Gradiconus regularis* (Panamic)--*Gradiconus cingulatus* (Caribbean); *Dauciconus virgatus* (Panamic)-- *Dauciconus daucus* (Caribbean); *Tenorioconus archon* (Panamic)--*Tenorioconus cedonulli* (Caribbean); *Gladioconus gladiator* (Panamic)--*Gladioconus mus* (Caribbean); *Perplexiconus perplexus* (Panamic)--*Perplexiconus puncticulatus* (Caribbean)
11. (Terebridae) *Pristiterebra glauca* (Panamic)--*Pristiterebra petiveriana* (Caribbean)
12. (Pseudomelatomidae) *Knefastia olivacea* (Panamic)--*Knefastia hilli* (Caribbean); *Hindsiclava militaris* (Panamic)--*Hindsiclava tippetti* (Caribbean)
13. (Drilliidae) *Clathrodrillia berryi* (Panamic)--*Clathrodrillia petuchi* (Caribbean)
14. (Turridae) *Polystira nobilis* (Panamic)--*Polystira coltrorum* (Brazilian)

A review of the paleontology of the Pliocene Caribbean fossil shell beds shows that the faunal composition of the modern Panamic Province is actually much more similar to that of the ancient Gatunian Paleoprovince than it is to that of the Recent Caribbean Province (see Petuch, 2004 for illustrations of Gatunian index fossils). The entire Panamic Province is essentially a living version of the Caribbean Gatunian Paleoprovince and can be considered to be a relictual fauna. During the severe sea level drops of the Pleistocene, the Caribbean banks, seamounts, and coral reef systems were emergent and the shallow water environments, along with their resident molluscan faunas, were destroyed. On the other hand, the long, continuous coastline of the tropical Eastern Pacific area, with its narrow continental shelf and step drop-off, was not as affected by these Pleistocene sea level drops. The majority of the Gatunian molluscan fauna managed to survive into recent times as the malacofauna of the Panamic Province.

Magdalenan Subprovince

Named for Bahia Magdalena, Baja California Sur, Mexico, the Magdalenan Subprovince extends from near Punta Abreojos, Baja California Sur, around Cabo San Lucas, to La Paz, Baja California Sur. The northern boundary of the subprovince marks the first appearance of true eutropical marine faunas along the Eastern Pacific coast, including Panamic Province index species such as the cypraeids *Naria albuginosa* and *Pseudozonaria arabicula*, the strombids *Strombus gracilior* and *Titanostrombus galeatus*, the muricid *Phyllonotus erythrostomus*, and the conids *Miliariconus tiaratus* and *Lividiconus diadema*. The subprovince also houses the largest coral reef complex known from anywhere along the western Mexican mainland, represented by large zonated reef structures off Cabo Pulmo, dominated by

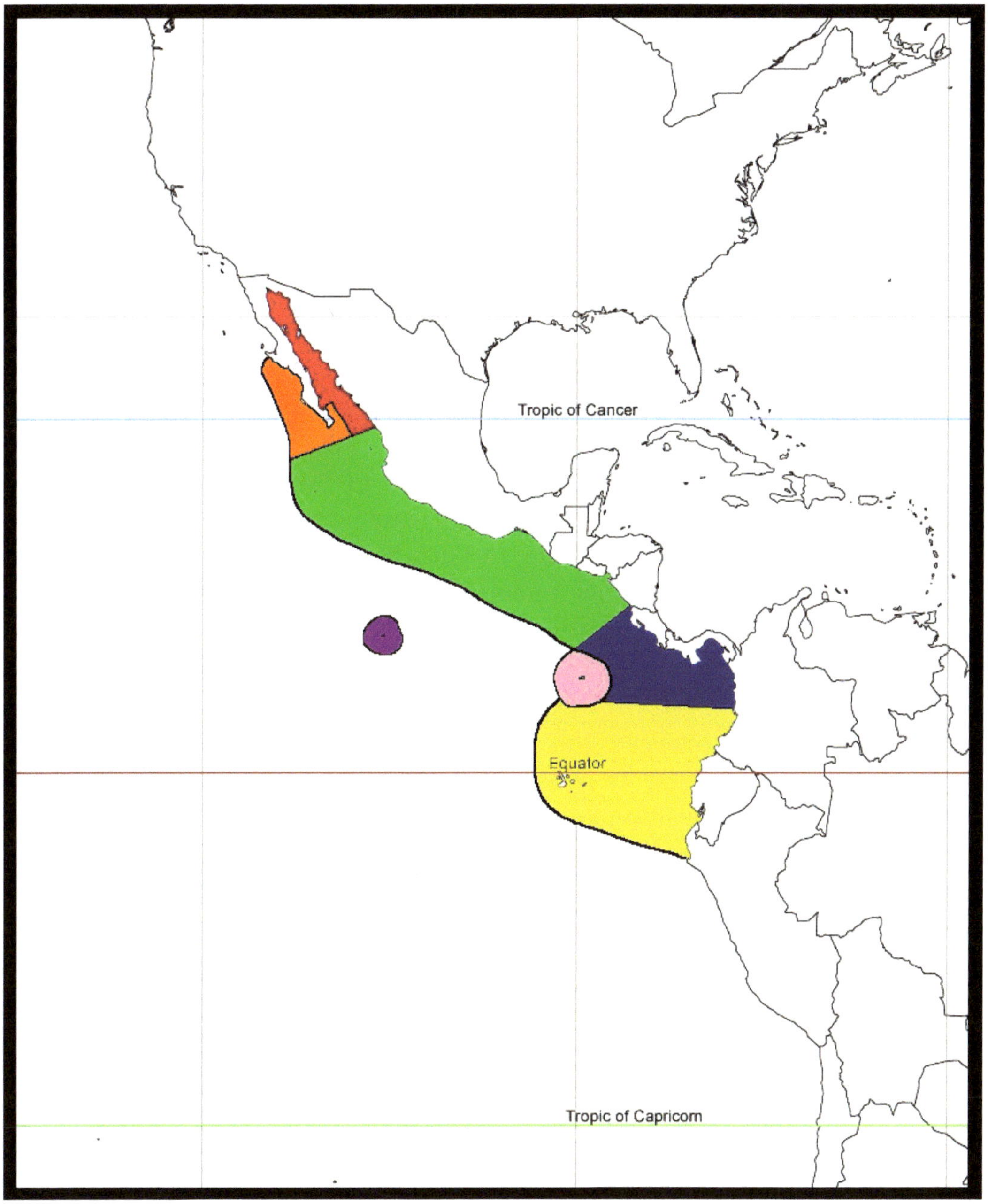

Figure 4.3 Map of the Panamic Molluscan Province, showing the areal extents of its subprovinces: the **Magdalenan Subprovince** (orange), the **Cortezian Subprovince** (red), the **Jaliscoan Subprovince** (green), the **Cocosian Subprovince** (light rose), the **Chiriquian Subprovince** (blue), and the **Ecuadorian Subprovince** (yellow), along with the **Clippertonian Infraprovince,** an isolated island component of the **Jaliscoan Subprovince** (purple), which has a unique mixed Polynesian-Panamic fauna.

Pocillopora corals. The Magdalenan Subprovince also contains a significant number of endemic gastropods, with the most noteworthy being the fasciolariid *Callifusus irregularis*, the muricid *Ceratostoma monoceros*, the small, high-spired volute *Enaeta cumingii pederseni*, the olivid *Americoliva hemphilli*, and the conid *Gradiconus magdalenensis* (all shown here on Figure 4.11). Of special interest within the Magdalenan Subprovince is a localized evolutionary hot spot that spans an area extending from Cabo San Lucas to Isla Cerralvo, at the southernmost tip of the Baja

California peninsula. Referred to here as the **Lucasian Infraprovince** (named for Cabo San Lucas), this center of speciation houses an interesting gastropod fauna that contains many endemic taxa, most of which have very small geographical ranges. Some of the more important of these include the olivids *Americoliva cumingi* and *Americoliva violacea*, and a small species radiation of the cone shell genus *Gradiconus*, including *G. dispar*, *G. skoglundae*, and *G. nybakkeni* (illustrated here on Figure 4.11). Some Lucasian species, such as the elongated olivid *Americoliva grovesi* (Figure 4.11G), are restricted to isolated offshore banks near the mouth of the Gulf of California, indicating that these flat-topped seamounts may house new and previously-unstudied ecosystems.

Cortezian Subprovince

Named for the "Sea of Cortez" (the Gulf of California), the Cortezian Subprovince encompasses the entire Gulf of California north of La Paz, including the major islands of Tiburon and Angel de la Guarda, and extends as far south as Topolobampo, Sinaloa State, along the mainland coast. In the Topolobampo area, the extensive brackish water estuary of the Rio Fuerte and the Saliaca, La Presa, Navachiste, Bacorehuis, and Ohuira Lagoon Systems, altogether, act as an ecological barrier to the southward dispersal of Cortezian endemic species. During the low sea level stands of the late Pleistocene, the northern third of the Gulf of California was oceanographically isolated from the southern two-thirds by a narrow emergent land bridge that connected Baja California with the Mexican mainland. This land barrier, formed by a narrow isthmus across the "Midriff Archipelago Region" of San Lorenzo, San Esteban, and Tiburon Islands, effectively cutting off the northern Gulf region from the Pacific Ocean. This isolated marine world, here named the **Sonoran Paleosea** (for the adjacent Sonora State, Mexico) became, essentially, a large salt water lake and its resident gastropod fauna was genetically separated from the main Panamic molluscan assemblages. Although existing in a desert region where its water should have evaporated away, the Sonoran Paleosea retained normal salinity and open oceanic conditions due to the large amount of freshwater effluent from the Colorado River.

After a million years of genetic isolation, many of the gastropods of the Sonoran Sea had evolved into new endemic species and these spread throughout the Gulf area when sea levels again rose and flooded the Midriff land bridge later in the Pleistocene. These Sonoran endemics have not dispersed beyond the present-day Gulf of California area and are a major component of the classic Cortezian Subprovince molluscan fauna. Some of the more important of these endemics include the cypraeid *Pseudozonaria annetteae* (erroneously shown as ranging outside of the Gulf region by Lorenz, 2017: map on page 400), the muricids *Vokesmurex ruthae*, *Mexacanthina angelica*, *Pteropurpura erinaceoides*, and *Muricanthus nigritus*, and a small species radiation of the cone shell genus *Gradiconus*, including *G. scalarissimus*, *G. gradatus*, and *G. regularis* (all illustrated here on Figure 4.12). Of special interest within the Cortezian Subprovince is a species radiation of the olivid genus *Americoliva*, the largest known from anywhere along the western coast of the Americas. At least ten endemic taxa are known from the Gulf of California, with several, such as *Americoliva harpularia*, *A. pindarina*, *A. subangulata melchersi*, and *A. intertincta*, being restricted to geographically-small areas and isolated bays. These Cortezian olives, along with other members of the endemic species complex, are shown here on Figure 4.14. Also of interest within the northern Gulf of California is a radiation of the endemic muricid genus *Zacatrophon*, with at least five known species. These spectacular deep water Cortezian endemics include *Z. beebei*, *Z. goliath*, *Z. skoglundae* (all three shown here on Figure 4.12), *Z. scotti*, and *Z. coani*. For a review of the genus *Zacatrophon*, see Houart and Hendrickx, 2020 and Houart and Löser, 2020.

The most important area, biologically, of the entire Cortezian Subprovince is found in the far northern Gulf of California, adjacent to the delta of the Colorado River. This small biotic region, centered on the deep Wagner Basin south of Puerto Peñasco, Sonora State, houses an evolutionary hot spot with a high level of endemicity. Referred to here as the **Felipean Infraprovince** (for San Felipe, Baja California), this infraprovincial area contains a highly-distinctive gastropod fauna with a pronounced relictual component. Several Californian Province-derived genera have managed to survive within the Wagner Basin area, remnants of a northward invasion into the Cortezian area during Pleistocene cold climatic times (Berschauer, Petuch, and Clark, 2018). Two of the more important of these relictual taxa include the muricid genus *Forreria* (*F. corteziana*) and the bursid genus *Crossata* (*C. californica sonorana*), both shown here on Figure 4.13. The large *Forreria* muricid is particularly interesting in that it is found only in deep water areas within the Wagner Basin, where it lives in association with petroleum seeps. Other Felipean endemics, which evolved within the Pleistocene Sonoran Paleosea, include the calliostomatid *Calliostoma palmeri*, the littorinid *Littorina albicarinata*, the large cowrie *Macrocypraea cervinetta californica*, the ranellid *Turritriton gibbosus adairense*, the muricid *Eupleura limata*, the fasciolariids *Hesperaptyxis seriatus*, *H. fredbakeri*, and *H. felipensis*, and the olivid *Americoliva corteziana* (most shown here on Figure 4.13). The geographically-small area of the Felipean Infraprovince and the Wagner Basin is also the habitat of the critically endangered Vaquita Dwarf Porpoise (*Phocoena sinus*), another relictual species that evolved within the Sonoroan Paleosea and never dispersed beyond its original home.

Jaliscoan Subprovince

Named for the Mexican State of Jalisco, the geographic center of the biotic unit, the Jaliscoan Subprovince encompasses the Mexican coastline between Topolobampo, Sinaloa State and Salina Cruz, Oaxaca State. Along the Oaxaca and Chiapas coasts, the extensive estuarine systems of the Rio Novillero, Rio Ostuta, Rio Los Perros, Rio Tehuantepec, and Rio Zanatenco, along with the brackish water lagoons of La Joya, Laguna Superior, Laguna Inferior, Laguna Xocotepec, and the Mar Muerto, altogether, act as a major ecological barrier for the northward dispersal of endemic species from the south. The Jaliscoan Subprovince, delineated by the Topolobampo lagoon system in the north and the Salina Cruz lagoon system in the south, has no areas of special evolution and houses molluscan assemblages containing only widespread Panamic taxa and few regional endemic species. The most prominent of these Jaliscoan endemics is the large muricid *Muricanthus ambiguus* (Figure 4.15A), which ranges from Mazatlan south to Salina Cruz. This conspicuous spiny species is part of a complex of muricids that have sharply-defined range limits within the Panamic Province: *Muricanthus nigritus* in the Cortezian Subprovince, *Muricanthus ambiguus* in the Jaliscoan Subprovince, *Muricanthus callidinus* in the Chiriquian Subprovince, and *Muricanthus radix* in the Chiriquian and Ecuadorian Subprovinces (the latter two species are discussed later in this chapter). Many Jaliscoan taxa, such as the olivid *Americoliva julieta* (Figure 4.15C), are most abundant along the Mexican coast but do range into the Chiriquian and Ecuadorian Subprovinces to the south, where they are far less common.

Located over 128 km southwest of Acapulco, Mexico, and the Jaliscoan Subprovince, Clipperton Island houses the westernmost faunal influence of the entire Panamic Province. This tiny, isolated coral island is considered to be the most remote atoll in the world and supports an impoverished mixed gastropod fauna of only 248 species, composed of Panamic, Indo-Pacific, and endemic taxa (see Kaiser, 2007 for a detailed overview of the Clipperton fauna). These Clipperton endemics demonstrate the presence of an evolutionary hot spot on the island, which we here refer to as the **Clippertonian Infraprovince**. Some of the Clippertonian infraprovincial endemics include the turbinid *Homalopoma clippertonense*, the littorinid *Protolittorina pintado schmitti*, the cowrie *Talostolida jacksoni*, the fasciolariid *Pustulatirus clippertonensis*, and the cone shell *Miliariconus roosevelti* (some shown here on Figure 4.15; see Daughenbaugh and Beals, 2016 for details on *Talostolida jacksoni*). Panamic Province faunal affinities are demonstrated by the presence of classic index taxa such as the cowries *Naria albuginosa* and *Luria isabellamexicana*, the cassid *Cypraecassis tenuis*, the muricids *Tribulus planospira* and *Plicopurpura pansa*, the mitrid *Nebularia rupicola*, and the cone shells *Lividiconus diadema*, *Chelyconus purpurascens*, *Harmoniconus nux*, and *Tesselliconus edaphus*. The classic Panamic taxa, along with the Clippertonian endemics, live together with a large number of Indo-Pacific and Polynesian taxa on the zonated coral reef systems that surround the brackish water central lagoon of the atoll. Some of these include the cowries *Monetaria caputserpentis*, *Monetaria moneta*, *Naria helvola callista*, *Mauritia depressa*, *Mauritia scurra*, and *Talparia talpa*, the muricids *Pterynotus tripterus*, *Drupa ricinus*, and *Nassa serta*, the harpid *Harpa gracilis*, the mitrid *Mitra papalis*, the conids *Virroconus ebraeus* and *Virroconus chaldeus*, and the terebrid *Terebra crenulata* (many of these are illustrated in Chapters 5 and 6). This bizarre mixture of faunal elements is unique in the Pacific Ocean and the Clipperton marine ecosystem stands out as one of the strangest known from anywhere on Earth.

Cocosian Subprovince

Located 550 km southwest of the Costa Rican mainland, Isla del Coco (Cocos Island) contains the only insular subprovince of the Panamic Province. This small island actually represents the only emergent section of a line of huge seamounts that have formed along the submerged Cocos Ridge. Being separated by such a great distance from the Central American mainland, the gastropod fauna of Cocos Island has been genetically isolated since the island first formed in the late Pliocene. This spatial separation has allowed allopatric speciation to take place among nonvagile species and has resulted in the evolution of a fauna with a high degree of endemism. Some of the more important Cocosian endemic taxa include the abalones *Haliotis (Padollus) roberti* and *Haliotis drogeni*, the caecid *Caecum cocosensis*, the cowrie *Talostolida sumeihoae*, the large muricid *Phyllonotus eversoni*, the olive shells *Americoliva deynzerae* and *Vullietoliva foxi*, and the cone shells *Dauciconus kaiserae*, *Kohniconus scariphus*, and *Kohniconus shaskyi*. These Cocosian endemics occur together with Indo-Pacific taxa such as the cowries *Monetaria moneta* and *Talparia talpa*, the mitrid *Mitra mitra*, the cone shells *Virroconus ebraeus* and *Virroconus chaldeus*, and the terebrid *Oxymeris maculatus*, and also a full complement of classic Panamic index species. The normally-rare Panamic cone shell, *Tesselliconus edaphus*, is common on coral reefs in protected bays all around Cocos Island. Some of the Cocosian endemic species are shown here on Figure 4.15; see Daughenbaugh and Beals, 2016 for details on *Talostolida sumeihoae*.

Chiriquian Subprovince

Named for the Golfo de Chiriqui of southwestern Panama, the Chiriquian Subprovince extends from Salina Cruz, Oaxaca State, Mexico, south to the Golfo Tortugas, near Buenaventura, Colombia. The subprovince also encompasses the islands in the Golfo de Nicoya of Costa Rica, the Perlas Islands in the Golfo de Panama, and numerous large islands like Coiba, Cebaco, Jicaron, Partida, Secas, and Taboga along the Panamanian coast. Being under the influence of the North Equatorial Counter Current, the Chiriquian Subprovince receives the warmest water of the entire Panamic Province and houses the largest coral reef complexes found anywhere along the west coast of the Americas. Containing at least 23 species of scleractinian corals, the Chiriquian Subprovince reef systems are the richest known from the Panamic Province and are dominated by four genera of stony corals, including *Pocillopora* (5 species), *Pavona* (7 species), *Psammocora* (4 species), and *Porites* (5 species), along with the hydrocorals *Millepora* and *Distichopora*. These extensive and species-rich coralline environments also support the most diverse molluscan fauna found anywhere within the boundaries of the Panamic Province. Besides supporting the full complement of widespread Panamic Province index species, the Chiriquian Subprovince has evolved a high percentage of endemic gastropod taxa, including the cowries *Pseudozonaria aequinoctialis* and *Talostolida panamensis*, the cassid *Cypraecassis wilmae*, the muricids *Muricanthus radix* and *Muricanthus callidinus* (northern part of the subprovince only), the fasciolariid *Leucozonia rudis*, the olivids *Americoliva olssoni*, *Americoliva spicata*, and *Americoliva truncata*, and the cone shells *Gradiconus scalaris*, *Globiconus baccatus*, and *Ximeniconus gubernatrix*. All of these Chiriquian endemics are shown here on Figure 4.16.

Ecuadorian Subprovince

Named for the country of Ecuador, the Ecuadorian Subprovince extends from the Golfo Tortugas, Colombia, south to Isla Lobos de Tierra, Paita District, Peru, and includes the nearshore islands of Puna, La Plata, Salango, and Gorgonilla as well as the offshore Galapagos Islands. The southern part of the subprovince is under the influence of the cold Humboldt Current, which comes close to shore south of Ecuador. This results in many of the classic Panamic eutropical gastropods disappearing from the local ecosystems and producing locally-impoverished malacofaunas. Although having a less species-rich gastropod fauna than the Chiriquian Subprovince, the Ecuadorian Subprovince still has evolved a large number of endemic species, some of which include the cowrie *Pseudozonaria nigropunctata* (also found in the Galapagos Islands), numerous muricids such as *Vokesimurex elenensis*, *Maxwellia angermyerae* (also found on the Galapagos Islands), *Homalocantha tortua*, and *Homalocantha tortua multicrispata*, and the endemic olivids *Americoliva mcleani*, *Americoliva polpasta radix*, *Felicioliva kaleontina chimu*, and *Strephonella undatella ecuadoriana* (most shown here on Figure 4.17). Of special interest within the Ecuadorian Subprovince is the presence of a living member of the muricid genus *Pterorhytis*, an archaic group that flourished in Pliocene eastern North America (Caloosahatchian Paleoprovince; see Petuch, 2004) but became extinct in the Atlantic Ocean during the Pleistocene. The last-living species of the genus, *Pterorhytis hamatus*, has somehow managed to find a refugium along the Ecuador coast and its presence there is quite problematic.

Most noteworthy within the Ecuadorian Subprovince are the molluscan faunas of the offshore Galapagos Islands. Although these fabled islands are famous for their exotic terrestrial faunas, the marine molluscan communities around the islands are not comparable in their levels of endemicity and contain, essentially, a standard Panamic Province malacofauna. Being bathed in the cold water of the Humboldt Current, the marine faunas of the islands have a slightly impoverished aspect and are missing many of the classic Panamic tropical index species that are abundant along the adjacent coast of South America. On the other hand, the cooler water conditions of the archipelago, as attested by the presence of Sea Lions and Galapagos Penguins, have allowed for the development of a distinctive evolutionary hot spot, here referred to as the **Galapagan Infraprovince**. Only a small number of infraprovincial endemics have evolved here, with the most noteworthy being the muricids *Babelomurex santacruzensis* and *Neorapana grandis*, the fasciolariid *Pustulatirus sanguineus*, the columbellid *Costanachis atramentaria*, the cancellariid *Sveltia gladiator*, and the pseudomelatomid *Hindsiclava hertleini*. Some of these are shown here on Figure 4.17. There are no endemic Galapagos species of Cypraeidae, Strombidae, and Conidae, and the family Volutidae is completely absent from Galapagan ecosystems.

ICONOGRAPHY OF GASTROPODS OF THE EASTERN PACIFIC REGION

(Principal Index Gastropods are shown on Figures 4.4 to 4.17)

Figure 4.4 Widespread Index Gastropods of the Eastern Pacific Region, occurring in the Aleutian, Oregonian, and Californian Provinces.

A= *Haliotis kamtschatkana* Jonas, 1845, length 146 mm, on rocks, 2 m depth off Santa Barbara, California; **B=** *Haliotis walallensis* Stearns, 1899, length 113 mm, on rocks, 1 m depth off Santa Cruz, California; **C=** *Bathybembix bairdii* (Dall, 1889), height 44 mm, 800 m depth off Port Orford, Curry County, Oregon; **D=** *Cidarina cidaris* (Carpenter, 1864), height 30 mm, 500 m depth off Crescent City, California; **E=** *Calliostoma annulatum* (Lightfoot, 1786), height 32 mm, found on sponges, 10 m depth off Monterey, California; **F=** *Pomaulax gibberosus* (Dillwyn, 1817), width 58 mm, on coralline algae, 5 m depth off Crescent City, California; **G=** *Ceratostoma foliatum* (Gmelin, 1791), length 82 mm, 10 m depth off La Jolla, California; **H=** *Nucella emarginata* (Deshayes, 1839), length 30 mm, on rocks exposed at low tide, Monterey, California; **I=** *Paciocinebrina lurida* (Middendorf, 1848), length 16 mm, low tide, Whitbey Island, Puget Sound, Washington; **J=** *Neptunea (Sulcosipho) tabulata* (Baird, 1863), length 86 mm, trawled from 150 m depth off Monterey, California; **K=** *Callianax biplicata* (Sowerby I, 1825), length 22 mm, low tide on mud flats, Monterrey Bay, Monterey, California.

Figure 4.5 Widespread Californian Province Gastropods.
A= *Haliotis corrugata* Wood, 1828, length 160 mm, 2 m depth off Encinitas, San Diego County, California; **B=** *Haliotis cracherodi* Leach, 1814, length 142 mm, 1 m depth in rock crevices in the surf zone, Palos Verdes, Los Angeles County, California; **C=** *Haliotis rufescens* Swainson, 1822, length 210 mm, 2 m depth on rocks and kelp, off Ventura, Ventura County, California; **D=** *Neobernaya spadicea* (Swainson, 1823), length 55 mm, under rock rubble, 5 m depth off Dana Point, Orange County, California; **E=** *Pteropurpura festiva* (Hinds, 1844), length 48 mm, 2 m depth on Mission Bay boat basin jetty, San Diego, San Diego County, California; **F=** *Pteropurpura macroptera* (Deshayes, 1839), length 62 mm, on exposed rock wall, 10 m depth off Dana Point, Orange County, California; **G=** *Atrimitra idae* (Melville, 1893), length 46 mm, 2 m depth off Two Harbors, Catalina Island, California; **H=** *Cancellaria cooperii* (Gabb, 1865), length 75 mm, 10 m depth off Santa Barbara, Santa Barbara County, California; **I=** *Californiconus californicus* (Reeve, 1844), length 32 mm, in a large tide pool, La Jolla, San Diego County, California; **J=** *Megasurcula carpenteriana* (Gabb, 1865), length 83 mm, 10 m depth off Santa Barbara, Santa Barbara County, California; **K=** *Harfordia mcleani* Callomon and Snyder, 2017, length 57 mm, 25 m depth off Cedros Island, Baja California del Sur State, Mexico; **(Northern Californian Provinciatone)** L= *Tegula montereyi* (Kiener, 1850), diameter 26 mm, 1 m depth on rocks, Carmel Point, Carmel, Monterey County, California.

Figure 4.6 Widespread Californian Province Gastropods.
A= *Haliotis fulgens* Philippi, 1845, length 178 mm, 2 m depth, off Dana Point, Orange County, California; **B**= *Haliotis sorenseni* Bartsch, 1940, length 165 mm, 8 m depth off Carlsbad, San Diego County, California; **C**= *Megastraea undosa* (Wood, 1828), diameter 95 mm, on rocks in 10 m depth off Point Loma, San Diego County, California; **D**= *Austrotrophon catalinensis* (Oldroyd, 1927), length 82 mm, trawled from 200 m depth off Avalon, Catalina Island, California; **E**= *Forreria belcheri* (Hinds, 1844), length 115 mm, 20 m depth off Seal Beach, Los Angeles County, California; **F**= *Maxwellia gemma* (Sowerby, II, 1879), length 32 mm, 2 m depth off La Jolla Cove, La Jolla, San Diego County, California; **G**= *Pteropurpura trialata* (Sowerby II, 1834), length 77 mm, 10 m depth off Dana Point, San Diego County, California; **H**= *Pteropurpura vokesae* (Emerson, 1964), length 52 mm, 20 m depth off Encinitas, San Diego County, California; **I**= *Kelletia kelleti* (Forbes, 1852), length 110 mm, 20 m depth off Imperial Beach, San Diego County, California; **J**= *Megasurcula stearnsiana* (Raymond, 1904), length 28 mm, in sand, 20 m depth off San Clemente Island, Channel Islands, California; **K**= *Crossata californica* (Hinds, 1843), length 96 mm, 20 m depth off Encinitas, San Diego County, California; **L**= *Cancellaria crawfordiana* (Dall, 1891), length 46 mm, 10 m depth off Dana Point, San Diego County, California.

Figure 4.7 Widespread Californian Province and Diegan Subprovince Gastropods.
A= *Maxwellia santarosana* (Dall, 1905), length 37 mm, 20 m depth off Encinitas, San Diego County, California; **B=** *Roperia poulsoni* (Carpenter, 1864, length 46 mm, 2 m depth off La Jolla Cove, La Jolla, San Diego County, California; **C=** *Babelomurex oldroydi* (Oldroyd, 1929), length 25 mm, 20 m depth off Avalon, Catalina Island, California; **D=** *Hesperaptyxis luteopictus* (Dall, 1877), length 22 mm, 25 m depth off San Benito Island, Baja California Sur State, Mexico; **E=** *Atrimitra catalinae* (Dall, 1919), length 24 mm, 20 m depth off Encinitas, San Diego County, California; **F=** *Acanthinucella spirata* (Blainville, 1832), length 26 mm, on rocks in 1 m depth, Mission Bay Yacht Basin, Mission Bay, San Diego County, California; **G=** *Paciocinebrina circumtexta* (Stearns, 1871), length 22 mm, 1 m depth off Point Loma, San Diego County, California; **H=** *Macron lividus* (A. Adams, 1855), length 18 mm, low tide on mud flats, Mission Bay, San Diego County, California; **I=** *Ceratostoma nuttalli* (Conrad, 1837), length 31 mm, 1 m depth off Dana Point, Orange County, California; **(Diegan Subprovince) J=** *Harfordia chucksnelli* Callomon and Snyder, 2017, length 44 mm, 20 m depth off Santa Barbara, Santa Barbara County, California; **K=** *Tegula (Stearnsia) regina* (Stearns, 1892), diameter 42 mm, 10 m depth off La Jolla, San Diego County, California; **L=** *Callianax diegensis* (Oldroyd, 1921), length 19 mm, off San Diego, San Diego County, California.

Figure 4.8 Index Gastropods of the Cedrosian and Guadalupean Subprovinces, Californian Province.
(Cedrosian Subprovince) **A**= *Haliotis fulgens turveri* Bartsch, 1942, length 140 mm, 2 m depth, off Santa Rosalita, Baja California State, Mexico; **B**= *Macron aethiops* (Reeve, 1847), length 80 mm, 50 m depth off Punta Colonet, Baja California State, Mexico; **C**= *Macron aethiops* form *kelletii* (A. Adams, 1855), length 62 mm, low tide near San Quintin, Baja California State, Mexico; **(Southern Diegan and Cedrosian Subprovinces)** **D**= *Megastraea turbanica* (Dall, 1910), length 160 mm, 10 m depth off Ensenada, Baja California State, Mexico; **E**= *Calcitrapessa leeana* (Dall, 1890), length 78 mm, 50 m depth due east of Cedros Island, Baja California State, Mexico; **(Guadalupean Subprovince)** **F**= *Haliotis cracherodi californiensis* Swainson, 1822, length 104 mm, 5 m depth along the southeastern side of Guadalupe Island, off Baja California State, Mexico; **G**= *Haliotis corrugata oweni* Talmadge, 1966, length 140 mm, 3 m depth along the southeastern side of Guadalupe Island, off Baja California State, Mexico; **H**= *Haliotis fulgens guadalupensis* Talmadge, 1964, length 142 mm, 2 m depth on open rock ledges, along the southeastern side of Guadalupe Island, off Baja California State, Mexico; **I**= *Tegula gallina multifilosa* (Stearns, 1892), diameter 24 mm, on exposed rocks at the base of a cliff along the southeastern side of Guadalupe Island, off Baja California State, Mexico; **J**= *Barbarofusus guadalupensis* Callomon and Snyder, 2017, holotype, length 51 mm, Guadalupe Island, off Baja California State, Mexico; **K**= *Dendropoma* cf. *lituella* (Moerch, 1861), numerous 8-10 mm diameter specimens growing on the back of the abalone *Haliotis corrugata oweni*, from 2 m depth off the southern side of Guadalupe Island (may be an undescribed endemic species).

Figure 4.9 Widespread Panamic Province Gastropods.
A= *Pseudozonaria arabicula* (Lamarck, 1811), length 25 mm, 3 m depth, Puerto Marques Bay, Acapulco, Guerrero State, Mexico; **B=** *Jenneria pustulata* (Lightfoot, 1786), length 24 mm, 2 m depth off El Islote, Rincon de Guayabitos, Nayarit State, Mexico; **C=** *Malea ringens* (Swainson, 1822), length 93 mm, low tide, Gubernadora Island, Golfo de Montijo, Panama; **D=** *Opeatostoma pseudodon* (Burrow, 1815), length 46 mm, 1 m depth in Puerto Marques Bay, Acapulco, Guerrero State, Mexico; **E=** *Enaeta cumingii* (Broderip, 1832), length 32 mm, low tide in San Carlos Bay, Guaymas, Sonora State, Mexico; **F=** *Harpa crenata* Swainson, 1822, length 80 mm, 100 m depth, off Mazatlan, Sinaloa State, Mexico; **G=** *Pleuroploca granosa* (Broderip, 1832), length 145 mm, 5 m depth off San Felipe, Baja California State, Mexico; **H=** *Porphyria porphyria* (Linnaeus, 1758), length 81 mm, 3 m depth off Cerralvo Island, Baja California Sur State, Mexico; **I=** *Cylinder dalli* (Stearns, 1873), length 50 mm, 3 m depth off Cebaco Island, Panama; **J=** *Ductoconus princeps* (Linnaeus, 1758), length 53 mm, 1 m depth in Puerto Marquez Bay, Acapulco, Guerrero State, Mexico; **K=** *Ximeniconus ximenes* (Gray, 1839), length 41 mm, 2 m depth in San Carlos Bay, Guaymas, Sonora State, Mexico; **L=** *Gradiconus monilifer* (Broderip and Sowerby I, 1833), length 41 mm, trawled from 35 m depth off Mazatlan, Sinaloa State, Mexico.

Figure 4.10 Widespread Panamic Province Gastropods.
A= *Luria isabellamexicana* (Stearns, 1893), length 31 mm, 2 m depth off El Islote, Rincon de Guayabitos, Nayarit State, Mexico; **B=** *Macrocypraea cervinetta* (Kiener, 1843), length 59 mm, low tide, Gubernadora Island, Panama; **C=** *Naria albuginosa* (Gray, 1825), length 29 mm, 2 m depth off El Islote, Rincon de Guayabitos, Nayarit State, Mexico; **D=** *Pseudozonaria robertsi* (Hidalgo, 1906), length 23 mm, low tide, Jesusita Island, Golfo de Nicoya, Costa Rica; **E=** *Cypraecassis tenuis* (Wood, 1828), length 130 mm, 5 m depth off Cerralvo Island, Baja California Sur, Mexico; **F=** *Homalocantha oxyacantha* (Broderip, 1833), length 49 mm, 3 m depth in Puerto Marques Bay, Acapulco, Guerrero State, Mexico; **G=** *Vullietoliva splendidula* (Sowerby I, 1825), length 43 mm, 5 m depth off Cerralvo Island, Baja California Sur State, Mexico; **H=** *Miliariconus tiaratus* (Sowerby I, 1833), length 23 mm, on a rock platform at low tide, San Jose del Cabo, Baja California Sur State, Mexico; **I=** *Perplexiconus lucidus* (Wood, 1828), length 25 mm, under coral rubble, Gubernadora Island, Panama; **J=** *Purpuriconus vittatus* (Hwass, 1792), length 28 mm, 3 m depth in Puerto Marquez Bay, Acapulco, Guerrero State, Mexico; **K=** *Stephanoconus bartschi* (Hanna and Strong, 1949), length 32 mm, 20 m depth off Ceralvo Island, Baja California Sur State, Mexico; **L=** *Titanostrombus galeatus* (Swainson, 1923), length 162 mm, 10 m depth off Punta Mita, Nayarit State, Mexico.

Figure 4.11 Index Gastropods of the Magdalenan Subprovince and Lucasian Infraprovince, Panamic Province. **(Magdalenan Subprovince)** **A=** *Americoliva hemphilli* (Ford, 1915), length 44 mm, 2 m depth north of Puerto Chale, Magdalena Bay region, Baja California Sur State, Mexico; **B=** *Callifusus irregularis* Grabau, 1904, length 136 mm, trawled from 30 m depth in Magdalena Bay, Baja California Sur State, Mexico; **C=** *Gradiconus magdalenensis* (Bartsch and Rehder, 1939), length 23 mm, 30 m depth in Magdalena Bay, Baja California Sur State, Mexico; **D=** *Ceratostoma monoceros* (Sowerby II, 1841), length 32 mm, on exposed rocks, 4 m depth in Bahia Magdalena, Baja California Sur, Mexico; **E, F=** *Enaeta cumingii pederseni* Verrill, 1870, length 26 mm, low tide south of Todos Santos, Baja California Sur State, Mexico; **(Lucasian Infraprovince)** **G=** *Americoliva cumingii* (Reeve, 1850), length 48 mm, 2 m depth off San Jose del Cabo, Baja California Sur State, Mexico; **H=** *Americoliva violacea* (Marrat, 1867), length 46 mm, low tide off El Cajete, La Paz, Baja California Sur State, Mexico (the interior of the aperture is deep violet in color); **I** = *Americoliva grovesi* (Petuch and Myers, 2014), length 44 mm, 128 m depth on the Outer Gorda Bank, southern Gulf of California, Baja California Sur State, Mexico; **J=** *Gradiconus nybakkeni* (Tenorio, Tucker, and Chaney, 2012), length 25 mm, 100 m depth off Cabo Pulmo, Baja California Sur State, Mexico; **K=** *Gradiconus dispar* (Sowerby I, 1833), length 37 mm, trawled from 30 m depth southeast of Cerralvo Island, Baja California Sur State, Mexico; **L=** *Gradiconus skoglundae* Tenorio, Tucker, and Chaney, 2012, length 20.6 mm, trawled from 60 m depth off Los Frailes, Baja California Sur State, Mexico.

Figure 4.12 Index Gastropods of the Cortezian Subprovince, Panamic Province.
A= *Pseudozonaria annetteae* (Dall, 1909), length 47 mm, low tide, San Carlos Bay, Guaymas, Sonora State, Mexico; **B**= *Muricanthus nigritus* (Philippi, 1845), length 85 mm, low tide off San Felipe, Baja California State, Mexico; **C**= *Mexacanthina angelica* (Oldroyd, 1918), length 40 mm, low tide, San Carlos Bay, Guaymas, Sonora State, Mexico; **D**= *Pteropurpura erinaceoides* (Valenciennes, 1832), length 39 mm, 1 m depth in San Carlos Bay, Guaymas, Sonora State, Mexico; **E**= *Vokesimurex ruthae* (E. Vokes, 1988), length 64 mm, 50 m depth off Guaymas, Sonora State, Mexico; **F**= *Zacatrophon beebei* (Hertlein and Strong, 1948), length 36 mm, 100 m depth, off Puerto Penasco, Sonora State, Mexico; **G**= *Gradiconus* cf. *recurvus* (Broderip, 1833) subspecies?, length 25 mm, 35 m depth off Cerralvo Island, Baja California Sur State, Mexico; **H**= *Gradiconus gradatus* (Wood, 1828), length 49 mm, 7 m depth in Bahia San Francisquito, Baja California Sur State, Mexico; **I**= *Gradiconus regularis* (Sowerby II, 1833), length 56 mm, low tide, 30 km south of San Felipe, Baja California State, Mexico; **J**= *Gradiconus scalarissimus* (da Motta, 1988), length 45 mm, by shrimpers, 30 m depth off Cabo Pulmo, Baja California Sur State, Mexico; **K**= *Zacatrophon skoglundae* Houart, 2010, length 41.9 mm, trawled from 90 m depth off Punta San Antonio, Sonora State, Mexico; **L**= *Zacatrophon goliath* Houart and Löser, 2020, paratype, length 79.5 mm, trawled by shrimpers in the Gulf of California, between Puerto Penasco and Guaymas, Sonora State, Mexico.

Figure 4.13 Index Gastropods of the Felipean Infraprovince, Panamic Province.
A, B= *Macrocypraea cervinetta californica* Lorenz, 2017, length 89 mm, 5 m depth off Puerto Penasco, Sonora State, Mexico; **C**= *Calliostoma palmeri* Dall, 1871, diameter 27 mm, 2 m depth off Playa Encanto, south of Puerto Penasco, Sonora State, Mexico; **D**= *Turritriton gibbosus adairense* Dall, 1910, length 30 mm, low tide, Bahia Adair, north of Puerto Penasco, Sonora State, Mexico; **E**= *Hesperaptyxis seriatus* (Callomon and Snyder, 2017), length 28 mm, trawled from 50 m depth east of Puertocitos, Baja California State, Mexico; **F**= *Forreria corteziana* Berschauer, Petuch, and Clark, 2018, low-spired form, length 82 mm, trawled from 140 m depth, on a petroleum seep in the Wagner Basin, 65 km south of Puerto Penasco, Sonora State, Mexico; **G**= *Forreria corteziana* Berschauer, Petuch, and Clark, 2018, length 94 mm, trawled from 140 m depth, on a petroleum seep in the Wagner Basin, 65 km south of Puerto Penasco, Sonora State, Mexico; **H**= *Hesperaptyxis felipensis* (Lowe, 1935), length 23 mm, low tide on rocks, San Felipe, Baja California State, Mexico; **I**= *Hesperaptyxis fredbakeri* (Lowe, 1935), length 52 mm, 50 m depth off Puerto Penasco, Sonora State, Mexico; **J**= *Americoliva corteziana* (Petuch and Sargent, 1986), length 42 mm, low tide, Cholla Bay, north of Puerto Penasco, Sonora State, Mexico; **K**= *Crossata californica sonorana* (Berry, 1960), length 124 mm, trawled by shrimpers from 50 m depth off Bahia Kino, Sonora State, Mexico.

Figure 4.14 The *Americoliva* Species Radiation of the Cortezian Subprovince, Panamic Province.
A= *Americoliva harpularia* (Lamarck, 1811), length 33 mm, 2 m depth, Bahia de la Ventana, opposite Cerralvo Island, Baja California Sur State, Mexico (note the characteristic blue zig-zag lines on the body whorl); **B**= *Americoliva intertincta* (Carpenter, 1857), length 33 mm, low tide, San Jose del Cabo, Baja California Sur State, Mexico; **C**= *Americoliva incrassata* form *burchorum* Ziegler, 1969, length 38 mm, low tide, San Felipe, Baja California State, Mexico; **D**= *Americoliva kerstitchi* (da Motta, 1985), length 27 mm, low tide, El Cajete, north of La Paz, Baja California Sur State, Mexico; **E**= *Americoliva subangulata melchersi* (Menke, 1851), length 43 mm, 2 m depth in San Carlos Bay, Guaymas, Sonora State, Mexico; **F**= *Americoliva pindarina* (Duclos, 1840), length 47 mm, 20 m depth off Isla Carmen, Baja California Sur State, Mexico (interior of aperture violet in color; *ionopsis* Berry, 1969 is a synonym); **G**= *Americoliva polpasta davisae* (Durham, 1950), length 45 mm, 30 m depth off Guaymas, Sonora State, Mexico (originally described as a late Pleistocene fossil); **H**= *Americoliva rejecta* (Burch and Burch, 1962), length 42 mm, low tide, Loreto, Baja California Sur State, Mexico; **I**= *Americoliva subangulata* (Philippi, 1848), length 42 mm, 5 m depth in sand, off Los Barriles, Baja California Sur State, Mexico; **J**= *Americoliva venulata* (Lamarck, 1811), length 51 mm, low tide, El Cajete, Baja California Sur State, Mexico; **K**= *Americoliva venulata punctata* (Marrat, 1870), length 45 mm, in sand at extreme low tide, Puerto Libertad, Sonora, Mexico.

Figure 4.15 Index Gastropods of the Jaliscoan and Cocosian Subprovinces and the Clippertonian Infraprovince, Panamic Province.

(Jaliscoan Subprovince) A= *Muricanthus ambiguus* (Reeve, 1845), length 41 mm, on exposed rock face in 2 m depth, Puerto Marques, Acapulco, Guerrero State, Mexico; **B=** *Americoliva julieta* (Duclos, 1835), length 59 mm, low tide, Playa de Jaco, Puntarenas Province, at the mouth of the Golfo de Nicoya, Costa Rica (also Chiriquian and Ecuadorian Subprovinces); **(Cocosian Subprovince) C=** *Tesselliconus edaphus* (Dall, 1910), length 46 mm, 10 m depth under coral, Chatham Bay, Cocos Island, Costa Rica (Cocos Island form); **D, E=** *Talostolida sumeihoae* Daughenbaugh and Beals, 2013, length 38 mm, 22 m depth off Manta Point, Cocos Island, Costa Rica (Paratype 8 from the type lot); **F=** *Americoliva deynzerae* (Petuch and Sargent, 1986), length 47 mm, 10 m depth on open sand, Chatham Bay, Cocos Island, Costa Rica; **G=** *Vullietoliva foxi* (Stingley, 1984), length 21 mm, 10 m depth on open sand, Chatham Bay, Cocos Island, Costa Rica; **(Clippertonian Infraprovince) H, I=** *Talostolida jacksoni* Daughenbaugh and Beals, 2013, length 32 mm, 15 m depth, off northeastern side of Clipperton Island, French Overseas Territory (Paratype 2 from the type lot); **J, K=** *Miliariconus roosevelti* (Bartsch and Rehder, 1939), length 20 mm, 1 m depth off the landing strip on the northwestern side of Clipperton Island, French Overseas Territory.

Figure 4.16 Index Gastropods of the Chiriquian Subprovince, Panamic Province.
A= *Talostolida panamensis* (Lorenz, 2002), length 28 mm, in living coral heads, 2 m depth off Cebaco Island, Panama; **B**= *Pseudozonaria aequinoctialis* Schilder, 1933, length 41 mm, 3 m depth, on a coral reef off Cebaco Island, Golfo de Montijo, Panama; **C**= *Cypraecassis wilmae* Kreipl and Alf, 2000, length 53 mm, low tide, northeastern coast of Taboga Island, Panama; **D**= *Leucozonia rudis* (Reeve, 1847), length 40 mm, low tide, Cebaco Island, Golfo de Montijo, Panama; **E**= *Muricanthus radix* (Gmelin, 1791), length 60 mm, 4 m depth off Venado Island, Panama; **F**= *Americoliva olssoni* (Petuch and Sargent, 1986), length 33 mm, dredged from 30 m depth, off Puerto Armuelles, Bahia de Charco Azul, Panama; **G**= *Muricanthus callidinus* (Berry, 1958), length 50 mm, low tide, off La Boquita, Carazo Department, Nicaragua; **H**= *Americoliva truncata* (Marrat, 1867), length 39 mm, 30 m depth, off Puerto Armuelles, Bahia de Charco Azul, Panama; **I**= *Globiconus baccatus* (Sowerby III, 1877), length 23 mm, low tide off Manzanillo, Cebaco Island, Golfo de Montijo, Panama; **J**= *Gradiconus scalaris* (Valenciennes, 1832), length 36 mm, 30 m depth, off Puerto Armuelles, Bahia de Charco Azul, Panama; **K**= *Ximeniconus gubernatrix* Petuch and Berschauer, 2018, length 38 mm, low tide, southeastern side of Gobernadora Island, Golfo de Montijo, Panama (common in the Perlas Islands); **L**= *Americoliva spicata* (Röding, 1798), length 47 mm, low tide off Pedro Gonzales Island, Perlas Islands, Panama (also Ecuadorian Subprovince).

Figure 4.17 Index Gastropods of the Ecuadorian Subprovince and the Galapagan Infraprovince, Panamic Province. **(Ecuadorian Subprovince) A=** *Pseudozonaria nigropunctata* (Gray, 1828), length 32 mm, 2 m depth off Puerto Villamil, Isabela Island, Galapagos Islands; **B=** *Homalocantha tortua* (Broderip, 1834), length 41 mm, 15 m depth off Isla Lobos de Afuera, Piura District, Peru; **C=** *Homalocantha tortua multicrispata* (Dunker, 1869), length 62 mm, 20 m depth off Salango Island, Manabi Province, Ecuador; **D=** *Americoliva mcleani* Petuch and Myers, 2014, length 23 mm, 30 m depth off Salango Island, Manabi Province, Ecuador; **E=** *Americoliva polpasta radix* (Petuch and Sargent, 1986), length 37 mm, 20 m depth off Salango Island, Manabi Province, Ecuador; **F=** *Felicioliva kaleontina chimu* (Petuch and Berschauer, 2017), length 37 mm, 35 m depth off Isla Lobos de Afuera, Piura District, Peru; **G=** *Strephonella undatella ecuadoriana* (Petuch and Sargent, 1986), length 19 mm, low tide in sand, off Salango, Manabi Province, Ecuador; **(Galapagan Infraprovince) H=** *Neorapana grandis* (Sowerby I, 1835), length 60 mm, 2 m depth off Puerto Villamil, Isabela Island, Galapagos Islands; **I=** *Costanachis atramentaria* (Sowerby II, 1844), length 8 mm, low tide, Puerto Villamil, Isabela Island, Galapagos Islands; **J=** *Pustulatirus sanguineus* (Wood, 1828), length 48 mm, 200 m depth off Academy Bay, Isla Santa Cruz, Galapagos Islands; **K=** *Babelomurex santacruzensis* (Emerson and D'Attilio, 1970), length 42 mm, 200 m depth off Academy Bay, Isla Santa Cruz, Galapagos Islands.

CHAPTER 5. Indo-Pacific Super-Region and the Central Pacific Tropical Region

The Indo-Pacific Super-Region is the largest single biogeographical unit in the world's oceans, and encompasses the entire Indian Ocean, Red Sea and Persian Gulf, the Western and Central Pacific Ocean, from southern Japan to northern Australia and east to Easter Island. Although characterized as having many localized pockets of endemism and species radiations, the Indo-Pacific Super-Region is broadly defined by numerous widely-distributed index species, all of which are known to have extremely long-lived planktotrophic veliger larvae. Some of these wide-ranging Indo-Pacific index taxa include the cowries *Mauritia mauritiana*, *Monetaria annulus*, *Monetaria moneta*, and *Luria isabella*, the charoniid *Charonia tritonis*, and the cone shells *Virroconus ebraeus*, *Virroconus chaldeus*, *Tesselliconus tessulatus*, *Conus bandanus*, *Textilia bullata*, *Pionoconus catus*, and *Calamiconus quercinus*. These taxa, and several others that occur in virtually every shallow water area of the Indo-Pacific, are shown here on Figure 5.6. This immense biogeographical unit contains four Regions, (the **Central Pacific Tropical Region**, **Western Pacific Tropical Region**, the **Indian Tropical Region,** and the **North Australian Tropical Region**), 10 Provinces (the **Hawaiian**, **Marquesan**, **Rapanuian**, **Polynesian**, **Indo-Malaysian**, **Japonic**, **Lemurian**, **Eritrean**, **Solanderian**, and **Dampierian**), and 25 Subprovinces (the **Tahitian**, **Micronesian**, **Indonesian**, **Melanesian**, **Neocaledonian**, **Philippinian**, **Shikokuan**, **Ryukyuan**, **South China**, **Andamanian**, **Bengalian**, **Malabaran**, **Mascarenean**, **Madagascan**, **Mozambican**, **Aqaban**, **Dahlakian**, **Omanian**, **Somalian**, **Moretonian**, **Cairnsian**, **Coralian**, **Perthian**, **Exmouthian**, and **Carpentarian**), making it the most complex large-scale biotic feature on Earth.

Central Pacific Tropical Region

Extending as far west as Palau, as far north as Midway Island, Hawaiian Islands, as far south as Easter Island (Rapa Nui), and as far east as Salas-y-Gomez Island, Chile, this vast region also encompasses all of the islands in French Polynesia, the Cook Islands, Samoa, Wallis and Futuna, Kiribati, Tonga, the Marshall Islands, and hundreds of others. Within this vast insular area, four distinct provinces have evolved; the **Hawaiian Province**, the **Marquesan Province**, the **Rapanuian Province**, and the **Polynesian Province**. Unlike the other two tropical regions, the Central Pacific Tropical Region and its four provinces are composed entirely of widely-dispersed and isolated islands, with no adjacent continental component. Because of this absence of a continental influence, the entire region lacks nutrient input from large continental rivers and has only near-sterile, nutrient-poor (oligotrophic) open oceanic water conditions. The mollusks that have evolved in these nutrient-poor ecosystems are often very small when compared to their congeners from continental coastlines and large continental-type islands. As a result of the difficulty for larval dispersal between the widely-separated island groups, many of the insular faunas are impoverished, often lacking entire genera and families that are abundant farther west. A classic example of this impoverishment can be seen in the complete absence of the family Volutidae from the entire Central Pacific Tropical Region. Typically having direct

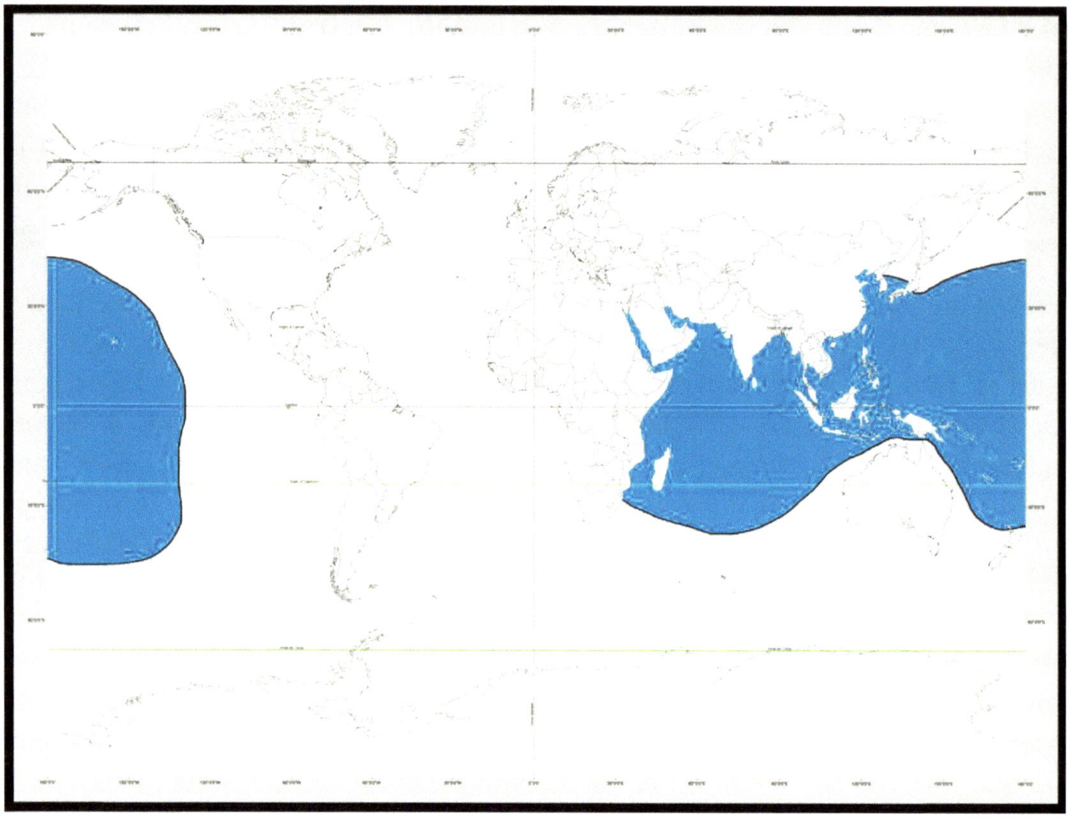

Figure 5.1 Map of the **Indo-Pacific Super-Region**, showing the areal extents of the combined tropical and paratropical subregions (blue). Included are the **Central Pacific Tropical Region**, the **Western Pacific Tropical Region**, and the **Indian Tropical Region**.

development, volutes cannot disperse across the wide, deep stretches of ocean between island groups, which have acted as faunal barriers. Similarly, no members of the Cancellariidae, and most of the Olivinae genera (Olividae), are also absent from the region.

Hawaiian Molluscan Province

Named for the Hawaiian Islands, the Hawaiian Province extends from Kure Atoll in the northwest to the island of Hawaii in the southeast and encompasses eight large volcanic islands, several small atolls, and numerous small islets (including Hawaii, Kahoolawe, Maui, Lanai, Oahu, Molokai, Kauai, and Niihau Islands as well as the northern atolls of Kure, Midway, Laysan, French Frigate Shoals, and others). There are no well-defined subprovinces, and the Hawaiian Province is here considered to be a single biotic area. The province contains a species-rich and highly endemic gastropod fauna, with combined percentages of endemism at nearly 80%. This very high level of allopatric speciation underscores the wide geographical separation from the other island systems in the region. During the late Pleistocene, the warm oceanic currents in the northern Pacific had an accelerated flow speed and many short-lived planktotrophic veligers were able to cross the wide expanse between the Hawaiian Islands and the islands to the west. Fossil beds dating from that time contain numerous Indo-Pacific genera and species that are now extinct on the island chain, most notably the scorpion conch *Harpago chiragra*. These same late Pleistocene fossil beds also contain extinct endemic taxa, such as the conid *Virroconus kahiko* and the strombid *Canarium ostergaardi*, demonstrating that rapid speciation was already taking place at that time. During the latest Pleistocene, these warm currents were deflected southward and the oceanic temperatures around Hawaii began to cool down to the present levels. Most of the Indo-Pacific species that had become established on the islands during the Pleistocene were now genetically isolated and began to undergo allopatric speciation, producing the high level of endemism now seen around the archipelago.

Because of the cooler water temperatures, the Hawaiian coral fauna is relatively impoverished, containing 57 species of scleractinians, but with only 25% of the total taxa making up the reef systems. These are primarily in the genera *Montipora* (*M. patula*, *M. capitata*, and *M. flabellata*), *Porites* (*P. compressa* and *P. lobata*), and *Pocillopora* (*P. meandrina*) and they form the main framework for all of the Hawaiian coral reef systems, along with massive coralline algal growth. These unusual carbonate environments house a very rich gastropod fauna, including distinctive endemics such as the muricids *Homalocantha pele*, *Vitularia sandwichensis*, and *Chicoreus insularum*, the strombids *Canarium helli* and *Euprotomus hawaiiensis*, the personiid *Distorsio burgessi*, the ranellid *Septa beui*, and the olivid *Omogymna ozodona sandwicensis* (all shown here on Figures 5.7 and 5.8, along with other Hawaiian endemics). Of primary interest on the Hawaiian coral reef systems is the single largest endemic cowrie fauna found anywhere on Earth, with over 22 species known from the island chain. Some of these include *Luria tessellata*, *Cypraea tigris schilderiana*, *Cribrarula gaskoini*, *Staphylaea semiplota*, *Talostolida rashleighana*, and *Talparia talpa rundorum*, along with localized species radiations of the genera *Lyncina* (*L. aliceae*, *L. sulcidentata*, and *L. leviathan*), *Pustularia* (*P. takahashii*, *P. wattsi*, and *P. mauiensis*), *Nucleolaria* (*N. granulata*, *N. hinuhinu*, and *N. pseudonucleus*), and *Naria* (*N. helvola hawaiiensis*, *N. ostergaardi*, *N. cernica marielae*) (most shown here on Figures 5.9 and 5.10). These occur along with a rich fauna of endemic reef-dwelling cone shells, such as *Virgiconus spiceri*, *Virgiconus peasei*, *Tesselliconus hawaiiensis*, *Miliariconus abbreviatus*, *Pionoconus striatus oahuensis*, and an endemic species radiation of the genus *Darioconus*, including *D. purus purus* (Oahu and Niihau Islands), *D. stellatus* (endemic to Kauai Island), *D. stellatus racemosus* (endemic to Hawaii Island), and *D. purus leviteni* (Oahu Island) (all of these are shown here on Figure 5.11).

The rocky cliffs, tidepools, and basaltic platforms of the eight large Hawaiian Islands house the richest fauna of endemic gastropods seen in any intertidal zone of the Indo-Pacific Super-Region. Some of these include the endemic nacellid limpets *Cellana sandwichensis, C. talcosa,* and *C. exarata* (the Hawaiian "Opihi"), the turbinid *Turbo sandwichensis*, the cypraeid *Monetaria caputophidii*, the muricids *Thais aperta* and *Neothais harpa*, and the cone shells *Harmoniconus paukstisi*, *Miliariconus abbreviatus*, and *Virgiconus peasei* (most shown here on Figures 5.7, 5.8, 5.9, and 5.11). The deeper water (> 100 m depths) sand-bottom sea floors adjacent to the Hawaiian coral reefs also house an interesting fauna of endemic gastropods. Some of these, which are collected primarily in deep water lobster traps, include the tonnid *Tonna sandwichensis*, the cassids *Semicassis umbilicata* and *Casmaria kalosmodix*, the fasciolariids *Fusinus michaelrogersi*, *Fusinus sandvichensis*, and *Cyrtulus mauiensis*, the harpids *Harpa goodwini* and *Harpa major conoidalis* (Hawaiian subspecies), the olivid *Omogymna richerti*, and the cone shell *Tesselliconus athenae* (some shown here on Figures 5.7 and 5.8). Many of the classic Central Pacific index cone shells, including *Textilia adamsoni*, *Darioconus auratinus*, and *Gastridium eldredi*, are absent from the Hawaiian Province.

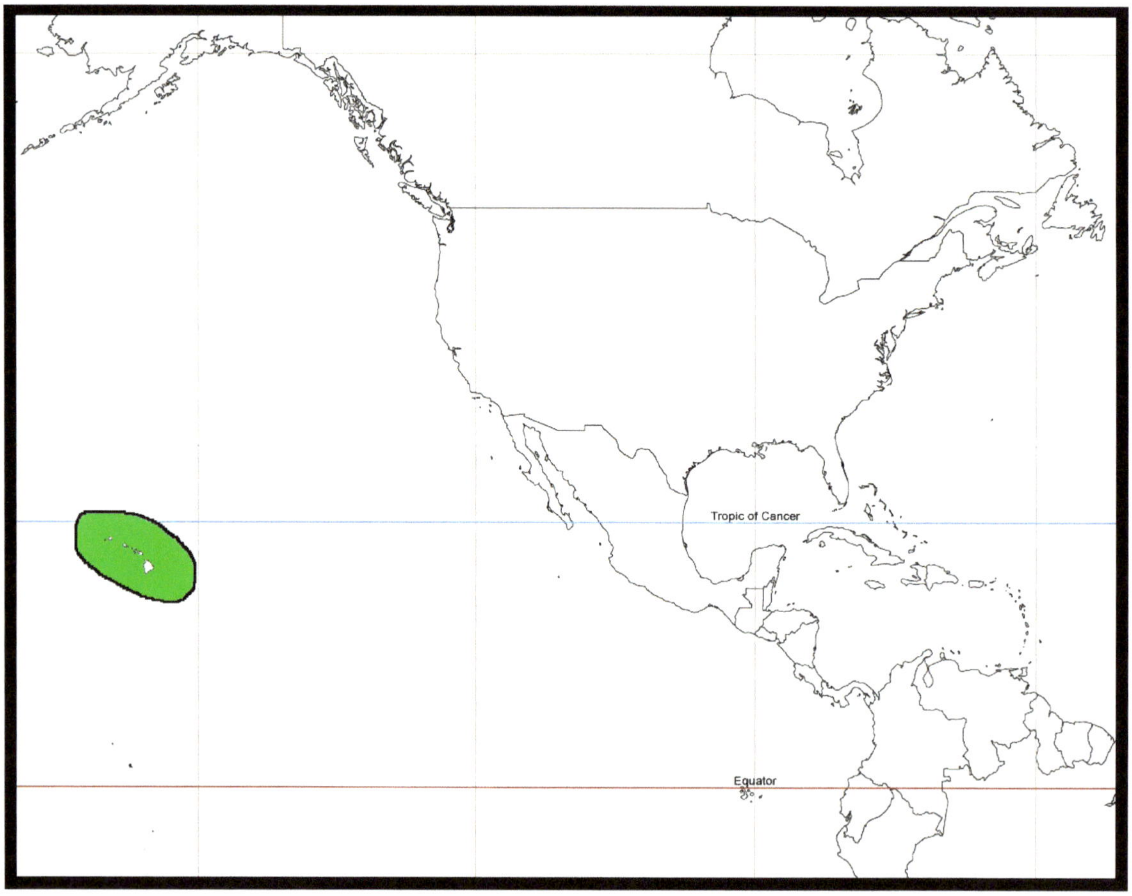

Figure 5.2 Map of the Hawaiian Archipelago, showing the areal extent of the **Hawaiian Molluscan Province** (green).

Marquesan Molluscan Province

Named for the Marquesas Islands of eastern French Polynesia, this isolated chain of nine large volcanic islands (Fatu Hiva, Hiva Oa, Motane, Tahuata, Nuku Hiva, Ua Pou, Vaipae'e, Eiao, and Hatutaa) and numerous smaller islets (such as Fatu Huku, Terihi, Matauapuna, Motu Iti, and Motu Nui), is almost as isolated from other land masses as are the Hawaiian Islands to the north. In similar fashion to the Hawaiian Islands, the Marquesas formed as a series of basaltic volcanoes that erupted above a mantle plume "hot spot". From the late Miocene (Messianian Age) to the mid-Pleistocene, these islands grew to over 4000 m above the Pacific Plate sea floor, forming high volcanic sea mounts that were separated from other island chains by great expanses of intervening deep water. The genetic isolation caused by these great distances and depths has led to the evolution of a highly endemic molluscan fauna with at least 70% endemism in the key tropical index faunas, supporting the provincial status of the island chain. During the late Pleistocene, the accelerated flow speed of warm currents from the west brought many low vagility Indo-Malaysian taxa to the Marquesas area. Some of these, which included the conid genus *Eugeniconus* and the strombid genus *Lambis* (*L. crocata* complex), evolved local endemic species but became extinct on the intervening islands between the Marquesas and the Western Pacific. Besides these extra-territorial faunal elements, the Marquesan Province also houses several endemic relictual taxa, primary among these being the fasciolariid *Cyrtulus serotinus* (Figure 5.12H), one of the last-living members of its normally Eocene-Oligocene genus. A preliminary survey of the molluscan faunas of the Marquesas Islands and the south-central Polynesian Area was given by Rehder and Wilson (1975).

When compared to the other island groups in French Polynesia, the Marquesas Islands stand out as having both the richest molluscan fauna and the largest number of endemic species and subspecies. Some of these Marquesan Province index taxa include the strombid *Lambis pilsbryi* (a member of the before-mentioned *Lambis crocata* complex), the muricids *Drupa iodostoma*, *Chicoreus thomasi*, *Chicoreus maurus steeriae*, and *Chicoreus lorenzi*, the harpids *Harpa ivogardai* and *Harpa kolaceki*, and the olivid *Acutoliva polita marquesana* (all shown here on Figure 5.12). Also like Hawaii, the isolated Marquesas Islands have evolved a highly-endemic cowrie fauna, one of the most species-rich known from the tropical Pacific. At least 17 endemic species and subspecies are now known to inhabit the Marquesan fringing coral reef systems, some of which include species such as *Naria thomasi*, *Nucleolaria cassiaui*, and *Cribrarula astaryi*, and subspecies such as *Leporicypraea mappa curvati*, *Lyncina propinqua pinguis*, *Ransoniella punctata proxima*, *Cypraea tigris lorenzi*, *Naria helvola bellatrix*, *Talparia talpa vivida*, *Luria isabella gauguini*, *Talostolida pellucens polynesiana*, *Mauritia maculifera martybealsi*, *Monetaria caputserpentis meae*, and *Palmadusta contaminata extrema* (all illustrated here on Figures 5.12 and 5.13). Although slightly outdated on the taxonomy, Daughenbaugh's 2015 work on the Marquesan cowries and their biogeographical distributions provides an excellent overview of this fauna.

Also of special interest in the Marquesan Province is the presence of the single largest cone shell fauna (Conidae) known from the Central Pacific Region. Besides containing classic Polynesian conids such as *Textilia adamsoni*, *Gastridium eldredi*, and *Darioconus auratinus*, and numerous wide-ranging Indo-Pacific taxa, the Marquesas Islands also house many endemic species such as *Cylinder textilinus*, *Miliariconus encaustus*, *Puncticulus vautieri*, *Rhizoconus taitensis*, *Pionoconus easoni*, *Lividiconus conco*, *Splinoconus hivanus*, *Splinoconus troendlei*, and *Rhombiconus pseudimperialis* (all shown here on Figure 5.14). Of special interest within the cone fauna of the Marquesan Province is the large and prominent *Pionoconus gauguini* (Figure 5.14G), an enigmatic species whose closest living relatives are *Pionoconus barthelemyi*, a species now confined to the central Indian Ocean, and *P. boutetorum* from Tahiti. Another enigmatic Marquesan species,

Eugeniconus marchionatus (Figure 5.14 C), belongs to a genus whose members are found only in the Philippines, Indonesia, and the Bay of Bengal and the presence of a lone example of *Eugeniconus* in far eastern Polynesia demonstrates that the genus was once far more widespread than it is today. Like the Eocene-survivor *Cyrtulus*, the conid genus *Eugeniconus* represents a relict taxon from an older time, when Polynesia contained an ancestral fauna that was more similar to Indonesia and the Bay of Bengal.

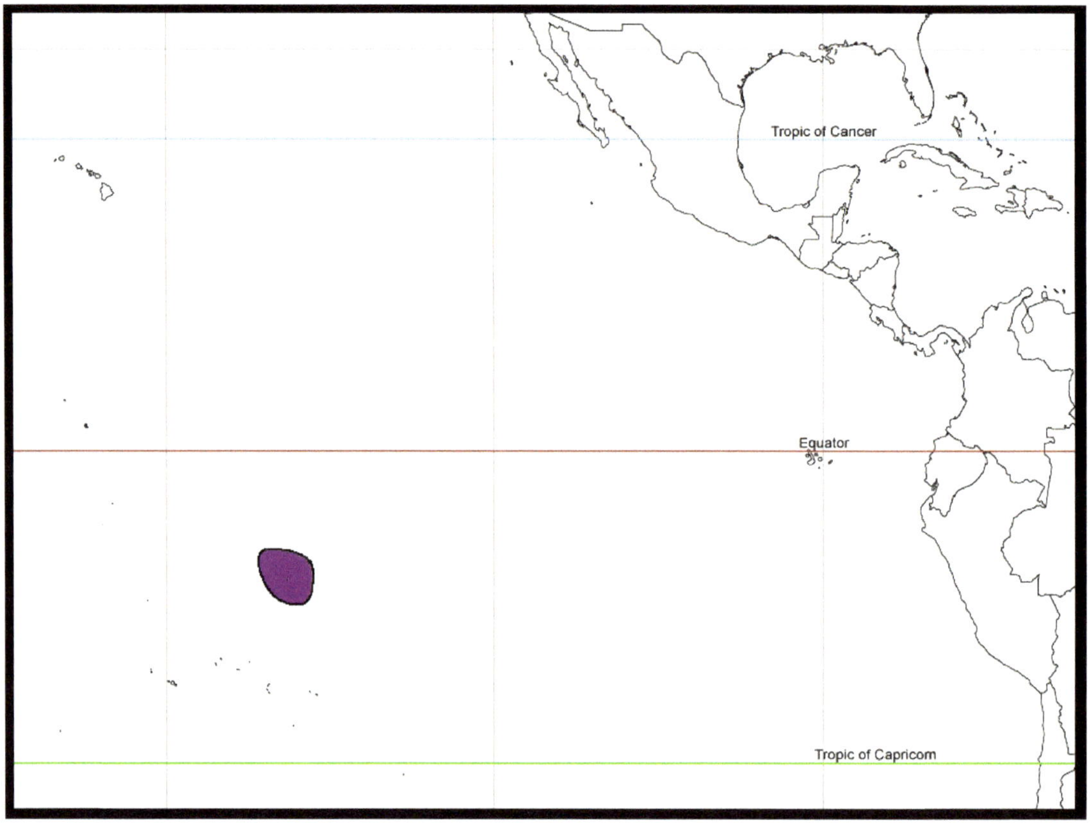

Figure 5.3 Map of the Marquesas Islands, showing the areal extent of the **Marquesan Molluscan Province** (purple).

Rapanuian Molluscan Province

This tiny insular province is named for Rapa Nui, also known as Easter Island or Isla de Pascua, on the extreme southeastern corner of the Polynesian Triangle. Comprising a single volcanic island and several adjacent islets and rocks, Rapa Nui is almost as far from its nearest neighbor, Pitcairn Island, as it is from Chile and the mainland of South America. Besides Rapa Nui Island itself, the molluscan province also includes the very tiny Salas y Gomez Island, 390 km to the east-northeast, which is considered to be the farthest-east point of Polynesia. The Rapanuian Province houses an impoverished Polynesian-Central Pacific molluscan fauna, with common shallow water species such as the cone shells *Harmoniconus nanus* and *Virroconus ebraeus*, the muricids *Drupa morum*, *Drupa ricinus*, *Morula uva*, and *Coralliophila violacea*, and the cowrie *Luria isabella controversa* being found in rocky tide pools all around the island (Rehder, 1980). A number of Rapanuian endemic gastropods also occur along with these widespread species and include the cowries *Monetaria caputdraconis*, *Cribrarula garciai*, *Naria englerti*, and *Naria leforti*, the strombid *Canarium rapanuense*, the muricid *Pascula citrina*, the mitrid *Strigatella flavocingulata*, and the cone shell *Miliariconus pascuensis*. The high intertidal areas and exposed basaltic cliffs also support several endemic taxa, including the planaxid *Hinea akuana*, the littorinid *Nodilittorina pascua*, and the nerite *Lisanerita lirellata* (shown here on Figure 5.15). The uninhabited tiny rock island of Salas y Gomez also has evolved several endemic species, and represents its own evolutionary hot spot, here referred to as the **Salas-Gomezian Infraprovince**. The most noteworthy infraprovincial endemic from this easternmost Polynesian outpost is the intertidal cypraeid, *Monetaria caputdraconis poppei* (Figure 5.15K), which has the most restricted range of any known cowrie shell on Earth. See Rehder, 1980 for a review of the molluscan faunas of Easter Island and Salas y Gomez Island.

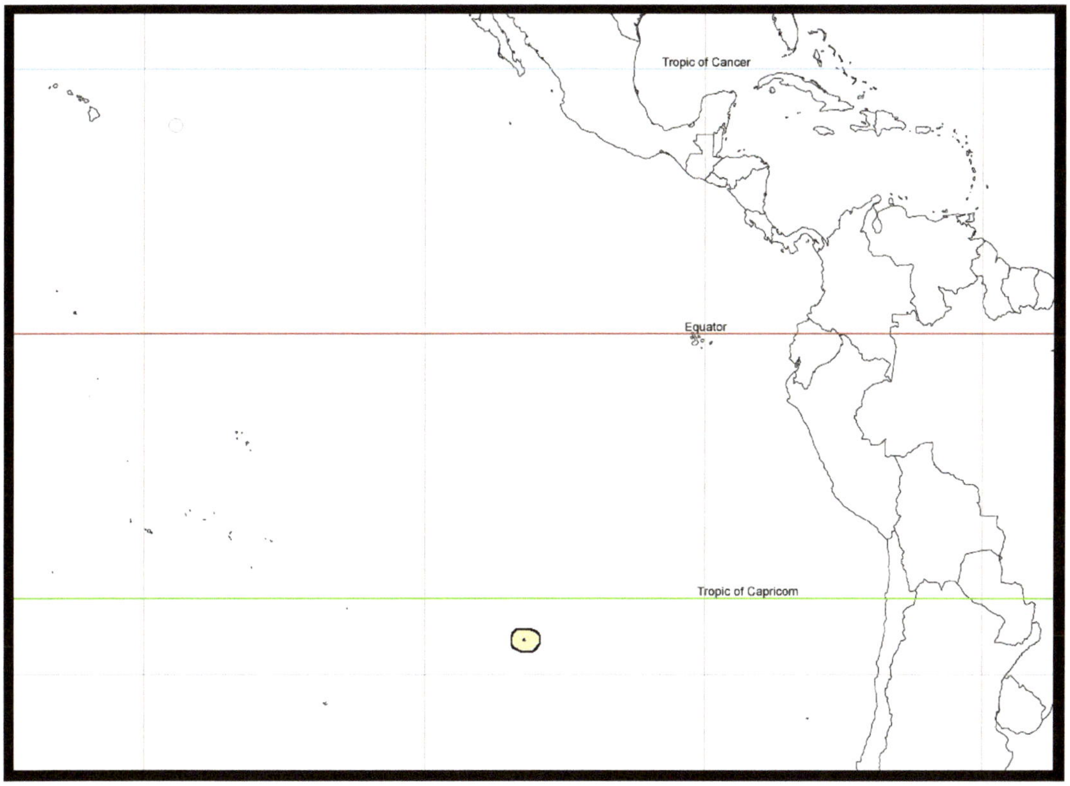

Figure 5.4 Map of Easter and Salas y Gomez Islands, showing the areal extent of the **Rapanuian Molluscan Province** (gold).

Polynesian Molluscan Province

Named for the cultural area of Polynesia ("Many Islands") in the Central Pacific, the Polynesian Molluscan Province extends from Palau in the west, to the Marianas and Marshall Islands in the north, to Tonga and Niue in the south, and to the Tuamotu and Austral Islands in the east and encompasses literally thousands of small volcanic islands and coral atolls. Within this vast area, the province contains two distinct subprovinces, the **Tahitian Subprovince** and the **Micronesian Subprovince**, along with three infraprovinces. Being so far from any continental mainlands and river effluents, the islands of the Polynesian Province are bathed in very sterile, nutrient-poor sea water, creating oligotrophic environmental conditions. For this reason, most of the Polynesian island faunas are not very species-rich and many of the resident species are dwarfed or of smaller-than-normal size for their genera. Some of these include the small abalone *Haliotis pulcherrima*, the dwarf cowrie *Naria irrorata*, the dwarf harpid *Harpa gracilis*, and the small vasid *Vasum armatum* (shown here on Figure 5.16). Other widespread Polynesian Province index taxa include the cowries *Cribrarula cumingi* and *Monetaria obvelata*, the muricid *Chicoreus maurus*, the olivid *Miniaceoliva efasciata*, and the cone shells *Gastridium eldredi*, *Darioconus auratinus*, and *Textilia adamsoni* (all shown here on Figure 5.16). Many classic and abundant Western Pacific gastropod taxa, such as the entire family Volutidae and the genera *Lyria*, *Plicolyria*, and *Cymbiola*, the entire family Cancellariidae, and the entire family Potamididae, are noticeably absent from the Polynesian fauna. The geologic and oceanographic history of Pitcairn Island and the South-Central Pacific was given by Spencer (1989) and Preece (1995).

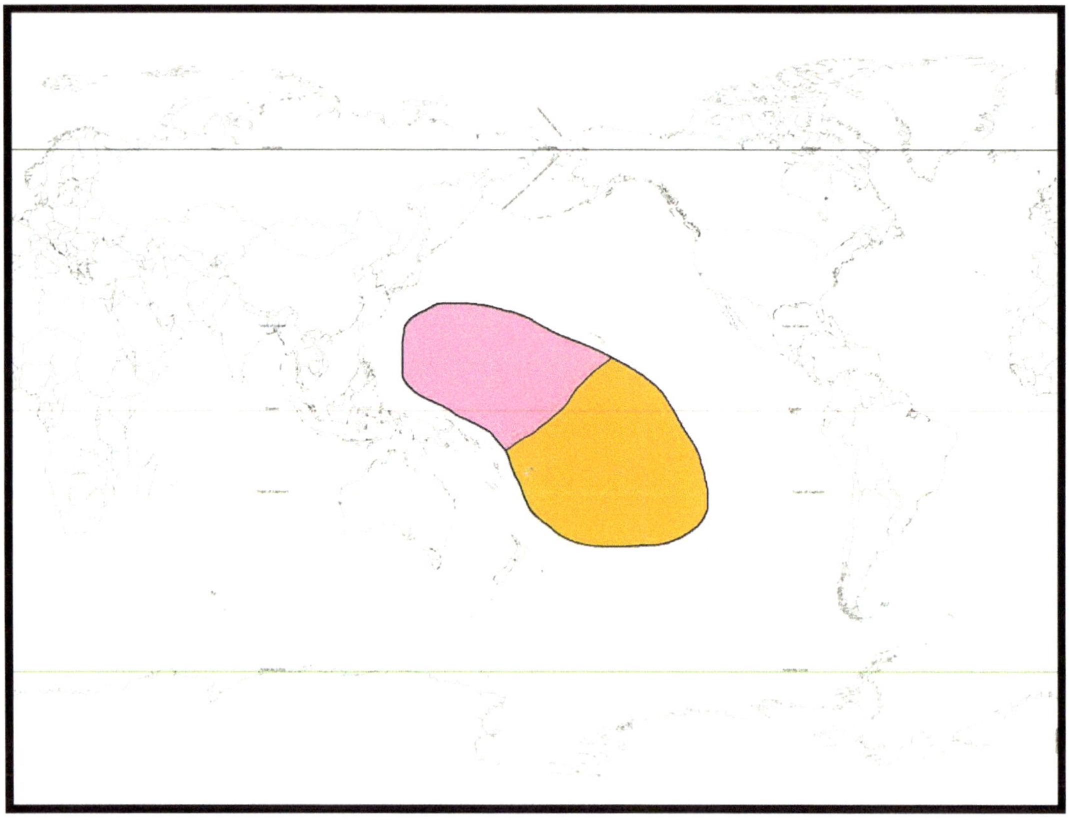

Figure 5.5 Map of the Polynesian Molluscan Province, showing the areal extents of its subprovinces: the **Tahitian Subprovince** (gold) and the **Micronesian Subprovince** (light rose).

Tahitian Subprovince

Although being centered on Tahiti and the Society Islands, the Tahitian Subprovince (named for the island of Tahiti) encompasses a wide area extending from Tokelau in the northwest to the Tuamotu Archipelago in the southeast. Although housing all of the classic wide-ranging Polynesian index species, the malacofauna of the Tahitian Subprovince also contains numerous endemic taxa such as the scorpion conch *Lambis robusta*, the cowries *Talparia talpa lutani*, *Talostolida violacincta*, *Talostolida teres janae*, *Ipsa childreni leforti*, and *Naria bernardi*, the olivid *Omogymna ozodona*, and the cone shells *Cylinder panniculus* and *Leporiconus pomareae* (most here on Figure 5.17). Tahiti, itself, houses a small number of endemic gastropods and these indicate that the island may have infraprovincial status. Some of these Tahitian island endemics include the septariid *Septaria taitana* (living in brackish and fresh water) and the cowries *Mauritia scurra mundula* and *Purpuradusta barbieri* (shown here on Figure 5.17), and the large cone shell *Pionoconus boutetorum*.

The eastern area of the Tahitian Subprovince, encompassing the hundreds of atolls and coral islands of the Tuamotu Archipelago ("Low Islands"), represents a major evolutionary hot spot, here referred to as the **Tuamotuan Infraprovince**. Most of the wisespread Polynesian and Tahitian gastropod index taxa are present throughout this myriad of atolls, but a large number of endemic species are also present. Of particular interest in this infraprovince is an unusually rich fauna of cowrie shells, many of which are restricted to the Tuamotu Archipelago. Some of these infraprovincial endemics include a small radiation of the genus *Monetaria*, with *M. obvelata perrieri*, *M. caputserpentis argentata*, and *M. caputserpentis nivalis*, and other cypraeids such as *Leporicypraea mappa admirabilis*, *Cribrarula compta*, *Pustularia tuamotensis*, *Talostolida subteres*, and the tiny *Purpuradusta oryzaeformis* (all shown here on Figure 5.18).

Micronesian Subprovince

Named for the cultural area of Micronesia ("Tiny Islands"), the widely-dispersed Micronesian Subprovince extends from Palau in the west, to the Marianas Islands in the northwest, to Samoa in the southeast, and also encompasses the Caroline, Marshall, and Tuvalu Islands and the island nations of Kiribati and Wallis and Futuna. Throughout this multitude of small volcanic islands and coral atolls, a classic Polynesian gastropod fauna is present, containing species such as the conids *Gastridium eldredi* and *Darioconus auratinus*, the muricid *Chicoreus maurus*, the harpid *Harpa gracilis*, and the olivid *Miniaceoliva efasciata*. These widespread Polynesian species also occur together with distinctive endemic Micronesian taxa such as the strombid *Sinustrombus taurus* and the cypraeid *Cribrarula taitae* (both shown here on Figure 5.19), forming a richer mixed fauna than those seen in the Tahitian Subprovince. Individual island endemism is prevalent throughout the vast area of the subprovince, the result of the great distances between many of the island chains. Several examples of this allopatric speciation pattern are seen in the Cypraeidae, where *Erronea ovum palauensis* is confined to the Palau Islands, *Leporicypraea mappa guamensis* is restricted to Guam, and *Blasicrura summersi* is confined to Tonga (and extraterritorially in the Fijis). Similarly, the olivids of the Micronesian Subprovince are often endemic to one island or small island group, such as Tarawa and its endemic olive shell *Miniaceoliva efasciata thierryi*, the Vava'u

Islands and their endemic species *Acutoliva hilli*, and Wallis and Futuna and their endemic species *Miniaceoliva lamberti chloeae* (the endemic Micronesian cowries and olives are shown here on Figure 5.19).

The island of Samoa, and its neighboring islands, also has evolved a large number of endemic gastropods and has become an evolutionary hot spot within the Micronesian Subprovince, here referred to as the **Samoan Infraprovince**. Some of the classic Samoan infraprovincial endemic taxa include the cowries *Blasicrura pallidula vivia*, *Monetaria sublittorea*, and *Erronea caurica samoensis* and the cone shell *Conus nigrescens* (all shown here on Figure 5.19). The Marshall Islands also have evolved numerous endemic taxa, and this is especially noteworthy on the giant atoll of Kwajalein, the largest atoll reef system on Earth. Referred to here as the **Kwajaleinian Infraprovince,** this Marshallese evolutionary hot spot contains numerous highly-restricted gastropod taxa, including the cowries *Bistolida stolida kwajaleinensis*, *Cribrarula cribraria oceanica*, *Cribrarula gaspardi*, *Palmadusta johnsonorum*, and *Leporicypraea mappa eluceta*, and the olive shells *Galeola carneola kwajaleinensis* and *Miniaceoliva efasciata berti*. These infraprovincial endemics are all illustrated here on Figure 5.20.

ICONOGRAPHY OF GASTROPODS OF THE INDO-PACIFIC SUPER-REGION AND THE CENTRAL TROPICAL PACIFIC REGION

(Principal Index Gastropods are shown on Figures 5.6 to 5.20)

Figure 5.6 Widespread Index Gastropods of the Indo-Pacific Super-Region, occurring in the Hawaiian, Marquesan, Polynesian, Indo-Malaysian, Japonic, Lemurian, Eritrean, Exmouthian, Carpenterian, and Solanderian Provinces.
A= *Mauritia mauritiana* (Linnaeus, 1758), length 95 mm, 1 m depth off Makaha, Oahu Island, Hawaii; **B=** *Charonia tritonis* (Linnaeus, 1758), length 380 mm, 5 m depth off Coron, Palawan, Philippines; **C=** *Calamiconus quercinus* (Lightfoot, 1786), length 52 mm, 2 m depth in the main lagoon of Rossel Island, Louisiade Archipelago, Papua-New Guinea; **D=** *Darioconus magnificus* (Reeve, 1843), length 63 mm, low tide off Samarai Island, Milne Bay, eastern Papua New Guinea; **E=** *Conus bandanus* Hwass, 1792, length 68 mm, 3 m depth off Vasini Island, Mombasa, Kenya; **F=** *Pionoconus catus* (Hwass, 1792), length 38 mm, under coral at low tide, Rossel Island, Louisiade Archipelago, Papua-New Guinea; **G=** *Tesselliconus tessulatus* (Born, 1778), length 50 mm, 2 m depth off Sykes Reef, Swains Reef Group, Queensland, Australia; **H=** *Textilia bullata* (Linnaeus, 1758), length 55 mm, 2 m depth off Balabac Island, Palawan, Philippines; **I=** *Virroconus ebraeus* (Linnaeus, 1758), length 38 mm, on an exposed coral reef, Nimoa Island, Calvados Chain, Louisiade Archipelago, Papua-New Guinea; **J=** *Virroconus chaldeus* (Röding, 1798), length 35 mm, under rocks at low tide, Faaa, Tahiti, Society Islands; **K=** *Monetaria moneta* (Linnaeus, 1758), length 30 mm, low tide on rocks, Nimoa Island, Calvados Chain, Louisiade Archipelago, Papua-New Guinea.

Figure 5.7 Widespread Hawaiian Province Index Gastropods.
A= *Turbo sandwicensis* Pease, 1861, height 45 mm, low tide, Haleiwa, Oahu, Hawaii; **B=** *Tonna hawaiiensis* Vos, 2007, length 92 mm, 100 m depth off Haleiwa, Oahu, Hawaii; **C=** *Semicassis umbilicata* (Pease, 1861), length 60 mm, 100 m depth, Waialua Bay, Haleiwa, Oahu, Hawaii; **D=** *Canarium helli* (Kiener, 1843), length 18 mm, 5 m depth off Mokuleia, Oahu, Hawaii; **E=** *Euprotomus hawaiensis* (Pilsbry, 1917), length 83 mm, 20 m depth in Pokai Bay, Waianae, Oahu, Hawaii; **F=** *Chicoreus insularum* (Pilsbry, 1921), length 71 mm, 3 m depth on Ahua Reef, Fort Kamehameha, Oahu, Hawaii; **G=** *Homalocantha pele* (Pilsbry, 1918), length 48 mm, 5 m depth on Tripod Reef, off Ewa Beach, Oahu, Hawaii; **H=** *Fusinus michaelrogersi* Goodwin, 2001, length 140 mm, 229 m depth off Tern Island, French Frigate Shoals, northwestern Hawaiian Islands; **I=** *Neocancilla langfordiana* (Cate, 1962), length 30 mm, 10 m depth on Tripod Reef, Ewa Beach, Oahu, Hawaii; **J=** *Scabricola newcombii* (Pease, 1869), length 27 mm, 5 m depth in Pokai Bay, Waianae, Oahu, Hawaii; **K=** *Omogymna ozodona sandwicensis* (Pease, 1860), length 27 mm, 20 m depth off Tripod Reef, Ewa Beach, Oahu, Hawaii.

Figure 5.8 Widespread Hawaiian Province Index Gastropods.
A= *Cellana sandwicensis* (Pease, 1861), length 47 mm, on exposed rocks in the surf zone, Waianae, Oahu, Hawaii; **B=** *Cellana talcosa* (Gould, 1846), length 62 mm, on exposed rocks in the surf zone, Waianae, Oahu, Hawaii; **C=** *Septa beui* Garcia-Talavera, 1985, length 26 mm, 5 m depth off Mokuleia, Oahu, Hawaii; **D=** *Neothais harpa* (Conrad, 1837), length 30 mm, on exposed rocks in the surf zone, Waianae, Oahu, Hawaii; **E=** *Thais aperta* (Blainville, 1832), length 34 mm, on exposed rocks in the surf zone, Waianae, Oahu, Hawaii; **F=** *Vitularia sandwichensis* (Pease, 1861), length 25 mm, 3 m depth on Ahua Reef, Fort Kamehameha, Oahu, Hawaii; **G=** *Fusinus sandvichensis* (Sowerby II, 1880), length 88 mm, 200 m depth, off Tern Island, French Frigate Shoals, Hawaii; **H=** *Distorsio burgessi* Lewis, 1972, length 47 mm, 25 m depth in Pokai Bay, Waianae, Oahu, Hawaii; **I=** *Cyrtulus mauiensis* (Callomon and Snyder, 2006), length 147 mm, 50 m depth, Maalaea Bay, Maui, Hawaii; **J=** *Hastula lanceata oahuensis* (Pilsbry, 1921), length 37 mm, 5 m depth off Kahaluu, Kaneohe Bay, Oahu, Hawaii; **K=** *Terebra nodularis* Deshayes, 1859, length 28 mm, 5 m depth off Kahaluu, Kaneohe Bay, Oahu, Hawaii.

Figure 5.9 Cowrie Species Radiations of the Hawaiian Province.
A= *Cribrarula gaskoini* (Reeve, 1846), length 25 mm, 5 m depth off Kahaluu, Kaneohe Bay, Oahu, Hawaii; **B=** *Cypraea tigris schilderiana* Cate, 1961, length 120 mm, 20 m depth off Makaha, Oahu, Hawaii; **C=** *Luria tessellata* (Swainson, 1822), length 34 mm, 10 m depth on Tripod Reef, off Ewa Beach, Oahu, Hawaii; **D=** *Lyncina aliceae* Lum, 2013, length 49 mm, 10 m depth on Ahua Reef, Fort Kamehameha, Oahu, Hawaii; **E=** *Lyncina leviathan* (Schilder and Schilder, 1937), length 75 mm, 2 m depth off Waianae, Oahu, Hawaii; **F=** *Lyncina sulcidentata* (Gray, 1824), length 40 mm, 10 m depth on Tripod Reef, off Ewa Beach, Oahu, Hawaii; **G=** *Monetaria caputophidii* (Schilder, 1927), length 30 mm, low tide, Waianae, Oahu, Hawaii; **H=** *Naria cernica marielae* (Cate, 1960), length 27 mm, 40 m depth, off Tern Island, French Frigate Shoals, Hawaii; **I=** *Naria helvola hawaiiensis* (Melvill, 1888), length 23 mm, 10 m depth, on Tripod Reef, off Ewa Beach, Oahu; **J=** *Naria ostergaardi* (Dall, 1921), length 20 mm, 20 m depth off Makaha, Oahu, Hawaii; **K=** *Cribrarula gaskoini fischeri* (Vayssiere, 1910), length 18 mm, 30 m depth off Hanalei, Kauai, Hawaii.

Figure 5.10 Cowrie Species Radiations of the Hawaiian Province.
A= *Nucleolaria granulata* (Pease, 1862), length 31 mm, 10 m depth off Kahaluu, Kaneohe Bay, Oahu, Hawaii; **B**= *Nucleolaria pseudonucleus* Moretzsohn, 2011, length 25 mm, 10 m depth off Makaha, Oahu, Hawaii; **C**= *Nucleolaria nucleus* cf. *gemmosa* (Perry, 1811), length 23 mm, 20 m depth, on Tripod Reef, Ewa Beach, Oahu, Hawaii; **D**= *Purpuradusta unifasciata* (Mighels, 1845), length 13 mm, 5 m depth off Kahaluu, Kaneohe Bay, Oahu, Hawaii; **E**= *Pustularia takahashii* Moretzsohn, 2011, length 14 mm, 3 m depth on Ahua Reef, Fort Kamehameha, Oahu, Hawaii; **F**= *Pustularia mauiensis* (Burgess, 1967), length 14 mm, in bryozoan colonies and algae at low tide, Maalaea Bay, Maui, Hawaii; **G**= *Pustularia wattsi* Lorenz, 2000, length 14 mm, 3 m depth Kahakuloa Bay, Maui, Hawaii; **H**= *Staphylaea semiplota* (Mighels, 1845), length 17 mm, Tripod Reef, Ewa Beach, Oahu, Hawaii; **I**= *Talparia talpa rundorum* Bridges, 2018, length 72 mm, 10 m depth off Hanauma Bay, Hawaii Kai, Oahu, Hawaii; **J**= *Talostolida alisonae* (Burgess, 1983), length 33 mm, 6 m depth off Makua, Oahu, Hawaii; **K**= *Talostolida rashleighana* (Melvill, 1880), length 19 mm, 20 m depth, off Makaha, Oahu, Hawaii.

Figure 5.11 Cone Shell Species Radiations of the Hawaiian Province.
A= *Darioconus purus leviteni* Tucker, Tenorio, and Chaney, 2011, length 41 mm, low tide under rocks, off Hauula, Oahu, Hawaii; **B=** *Darioconus stellatus racemosus* (Sowerby III, 1874), length 42 mm, 2 m depth off Hanalei Pier, Hanalei, Kauai, Hawaii; **C=** *Darioconus stellatus* (Kiener, 1847), length 32 mm, 1 m depth in Waimea Bay, Waimea, Kauai, Hawaii (endemic to Kauai); **D=** *Harmoniconus paukstisi* Tucker, Tenorio, and Chaney, 2011, length 18 mm, exposed at low tide, off Waianae, Oahu, Hawaii; **E=** *Miliariconus abbreviatus* (Reeve, 1843), length 39 mm, 1 m depth off Kahaluu, Kaneohe Bay, Oahu, Hawaii; **F=** *Pionoconus striatus oahuensis* Tucker, Tenorio, and Chaney, 2011, length 82 mm, 15 m depth off Waianae, Oahu, Hawaii; **G=** *Tesselliconus sandwichensis* (Walls, 1978), length 28 mm, 3 m depth on Tripod Reef, off Ewa Beach, Oahu, Hawaii; **H=** *Virgiconus peasei* (Brazier, 1877), length 41 mm, 20 m depth off Maile Sands, Waianae, Oahu, Hawaii; **I=** *Virgiconus spiceri* (Bartsch and Rehder, 1943), length 109 mm, 30 m depth in the Honolulu Channel, off Sand Island Beach, Oahu, Hawaii; **J=** *Vituliconus circumactus hammatus* (Bartsch and Rehder, 1943), length 21 mm, 100 m depth off Haleiwa, Oahu, Hawaii; **K=** *Darioconus purus* (Pease, 1863), length 42.6 mm, 2 m depth in sand and rubble, Nanakuli, Oahu Island, Hawaii.

Figure 5.12 Widespread Marquesan Province Index Gastropods.
A= *Lyncina propinqua pinguis* Lorenz, 2017, length 38 mm, 5 m depth, in Taiohae Bay, Nuku Hiva Island, Marquesas Islands; **B=** *Chicoreus thomasi* (Crosee, 1872), length 40 mm, 3 m depth in Taiohae Bay, Nuku Hiva Island, Marquesas Islands; **C=** *Ransoniella punctata proxima* Lorenz, 2017, length 9 mm, 5 m depth in Taiohae Bay, Nuku Hiva Island, Marquesas Islands; **D=** *Lambis pilsbryi* Abbott, 1961, female specimen, length 204 mm, 5 m depth off Cape Martin, Controller Bay, Nuku Hiva Island, Marquesas Islands; **E=** *Chicoreus maurus steeriae* (Reeve, 1845), length 80 mm, 3 m depth on coral rubble, Taiohae Bay, Nuku Hiva Island, Marquesas Islands; **F=** *Chicoreus lorenzi* Houart, 2009, length 57 mm, 5 m depth in West Bay, Nuku Hiva Island, Marquesas Islands; **G=** *Drupa iodostoma* (Lesson, 1840), length 32 mm, 1 m depth on exposed rock platform, Cape Martin, Nuku Hiva Island, Marquesas Islands; **H=** *Cyrtulus serotinus* (Hinds, 1844), length 67 mm, 35 m depth at night, off the Sentinal, entrance to Taiohae Bay, Nuku Hiva Island, Marquesas Islands; **I=** *Harpa ivojardai* Cossignani, 2013, length 34 mm, in sand, 20 m depth, Taiohae Bay, Nuku Hiva Island, Marquesas Islands; **J=** *Harpa kolaceki* Cossignani, 2011, length 38 mm, 20 m depth off Cape Martin, Nuku Hiva Island, Marquesas Islands; **K=** *Acutoliva polita marquesana* Petuch and Sargent, 1986, length 14 mm, 20 m depth in Taiohae Bay, Nuku Hiva Island, Marquesas Islands; **L=** *Cribrarula astaryi* Schilder, 1971, length 19 mm, 5 m depth, Colette Bay, Nuku Hiva Island, Marquesas Islands.

Figure 5.13 Cowrie Species Radiations of the Marquesan Province.
A= *Mauritia scurra hivaensis* Lorenz, 2017, length 49 mm, 30 m depth in Colette Bay, Taihohae, Nuku Hiva Island, Marquesas Islands; **B=** *Cypraea tigris lorenzi* Meyer and Tweedt, 2017, length 83 mm, 30 m depth, off the Sentinal, Taiohae Bay, Nuku Hiva, Marquesas Islands; **C=** *Leporicypraea mappa curvati* Beals and Lum, 2017, length 76.4 mm (Paratype #9; shell base with large dark patch), 25 m depth in Taiohae Bay, Nuku Hiva Island, Marquesas Islands; **D=** *Mauritia maculifera martybealsi* Lorenz, 2002, length 52 mm, 5 m depth in Taiohae Bay, Nuku Hiva Island, Marquesas Islands; **E=** *Monetaria caputserpentis meae* Lorenz, 2017, length 35 mm, under rocks at low tide, western side of Taiohae Bay, Nuku Hiva Island, Marquesas Islands; **F=** *Naria helvola bellatrix* Lorenz, 2009, length 23 mm, 3 m depth, Taiohae Bay, Nuku Hiva Island, Marquesas Islands; **G=** *Naria thomasi* (Crosse, 1865), length 15 mm, 10 m depth in Colette Bay, Nuku Hiva Island, Marquesas Islands; **H=** *Nucleolaria cassiaui* (Burgess, 1965), length 24 mm, 15 m depth in Taiohae Bay, Nuku Hiva Island, Marquesas Islands; **I=** *Talostolida pellucens polynesiana* Raybaudi, 1992, length 35 mm, 25 m depth in West Bay, Nuku Hiva Island, Marquesas Islands; **J=** *Talparia talpa vivida* Bridges, 2018, length 63 mm, 25 m depth, Matauapuna, Taiohae Bay, Nuku Hiva Island, Marquesas Islands; **K=** *Palmadusta contaminata extrema* Raybaudi, 1990, length 15 mm, 10 m depth, in Taiohae Bay, Nuku Hiva Island, Marquesas Islands; **L=** *Luria isabella gauguini* Lorenz, 2017, length 18 mm, 10 m depth in Colette Bay, Nuku Hiva Island, Marquesas Islands.

Figure 5.14 Cone Shell Radiations of the Marquesan Province.
A= *Splinoconus troendlei* (Moolenbeek, Zandbergen, and Bouchet, 2008), length 17 mm, 106 m depth off Nuku Hiva Island, Marquesas Islands; **B=** *Cylinder textilinus* (Kiener, 1847), length 53 mm, 30 m depth, bayside of Cape Martin, in Controller Bay, Nuku Hiva Island, Marquesas Islands; **C=** *Eugeniconus marchionatus* (Hinds, 1843), length 40 mm, 20 m depth off Motumano Point, Nuku Hiva Island, Marquesas Islands; **D=** *Lividiconus conco* (Puillandre, Stocklin, Favreau, *et al.*, 2014), length 51 mm, 30 m depth off the Sentinal, Taiohae Bay, Nuku Hiva Island, Marquesas Islands; **E=** *Miliariconus encaustus* (Kiener, 1845), length 28 mm, 5 m depth in Taiohae Bay, Nuku Hiva Island, Marquesas Islands; **F=** *Pionoconus easoni* Petuch and Berschauer, 2017, length 35 mm, 10 m depth off Motumano Point, Nuku Hiva Island, Marquesas Islands; **G=** *Pionoconus gauguini* (Richard and Salvat, 1973), length 78 mm, 30 m depth off the Sentinal, Taiohae Bay, Nuku Hiva Island, Marquesas Islands; **H=** *Puncticulus vautieri* (Kiener, 1845), length 42 mm, 25 m depth in Colette Bay, Nuku Hiva Island, Marquesas Islands; **I=** *Rhizoconus taitensis* (Hwass, 1792), length 41 mm, 5 m depth off Motumano Point, Nuku Hiva Island, Marquesas Islands; **J=** *Rhombiconus pseudimperialis* (Moolenbeek, Zandbergen, and Bouchet, 2008), length 31 mm, 30 m depth off Matauapuna, Taiohae Bay, Nuku Hiva Island, Marquesas Islands; **K=** *Splinoconus hivanus* (Moolenbeek, Zandbergen, and Bouchet, 2008), holotype, length 16 mm, 120 m depth off Nuku Hiva Island, Marquesas Islands.

Figure 5.15 Index Gastropods of the Rapanuian Province and Salas-Gomezian Infraprovince).
A= *Lisanerita lirellata* (Rehder, 1980), width 15 mm, low tide, Hanga Roa, Easter Island; **B=** *Hinea akuana* (Rehder, 1980), length 8 mm, under a rock at low tide, Hanga Roa, Easter Island; **C=** *Monetaria caputdraconis* (Melvill, 1888), length 32 mm, low tide off Ana Kakenga, Easter Island; **D=** *Naria leforti* (Senders and Martin, 1987), length 24 mm, 25 m depth off Motu Kao Kao Island, off southern Easter Island; **E=** *Naria englerti* (Summers and Burgess, 1965), length 18 mm, 10 m depth off Ahu Tahai, Easter Island; **F=** *Canarium rapanuense* Liverani, Wieneke, & Kronenberg, 2020, length 26 mm, low tide off Ana Kakenga, Easter Island; **G=** *Pascula citrina* Rehder, 1980, length 11 mm, off Hanga Roa, Easter Island; **H=** *Strigatella flavocingulata* (Lamy, 1938), length 25 mm, under rocks at low tide, Ana Kakenga, Easter Island; **I=** *Miliariconus pascuensis* (Rehder, 1980), length 27 mm, under rocks in large tidal pools, Ana Kakenga, Easter Island; **J=** *Nodilittorina pascua* (Rosewater, 1970), height 11 mm, high intertidal zone, Hanga Roa, Easter Island; **(Salas-Gomezian Infraprovince) K=** *Monetaria caputdraconis poppei* Martin, 1989, length 27 mm, in rock tidal pools, southern shore of Salas y Gomez Island, Chile.

Figure 5.16 Widespread Polynesian Province Index Gastropods.
A= *Haliotis pulcherrima* Gmelin, 1791, width 24 mm, low tide, Aitutaki Atoll, Cook Islands; **B**= *Cribrarula cumingi* (Sowerby II, 1832), length 18 mm, 5 m depth off Vaia'au, Raiatea Island, Society Islands; **C**= *Harpa gracilis* Broderip and Sowerby I, 1829, length 27 mm, in coral rubble, Mauke Island, Cook Islands; **D**= *Monetaria obvelata* (Lamarck, 1810), length 17 mm, low tide, Punaauia, Tahiti, Society Islands; **E**= *Gastridium eldredi* (Morrison, 1955), length 60 mm, found dead on a reef flat off Tiputa, Rangiroa Atoll, Tuamotu Islands; **F**= *Naria irrorata* (Gray, 1828), length 12 mm, 3 m depth off Vaia'au, Raiatea Island, Society Islands; **G**= *Chicoreus maurus* (Broderip, 1833), length 87 mm, 30 m depth off Christmas Island, Kiribati; **H**= *Vasum armatum* Broderip, 1833, length 27 mm, low tide on rocks, Punaauia, Tahiti, Society Islands; **I**= *Miniaceoliva efasciata* (Dautzenberg, 1927), length 57 mm, 2 m depth, off the Gateway Hotel, Majuro Atoll, Marshall Islands; **J**= *Darioconus auratinus* (da Motta, 1982), length 55 mm, 10 m depth in the main lagoon, Arutua Atoll, Tuamotu Islands; **K**= *Textilia adamsoni* (Broderip, 1836), length 50 mm, 25 m depth off Ututaotao, Colette Bay, Nuku Hiva Island, Marquesas Islands.

Figure 5.17 Index Gastropods of the Tahitian Subprovince, Polynesian Province.
A= *Septaria taitana* Mousson, 1869, length 27 mm, at the base of the waterfall at Punaauia, Tahiti, Society Islands; **B=** *Ipsa childreni leforti* Heiman, 2009, length 12 mm, 20 m depth off Faaa, Tahiti, Society Islands; **C=** *Mauritia scurra mundula* Lorenz, 2002, length 32 mm, 1 m depth off Tautira, Tahiti, Society Islands; **D=** *Talparia talpa lutani* Bridges, 2015, length 59 mm, 4 m depth on living coral, off Punaauia, Tahiti, Society Islands; **E=** *Talostolida teres janae* (Lorenz, 2002), length 25 mm, 1 m depth in coral rubble, Tautira, Tahiti, Society Islands; **F=** *Talostolida violacincta* (Lorenz, 2002), length 26 mm, 1 m depth, Tautira, Tahiti, Society Islands; **G=** *Lambis robusta* (Swainson, 1821), length 132 mm, 3 m depth off Punaauia, Tahiti, Society Islands; **H=** *Omogymna ozodona* (Duclos, 1835), length 15 mm, 3 m depth in Opunohu Bay, Moorea Island, Society Islands; **I=** *Cylinder panniculus* (Lamarck, 1810), length 46 mm, 1 m depth off Papiti, Huahine Island, Society Islands; **J=** *Leporiconus pomareae* Monnier and Limpalaër, 2014, length 26 mm, 5 m depth off Punaauia, Tahiti, Society Islands; **K=** *Purpuradusta barbieri* (Raybaudi, 1986), length 9 mm, in coral rubble, 10 m depth off Punaauia, Tahiti, Society Islands; **L=** *Naria bernardi* (Richard, 1974), length 14 mm, 25 m depth off Maharepa, Moorea Island, Society Islands.

Figure 5.18 Cowrie Species Radiations of the Tuamotuan Infraprovince, Polynesian Province.
A= *Cribrarula compta* (Pease, 1860), length 12 mm, 2 m depth off Rotoava, Fakarava Atoll, Tuamotu Islands; **B**= *Leporicypraea mappa admirabilis* Lorenz, 2002, length 77.6 mm, 15 m depth off Marae Takai, Takapoto Atoll, Tuamotu Islands; **C**= *Monetaria caputserpentis argentata* (Dautzenberg and Bouge, 1933), length 31 mm, low tide, Takume Atoll, Tuamotu Islands; **D, E**= *Monetaria caputserpentis nivalis* Lorenz and Vulliet, 2015, length 25 mm, in rock pools at low tide, Tiputa, Rangiroa Atoll, Tuamotu Islands; **F**= *Monetaria obvelata perrieri* (Rochebrune, 1884), length 13 mm, low tide, Roroava, Fakarava Atoll, Tuamotu Islands; **G**= *Purpuradusta oryzaeformis* Lorenz and Sterba, 1999, length 8 mm, 1 m depth off Marae Takai, Takapoto Atoll, Tuamotu Islands; **H**= *Pustularia tuamotensis* (Lorenz, 1999), length 14 mm, 3 m depth on the main reef of Tiputa, Rangiroa Atoll, Tuamotu Islands; **I**= *Talostolida subteres* (Weinkauff, 1881), length 20 mm, 2 m depth off Roroava, Fakarava Atoll, Tuamotu Islands; **J**= *Lyncina bouteti* (Burgess and Arnette, 1981), length 36 mm, 1 m depth in the main lagoon off Turipaoa, Manihi Atoll, Tuamotu Islands.

Figure 5.19 Index Gastropods of the Micronesian Subprovince and Samoan Infraprovince, Polynesian Province. **(Micronesian Subprovince) A**= *Blasicrura summersi* (Schilder, 1958), length 20 mm, under rocks at low tide, Vaitukakau Bay, Utu Vava'u Island, Vava'u Islands, Tonga; **B**= *Erronea ovum palauensis* Schilder and Schilder, 1938, length 34 mm, low tide, Karamadoo Bay, Babeldaob Island, Palau Islands; **C**= *Leporicypraea mappa guamensis* Bergonzoni and Passamonti, 2014, length 47 mm, 3 m depth in Apra Harbor, Guam; **D**= *Sinustrombus taurus* (Reeve, 1857), length 121 mm, 1 m depth off Vaiasala, Savaii Island, Western Samoa; **E**= *Acutoliva hilli* (Petuch and Sargent, 1986), length 9 mm, 1 m depth in Vaitukakau Bay, Utu Vava'u Island, Vava'u Islands, Tonga; **F**= *Miniaceoliva efasciata thierryi* Petuch and Myers, 2014, length 46 mm, 2 m depth off Buariki, Tarawa Atoll, Gilbert Islands, Republic of Kiribati; **(Samoan Infraprovince) G**= *Blasicrura pallidula vivia* Steadman and Cotton, 1953, length 16 mm, 1 m depth, Vaisala Lagoon, Savaii Island, Western Samoa; **H**= *Cribrarula taitae* (Burgess, 1993), length 14 mm, 3 m depth off Faiaai Savaii Island, Western Samoa; **I**= *Erronea caurica samoensis* Lorenz, 2002, length 30 mm, 2 m depth off Satuiatua, Savaii Island, Western Samoa; **J**= *Monetaria sublittorea* Lorenz, 1998, length 13 mm, 2 m depth, Pago Pago Harbor, Tutuila Island, American Samoa; **K**= *Conus nigrescens* Sowerby II, 1859, length 26 mm, 3 m depth in Faga'itua Bay, Tutuila Island, American Samoa; **(Wallis and Futuna) L**= *Miniaceoliva lamberti chloeae* Petuch and Myers, 2014, length 60 mm, 3 m depth off Pointe Tepako, Uvea, Wallis Island, Wallis and Futuna Islands.

Figure 5.20 Index Gastropods of the Kwajaleinian Infraprovince, Polynesian Province.
A, B= *Bistolida stolida kwajaleinensis* (Martin and Senders, 1983), length 23 mm, 15 m depth, R-Buoy Pinnacle, southern lagoon, Kwajalein Atoll, Marshall Islands; **C=** *Cribrarula cribraria oceanica* Raybaudi, 1993, length 20 mm, 10 m depth off the P-North Buoy, Kwajalein Atoll, Marshall Islands; **D=** *Cribrarula gaspardi* Biraghi and Nicolay, 1993, length 13 mm, 10 m depth on the R-Buoy Pinnacle, southern lagoon, Kwajalein Atoll, Marshall Islands; **E, F=** *Leporicypraea mappa eluceta* Lorenz and Hubert, 2000, length 67 mm, 30 m depth on Oceanside West Reef, Kwajalein Atoll, Marshall Islands; **G=** *Palmadusta johnsonorum* Lorenz, 2002, length 14 mm, 15 m depth at night, R-Buoy Pinnacle, southern lagoon, Kwajalein Atoll, Marshall Islands; **H=** *Galeola carneola kwajaleinensis* (da Motta, 1985), length 13 mm, 3 m depth off Emon Beach, Kwajalein Island, Kwajalein Atoll, Marshall Islands; **I=** *Miniaceoliva efasciata berti* (Terzer, 1986), length 47 mm, 3 m depth off Emon Beach, Kwajalein Island, Kwajalein Atoll, Marshall Islands; **J=** *Darioconus cathyae* Monnier, Limpalaër, and Prugnaud, 2020, length mm, 20 m depth off the G-Buoy, southern lagoon, Kwajalein Atoll, Marshall Islands.

Intertidal zone at Keawaula Beach along the northwest coast of Oahu, Hawaii, where the endemic nacellid limpets *Cellana sandwichensis, C. talcosa,* and *C. exarata* (the Hawaiian "Opihi"), and the endemic cypraeid *Monetaria caputophidii* can be found among the volcanic rocks. Photo by Morgan Taylor Photography, used with permission. ©

CHAPTER 6. Western Pacific Tropical Region

The Western Pacific Tropical Region spans a wide area that extends from central Japan in the north to New Guinea in the south, and eastward to New Caledonia and the Fiji Islands, and includes the immense archipelagoes of the Philippines, Indonesia, and Melanesia. This large marine region encompasses the tropical coastal areas of eastern Eurasia and is enriched by the heavy river effluent draining off the continent and from rivers on the adjacent large continental islands. Because of the high nutrient budget pouring into the marine environment and the perpetual warm water conditions, this region contains the richest molluscan fauna found anywhere on Earth. The Western Pacific Tropical Region is now known to contain two major molluscan biotic units, the **Indo-Malaysian Province** in the south and east and the **Japonic Province** in the north, and seven distinct subprovinces. Besides containing the widespread Indo-Pacific taxa listed in the previous chapter, the Western Pacific Tropical Region also houses a large number of characteristic regionally-endemic taxa that range from Japan to New Caledonia, including the cypraeids *Austrasiatica hirasei* and *Palmulacypraea musumea*, and the cone shells *Turriconus excelsus*, *Yeddoconus aphrodite*, *Yeddoconus boholensis* (some illustrated on Figure 6.3), *Afonsoconus kinoshitai*, *Graphiconus armadillo*, and *Kioconus hirasei* (all illustrated on Figure 6.23).

Indo-Malaysian Molluscan Province

Named for the Indonesian and Malaysian Archipelagoes, which represent the core area of the biotic unit, this province contains the richest-known marine molluscan fauna in the world. The province is essentially defined by the **Coral Triangle**, a vast series of coral reef complexes that extends from the Philippines in the north, the island of Sumatra in the west, and the Solomon Islands in the east. These reef systems, along with the adjacent mangrove forests, volcanic rocky coasts, and muddy brackish estuaries, offer a large number of habitats that support species-rich malacofaunas. Many of the islands within the Indo-Malaysian Province are separated by deep water trenches and these barriers to dispersal have produced a high level of endemism within low-vagility genera and families. These intraprovincial pockets of evolution have led to the formation of four distinct subprovinces, the **Indonesian**, the **Melanesian**, the **Neocaledonian**, and the **Philippinian** and at least five infraprovinces. A large number of wide-ranging conspicuous gastropods sharply demarcate the limits of the entire region, some of which include the cypraeids *Callistocypraea aurantium*, *Callistocypraea leucodon*, *Erronea fernandoi*, *Ransoniella martini*, and *Leporicypraea valentia*, the ovulid *Amonovula piriei*, the harpid *Morum (Herculea) ponderosum,* the olivid *Musteloliva xenos*, and the conid *Cylinder gloriamaris* (all shown here on Figure 6.3).

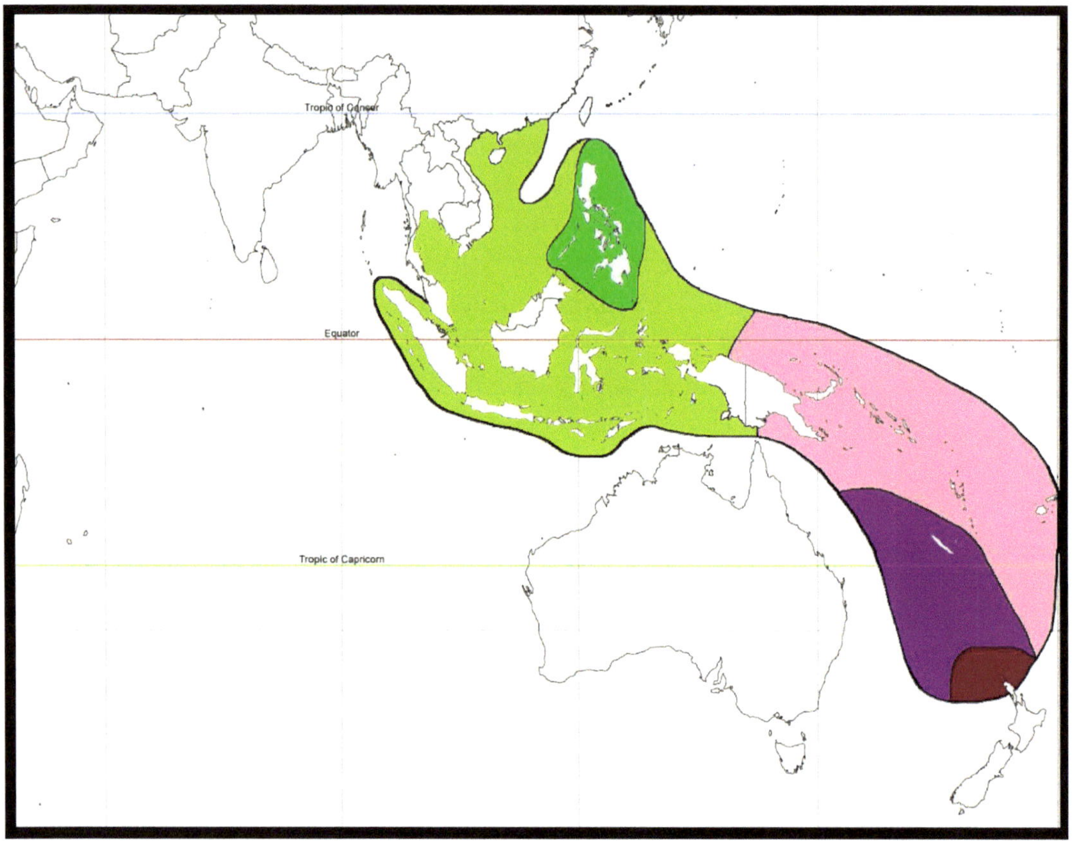

Figure 6.1 Map of the Indo-Malaysian Molluscan Province, showing the areal extents of its subprovinces: the **Indonesian Subprovince** (lime green), the **Melanesian Subprovince** (light rose), the **Neocaledonian Subprovince** (purple), and the **Philippinian Subprovince** (green), and also the **Auporian Subprovince** of the **Neozealandic Province** (dark rose).

Indonesian Subprovince

Named for the country of Indonesia and the widespread Indonesian Archipelago, this subprovincial area extends from Sumatra in the west to Irian Jaya (western New Guinea) in the east and also encompasses the southern part of the Sulu Sea of the Philippines. Although appearing to be a contiguous faunal subdivision, the Indonesian Subprovince actually is composed of three separate ecozones; the Asian, the Wallacean, and the Australian. These present-day separate faunal influences reflect the appearance of the Indonesian Archipelago during the late Pleistocene, when lowered sea levels exposed wide areas of shallow sea floors. Around 400,000 years B.P., the islands of Sumatra, Java, Borneo, Bali, and all the smaller islands in-between, were connected to mainland Asia and had fused into a single giant peninsular landmass referred to as **Sundaland**. Likewise, Australia, New Guinea, and the entire Arafura Sea area were combined into a single continental landmass referred to as **Sahulland**. The islands of central Indonesia, including the present-day area extending from Lomboc in the west, to Timor in the east, and northward to Sulawesi, represented a faunal transition zone that is referred to as **Wallacea**. Named for the 19th century biologist and explorer, Alfred Russel Wallace, this area remained as a series of separated islands during the late Pleistocene and acted as a barrier to gene flow from Sundaland to Sahulland. This faunal pattern is still seen today as

"Wallace's Line", a distinct break seen in the distribution of terrestrial organisms between the islands of Bali and Lombok. The old Sundaland-Wallacea-Sahulland faunal patterns are still reflected in the Recent Indonesian marine molluscan faunas, primarily by infraprovincial hot spots. These three separate faunal influences are superimposed upon a larger pattern of widespread Indonesian endemic species. Some of these general Indonesian endemics, which can be used to define the subprovince, include the muricids *Murex spinistreptos* and *Hexaplex bundharmai*, the buccinid *Babylonia semipicta*, the volutes *Cymbiola innexa* and *Cymbiola octagonalis*, the olivid *Galeola vicweei*, and the cone shells *Conus equestris*, *Pionoconus frauenfeldi*, and *Textilia cervus* (all shown here on Figure 6.4).

Deriving from the Sahulland Pleistocene molluscan fauna, the present-day area of the Maluku (Moluccas) Islands, Raja Ampat Archipelago, and western Irian Jaya constitutes an evolutionary hot spot, referred to here as the **Malukuan Infraprovince**. Of particular interest within this localized evolutionary center are a number of highly-restricted endemic cone shells, such as *Phasmoconus daphne*, *Conus nocturnus*, and *Conus nocturnus deburghiae* (the last two species are illustrated here on Figure 6.4). The Pleistocene Wallacean malacofaunas are also still present within the Indonesian Subprovince, as demonstrated by highly-restricted taxa that are confined to an evolutionary hot spot centered on the East and West Nusa Tenggarra Islands (Lesser Sunda Archipelago). Referred to here as the **Nusa Tenggaran Infraprovince**, this area extends from Lombok eastward to Wetar and Timor and contains the highest number of endemic taxa seen within the entire Indonesian Subprovince. Some of these Wallacean-derived Nusa Tenggaran species include the cypraeid *Erronea vredenbergi*, the volutids *Cymbiola chrysostoma* and *Cymbiola cymbiola*, and the conids *Cylinder abbas johnabbasi*, *Graphiconus wittigi*, and *Phasmoconus giorossii* (all illustrated here on Figure 6.5). Of special interest within the Nusa Tenggaran Infraprovince is a species radiation of the conid genus *Eugeniconus* that comprises at least four named subspecies, including *E. nobilis nobilis* (Flores Island), *E. nobilis abbai* (southern Lembata Island), *E. nobilis skinneri* (Bali Sea), and *E. nobilis victor* (northern Lembata Island). These strikingly-colored endemic gastropods, the most beautiful cone shells known from the Wallacean area, are shown here on Figure 6.5.

Of special interest within the Indonesian Subprovince is the northwestern area, which extends into the South China Sea along the shallow coastal waters of Vietnam. Being isolated from the main core of the subprovince, this extreme northern area houses a shallow water malacofauna that contains many endemic taxa and has only recently been recognized as constituting an evolutionary hot spot. This northern edge of the Indonesian Subprovince is here referred to as the **Vietnamese Infraprovince** (named for Vietnam) and it encompasses an unusually rich molluscan fauna that is different from those of both the Philippinean Subprovince to the east and the classic Indonesian Subprovince to the south. Some of the more typical shallow water Vietnamese endemic taxa include the angariids *Angaria petuchi*, *A. moolenbeeki*, and *A. guntheri*, the turbinid star shell *Astralium dekkersi*, the nassariid *Nassarius thachorum*, the large melongenid *Pugilina elongata nhatrangensis*, the columbellid

Pardanilops borroni, the moruminine harpid *Oniscidia kreipli*, and the large mitrid *Nebularia thachi*. For more detailed discussions of this relatively unknown malacofauna, see Thach (2018).

Melanesian Subprovince

Named for the cultural region of Melanesia, the Melanesian Subprovince extends from eastern New Guinea (Papua-New Guinea) eastward to the Fiji Islands, and encompasses New Britain Island and the Bismarck Archipelago, the Solomon Islands, the island nation of Vanuatu, New Caledonia, and the entire Coral Sea area east of the Great Barrier Reef islands. The large continental island of New Guinea, itself a remnant of the Pleistocene landmass of Sahulland, supplies the surrounding seas and enclosed basins with an immense volume of nutrient-rich river effluent. These high-productivity continental-type water conditions have allowed most of the wide-ranging Indo-Malaysian gastropods to become established around the island and within the Bismarck and Solomon Sea basins. Being isolated on the extreme eastern edge of the Coral Triangle, the Melanesian area has also evolved its own fauna of endemic gastropods and these can be used to demarcate the boundaries of the subprovince. Some of the more important and characteristic of these index taxa include the cowries *Bistolida stolida crossei*, *Contradusta bregeriana*, *Cribrarula catholicorum*, *Erronea (Adusta) melanesiae*, *Naria eburnea*, and *Talostolida pseudoteres*, the olivid *Viduoliva rubrolabiata*, the volute *Cymbiola rutila*, and the cone shells *Bathyconus fijiensis* and *Phasmoconus sutanorcum* (all shown here on Figure 6.6).

Within the Melanesian area, the Solomon Sea, and its multiple component island chains such as the Louisiades, Trobriands, and D'Entrecasteaux, along with the Solomon Islands, stands out as having the most species-rich molluscan fauna in the entire subprovince. Referred to here as the **Solomonian Infraprovince**, this evolutionary hot spot contains the largest number of endemic gastropods, including distinctive taxa such as the cypraeid *Contradusta bregeriana pervelata*, the strombid *Laevistrombus guidoi*, the olivid *Acutoliva buelowi* form *stoneorum*, the cone shell *Graphiconus gabryae*, a species radiation of the genus *Phasmoconus* including *P. solomonensis*, *P. nahoniaraensis*, and *P. zebra* (all illustrated here on Figure 6.7), and the muricid *Chicoreus paini*. Of special interest in the Solomonian Infraprovince is a highly localized species radiation of the cypraeid genus *Eclogavena*, with at least six species found on different isolated island chains. Some of these include *Eclogavena coxeni* (eastern Solomon Sea), *E. hesperina* (western Solomon Sea), *E. hesperina insolita* (New Britain), *E. pseudohesperina* (Solomon Islands) (all shown here on Figure 6.7), *E. steineri* (Trobriand Islands), and *E. coxeni hypercallosa* (Solomon Islands). Another particularly noteworthy Solomonian endemic species radiation is seen in the volutid genus *Cymbiola*, where at least five endemic species and subspecies have evolved on the widely-separated island chains. All of these large and colorful Solomonian volutes have evolved as localized island offshoots of the widespread Melanesian *Cymbiola rutila* and include *Cymbiola rueckeri* (endemic to the southern Solomon Islands), *C. ceraunia* (endemic to the northern Solomon Islands), *C. piperita* (endemic to eastern Papua and the eastern Louisiade

Archipelago), *C. macgillvrayi* (endemic to Rossel Island and the western Louisiade Archipelago)), and *C. norrisii* (endemic to the Trobriand Islands and northern D'Entrecasteaux Islands). These are all shown here on Figure 6.8.

Neocaledonian Subprovince

Named for the island of New Caledonia, the Neocaledonian Subprovince encompasses the area south of Vanuatu and includes New Caledonia itself, along with the Loyalty and Chesterfield Islands and the northern half of the Norfolk Ridge. The subprovince is situated on the northern tip of the narrow, mostly-submerged continent of Zealandia, which extends from New Caledonia and Lord Howe Island in the north to New Zealand in the south, and underlies much of the Tasman Sea. The long submarine mountain chain of the Norfolk Ridge, which extends along the eastern edge of Zealandia, from New Caledonia and Norfolk Island to New Zealand, represents the subduction zone where the smaller plate is sliding below the much larger Pacific Plate. New Caledonia, itself, has an odd geology, being composed of iron and nickel-rich crustal rocks that were uplifted when Zealandia collided with the Pacific Plate, south of what is now Vanuatu. These metallic-rich rocks are often present as submerged outcrops all along New Caledonia, producing high concentrations of nickel ions in the local marine environment. This anomalously-high nickel content is picked up by cowrie shells and incorporated into their shell enamel, producing the famous jet-black melanistic specimens. One of these classic New Caledonian melanistic cowries, *Mauritia eglantina* form *nigricans*, is shown here on Figure 6.9 G.

The slowly-moving Zealandia continental plate carries its own endemic fauna along with it, including some taxa that are probably remnants of faunas that had evolved on older, now-sunken predecessor islands. A classic example of one of these continental plate "hitchhikers" is the bizarre cerithiid *Gourmya gourmyi* (Figure 6.9A), which is found only around New Caledonia and a few islands in Vanuatu. Other non-dispersing taxa that are endemic to the Neocaledonian Subprovince include the volutids *Cymbiola rossiniana*, *Cymbiola deshayesi*, and *Cymbiolacca (Magnavictoria) thatcheri* (Chesterfield Islands) (all shown on Figure 6.9). the columbariid *Fustifusus pinicola* (with the genus *Fustifusus* being endemic to New Caledonia; Figure 6.9 F), and an entire fauna of cone shells, including species such as *Conus crosseanus*, *Conus suffusus*, *Conus pseudomarmoreus*, *Vituliconus swainsoni*, *Graphiconus lienardi*, and *Graphiconus richeri* (all shown here on Figure 6.10). Of special interest on New Caledonia is an endemic species radiation of the conid genus *Thoraconus*, with at least eight distinct species and subspecies being found in different isolated bays around the island. Some of these include *Thoraconus exiguus* (Ile Tenia), *T. bougei* (Prony Bay), *T. cabritii* (Nickel Bay), *T. plumbeus* (Vincer Bay), *T. optimus* (Bay Goro), and *T. vayssetianus* (Ile Ouen) (all illustrated here on Figure 6.10). The family Cypraeidae has also evolved a swarm of endemic species and subspecies within the Neocaledonian Subprovince, some of which include *Leporicypraea mappa montrouzieri*, *Leporicypraea mappa kanakinus*, *Austrasiatica langfordi cavatoensis*, *Ipsa childreni novaecaledoniae*, and *Ransoniella martini superstes* (all illustrated here on Figure 6.10).

The extremely long submarine mountain chain of the Norfolk Ridge begins at the Isle of Pines (Ile des Pins), New Caledonia and extends due south to Norfolk Island. There it makes a bend to the southeast and runs directly into the Three Kings Islands and Cape Reinga of North Island, New Zealand. The Ridge forms a submerged bridge 1300 km long that connects New Caledonia with New Zealand and demarcates the eastern edge of the Zealandia Continental Plate. Although averaging around 2000 m in depth, many of the Norfolk Ridge seamounts come within 400-500 m of the sea surface, such as the East and West Jumeaux Banks, Introuvable Bank, and Athos and Porthos Banks. Literally hundreds of isolated banks dot the length of the Norfolk Ridge, with each being essentially an island all to itself. These isolated seamounts support rich and diverse deep sea communities, with each mountain top having a different type of ecosystem. Many of these are named for the dominant organism of each community, such as Sponge Bank (named for the beds of Glass Sponges), Stylaster Bank (named for the reefs of delicate hydrocorals), and Brachiopod Bank (named for abundant encrusting brachiopods). The amazingly-rich seamount communities of the Norfolk Ridge, altogether, form one of the most spectacular evolutionary hot spots found on Earth, referred to here as the **Norfolkian Infraprovince** (named for the Norfolk Ridge). A large number of endemic gastropods have evolved within this infraprovince, some of which include the muricids *Chicoreus boucheti* and the incomparable *Babelomurex neocaledonicus*, the fasciolariid *Granulifusus bacciballus*, the siphonate buccinid *Serratifusus lineatus*, the olivid *Amalda aureomarginata*, the volutid *Plicolyria boucheti*, and the cone shells *Afonsoconus bruuni*, *Kioconus plinthus*, *Kioconus gondwanensis*, *Boucheticonus alisi*, *Yeddoconus boucheti*, and *Kurodaconus luciae*. These are all illustrated here on Figure 6.12.

Philippinian Subprovince

Located at the northern tip of the Coral Triangle, the Philippine Islands contain the richest single molluscan fauna found anywhere, both within the Indo-Malaysian Province and on the entire planet. Named for the Philippine Archipelago of over 8,000 islands, the Philippinian Subprovince extends from the Batanes Islands north of Luzon Island south to Balabac Island, Palawan Group and Siasi Island, Sulu Archipelago. This insular subprovince is rivalled only by the Indonesian Subprovince in species-richness, but out-competes its southern sister subprovince by containing many more endemic taxa. The multiple isolated island chains of the Philippines are separated from the each other by very deep trench-like basins, creating the perfect conditions for allopatric speciation in taxa with low vagility. Because of this, the Philippines have evolved a highly-characteristic molluscan fauna unlike any other in the Western Pacific. Some of the classic widespread subprovincial index gastropods include the cypraeid *Pustularia chiapponii*, the muricid *Homalocantha dondani*, the olivid *Acutoliva bathyalis*, and a large endemic cone shell fauna including *Turriconus beatrix*, *Turriconus rizali*, *Isoconus richardsae*, *Calamiconus escondidai*,

Strategiconus thalassiarchus, *Pionoconus leobottonii*, *Yeddoconus olangoensis*, *Yeddoconus nereis*, and *Yeddoconus gattegnoi* (all illustrated here on Figure 6.12). For a nearly-complete overview of the Philippines molluscan fauna, see the 10 volume set of iconographies entitled *Philippine Marine Mollusks* that was edited by Guido Poppe.

Of special interest within the Philippines is a large and unstudied species radiation of the conid genus *Calibanus*, possibly incorporating over 15 separate taxa. These interesting small cones, originally thought to be only color forms of a single species, *Calibanus furvus*, have now been found to inhabit many different marine habitats, including in the root masses of mangrove trees, intertidal mud flats, clean carbonate sand shallow sea floors, sea grass beds, and in coral rubble near living coral reefs. Many of the *Calibanus* species have also been found to live exclusively on separate island groups, with the majority of the known species occurring in the Sulu Archipelago, Palawan Islands, and around Cebu Island. Some of these species include *Calibanus furvus* (mangroves and mud flats, Palawan Islands), *C. aegrotus* (clean coral sand, Sulu Archipelago), *C. albus* (clean coral sand, Coron Island), *C. albicans* (coral rubble and sea grass beds, Sulu Archipelago), *C. cecilei* (coral rubble, Jolo Island), *C. crepusculum* (clean coral sand, Balabac Island), *C. granifer* (near living coral reefs, Mindanao Island), *C. neobuxeus* (sand and sea grass beds, Palawan Islands), *C. nivalis* (clean coral sand, southern Sulu Sea), *C. polygrammus* (mud flats, Cebu Island), and *C. turritinus* (deep reef areas, Palawan Islands). These taxa are all shown here on Figure 6.13 and we consider them to be full species, or at least subspecies, pending more research on their anatomy and biochemistry.

Another characteristic Philippinian species radiation is seen in the volutid genus *Cymbiola*, with at least 10 species and subspecies being distributed throughout the subprovince. Like the conid genus *Calibanus*, these large and colorful endemic volutes are often restricted to isolated islands or small island groups, primarily in the southern Philippines and Sulu Sea areas. Some of these include *Cymbiola aulica* (Sulu Sea), *C. alexisallaryi* (Pangutaran Island), *C. cathcartiae* (southern Palawan Island), *C. palawanica* (Palawan Island), *C. laminusa* (Sulu Archipelago), *C. malayensis* (southern Sulu Sea), *C. vespertilio augustinensis* (Mindanao Island), and *C. vespertilio matiensis* (Zamboanga area, Mindanao Island) (all shown here on Figure 6.15), along with the widespread Philippine *C. vespertilio mitis* and *C. imperialis*.

The southern half of the Sulu Sea, including the Zamboanga Peninsula of Mindanao Island, the Sulu Archipelago, and southern Palawan Island as well as the islands off northern Sabah, Malaysia, constitutes a faunal overlap zone between the Philippinian and Indonesian Subprovinces. The entire area has also been found to be an evolutionary hot spot with a very large number of endemic species being present within the confines of the Sulu Sea Basin. Referred to here as the **Suluan Infraprovince** (for the Sulu Sea), this area's largest faunal component is composed of Philippine endemic taxa, demonstrating that it can clearly be considered to be a subdivision of the Philippinian Subprovince. These endemics, however, are mixed together with many classic Indonesian taxa that are now known to range that far north, including desirable rarities such as the cone shells *Textilia cervus* and *Strategiconus thomae* (Figure 6.16 L). Of the known infraprovinces found around the world, the Suluan contains the single greatest pulse of marine gastropod evolution seen in any modern tropical sea.

Unlike most infraprovincial areas found elsewhere around the world, the Suluan Infraprovince houses an unusually-high number of endemic taxa, approaching 20% of the total fauna and reaching almost subprovincial status. The endemic Suluan taxa contained in the principal tropical gastropod index families are often present as species swarms, underscoring the rapid evolution that has taken place there since the Pleistocene. One of these swarms is seen in the Cypraeidae, which is represented by the Suluan endemics *Austrasiatica langfordi poppei*, *Ficadusta pulchella aliguayensis*, *Bistolida stolida aureliae*, *Eclogavena dani*, and *Eclogavena dayritiana mandejarorum* (all shown here on Figures 6.15, 6.16, and 6.17). A similar evolutionary pattern is seen in the Muricidae, with *Pterynotus mikoyoae* and *Chicoreus jessicae* and a large species radiation of the genus *Homalocantha*, including *H. granpoderi*, *H. ninae*, *H. pisori*, *H. vicdani*, *H. anomaliae* (all shown here on Figures 6.15), and *H. nivea*. The Conidae has evolved one of the largest clusters of endemic taxa seen within the Suluan Infraprovince, including species such as *Boucheticonus pseudokimioi*, *Fulgiconus cebuensis*, *Kioconus (Ongoconus) vanvilstereni*, *Rolaniconus dedonderi*, *Pionoconus robini*, *Darioconus viperinus*, *Calamiconus tethys*, *Phasmoconus balabacensis*, *Phasmoconus zapatosensis*, *Eugeniconus cordigera*, *Eugeniconus bitleri*, *Conus vidua*, and a radiation of the genus *Cylinder*, including *C. barbieri*, *C. telatus*, *C. scottjordani*, and *C. tagaroae* (all shown here on Figures 6.15, 6.16, and 6.17). These swarms of species occur along with numerous other important Suluan endemic gastropods such as the dwarf scorpion conch *Lambis adami*, the harpid *Oniscidia kurzi*, and the olivids *Recourtoliva poppei*, *Musteloliva boholensis*, *Viduoliva zamboangensis*, and *Viduoliva mindanaoensis* (all shown here on Figures 6.15, 6.16, and 6.17).

Japonic Molluscan Province

Named for the country of Japan and the Japanese Archipelago, the Japonic Province extends from the Izu Peninsula, Shizuoka Prefecture, Honshu Island, Japan, south through the entire Ryukyu Archipelago and eastern half of the East China Sea, to Taiwan and the deep Luzon Strait, and into the central deep basin of the South China Sea. The Japonic Province is very complex oceanographically, being distributed over three separate oceanic basins and including sections of the Sea of Japan, the East China Sea, and the South China Sea. Even as far north as the Izu Peninsula, the water temperatures are moderated by the warm, northward-flowing Kuroshio Current, and this has allowed for the evolution of an extremely rich endemic tropical malacofauna. In the southern part of the Japonic Province, water temperatures are even higher, supporting a much richer malacofauna that shares many taxa with the neighboring Philippinian Subprovince. Because of these diverse water temperatures and faunas, the

Japonic Province is now recognized as having three separate subprovinces; the **Shikokuan**, the **Ryukyuan**, and the **South China**. A large number of classic widespread Japonic index gastropods are found in two, or all three, of these subprovinces and these define the boundaries of the province. Some of the widespread Japonic taxa include the cypraeids *Perisserosa guttata azumai* and *Paradusta hungerfordi*, the charoniid *Charonia sauliae*, the muricid *Chicoreus asianus*, the babyloniid *Babylonia japonica*, and the cone shells *Pionoconus fulmen*, *Profundiconus profundorum*, and *Endemoconus sieboldi* (all illustrated here on Figure 6.18).

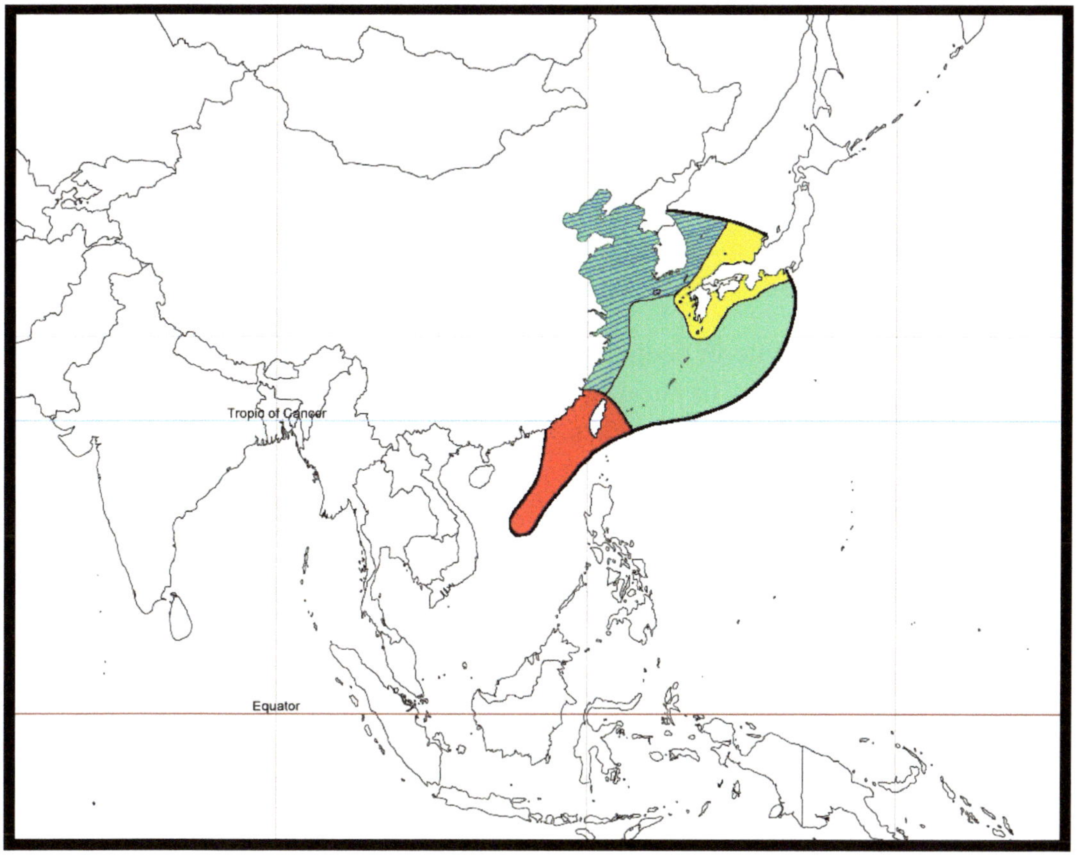

Figure 6.2 Map of the Japonic Molluscan Province, showing the areal extents of its subprovinces: the **Shikokuan Subprovince** (yellow), the **Ryukyuan Subprovince** (green), and the **South China Subprovince** (red). The cross-hatched green area demarcates environments with cooler water temperatures that support impoverished Japonic-type faunas; these may eventually prove to be a new subprovince.

Shikokuan Subprovince

Named for the central Japanese island of Shikoku, the Shikokuan Subprovince includes the southern half of Honshu Island south of the Izu Peninsula and the western side of Honshu south of the Kashimacho coast, the Japanese Inland Sea and its islands, and also the main islands of Shikoku and Kyushu and their numerous adjacent coastal islands. Although situated within a cool temperate climatic zone, the tropically-derived Kuroshio Current floods the Shikokuan coastal waters and creates warm temperate-to-eutropical oceanic conditions throughout the area. Conditions are even warm enough for over 100 species of scleractinian corals to flourish from

Kyushu to the Kii Peninsula and 25-40 species to grow within Tokyo Bay (Nishihira and Veron, 1995). Although present along the southern coastline of Japan, these corals never form large, complex reef systems and they are represented mostly by small bioherms that are scattered among volcanic rock formations. This unique mixture of tropical and temperate-type biotopes, not seen anywhere else in the North Pacific, has allowed for the evolution of a special tropical molluscan fauna with a high degree of endemism. Some of these Shikokuan endemics include the astraeine turbinid *Pomaulax japonicus*, the cypraeids *Naria cernica ogasawarensis*, *Purpuradusta japonica*, and *Palmadusta artufelli*, the strombid *Doxander japonicus*, the fasciolariid *Filifusus glaber*, and the cone shells *Miliariconus fulgetrum* and *Nitidiconus pauperculus* (all shown here on Figures 6.19 and 6.20). These eutropical gastropod groups are sympatric with a wide complement of cold water boreal and subarctic-derived groups such as the buccinoideans *Buccinum*, *Ancistrolepis*, *Neptunea*, and *Volutharpa*, and the muricids *Boreotrophon*, *Nipponotrophon*, *Scabrotrophon*, and a species radiation of the muricid genus *Pteropurpura* including *P. falcata*, *P. esycha*, and *P. stimpsoni* (shown here on Figure 6.19).

Below the thermocline in deeper and colder water, a very rich warm-temperate gastropod fauna has evolved in parallel with the coastal Japonic faunas, with many endemic and highly-restricted species. Some of these include the eucyclid *Bathybembix aeola*, the buccinids *Japelion adelphicus* and *Metajapelion pericochlion*, and a very large and remarkable fauna of fulgurariine volutids, containing 5 genera and at least 30 species. Some of these classic deep water Japanese volutes, which are commercially fished for food, include *Fulgoraria tosaensis*, *Fulgoraria rupestris*, *Fulgoraria hamillei*, *Musashia hirasei*, *Nipponomelon magna*, *Nipponomelon prevostiana*, *Psephaea concinna*, *Psephaea kaneko*, *Psephaea hayashii*, *Psephaea daviesi*, and *Saotomea delicata*. These are all illustrated on Figure 6.20. This cooler-water offshore area also houses numerous Japonic gastropod taxa that have extended their ranges southward into very deep water areas of the Indo-Malaysian Province. Species that would normally occur in 100 m depths off Japan have been collected in water as deep as 1000 m in the Philippines and Indonesia, exhibiting a pattern of submergence in lower latitude areas. Some of these Japonic panbathyal taxa, that follow the thermocline into the Bathyal Zone, include the cypraeids *Austrasiatica hirasei* and *Palmulacypraea musumea*, and the cone shells *Kurodaconus stupa*, *Kurodaconus stupella*, *Afonsoconus kinoshitai*, *Graphiconus kuroharai*, *Graphiconus armadillo*, *Kioconus hirasei*, *Profundiconus lani*, *Textilia dusaveli*, and *Turriconus miniexcelsus* (all illustrated here on Figure 6.23).

Ryukyuan Subprovince

Named for the Ryukyu Archipelago south of the Japanese main islands, the Ryukyuan Subprovince extends from Tanegashima Island in the north to Ishigaki and Hateruma Islands in the south, and encompasses the Amami Islands and Okinawa as well as the eastern half of the East China Sea. The Ryukyu Islands, being embedded within the warm Kuroshio Current, contain eutropical water conditions and house extensive and species-rich coral reef systems, with as many as 220 to 380 species of scleractinian corals. Although containing the full complement of Western Pacific Tropical Region gastropods, the reef-lined Ryukyu Islands and the adjacent East China Sea also house a rich gastropod fauna with a high level of endemism. Some of these shallow and deep water Ryukyuan endemics include the cypraeids *Austrasiatica langfordi langfordi*, *Austrasiatica sakurai*, *Lyncina kuroharai*, *Nesiocypraea teremachii*, *Raybaudia joyceae*, and *Eclogavena luchuana*, the muricid *Chicoreus ryukyuensis*, the babyloniids *Babylonia kirana* and *Babylonia magnifica*, the olivid *Viduoliva elegans hemiltona*, and the cone shell *Leptoconus kawamurai*. These are shown here on Figure 6.21.

South China Subprovince

This subprovince, named for the South China Sea, is an extreme southern outlier of the main Japonic Province. The entire subprovincial area is confined to the deep (below 200 m) central basin of the South China Sea and is bathymetrically connected to the East China Sea by the deep and narrow trench contained within the Luzon Strait. This trench-like feature has allowed gene flow between the deep water areas of the East China and South China Seas and has acted as a conduit for the migration of Japonic genera into the closed southern basin. Although the central area of the South China Sea houses a deep water pocket of Japonic taxa, the surrounding shallow water areas, along the coasts of southern China, Vietnam, and the Philippines, all have classic eutropical molluscan faunas and shorelines dominated by coral reefs. The extreme environmental and bathymetric isolation of the South China Sea deep water fauna has allowed for the evolution of a highly endemic molluscan fauna. Primary among these South China endemics is a large fauna of fulgurariine volutes, containing as many as 15 species. Some of these include *Musashia allaryi*, *Fulgoraria bailorum*, *F. ericarum*, *F. thachi*, *F. humerosa, Saotomea pratasensis*, and the endemic South China genus *Bondarevia*, with its single known species, *B. minima* (illustrated here on Figure 6.22). Other classic South China Subprovince endemic index taxa include the muricid *Chicoreus exuberans*, the buccinid *Ancistrolepis vietnamensis*, the babyloniid *Babylonia pieroangelai*, the giant marginellonid *Marginellona gigas*, the lyriine volutid *Canalilyria kurodai*, and the cone shell *Turriconus takahashii*. Most of these are illustrated on Figure 6.22.

Neozealandic Molluscan Province and Aupourian Subprovince

The Neozealandic Province, encompassing the North and South Islands of New Zealand and the numerous adjacent archipelagoes such as the Chatham, Stewart, Auckland, Antipodes, and Campbell Islands, contains one of the richest cold-temperate molluscan faunas known anywhere on Earth (with almost 2,000 species documented by Powell, 1979). This amazingly-rich subantarctic malacofauna houses numerous distinctive endemic genera and species complexes which have evolved in isolation around the landmasses of New Zealand. The two main islands and their ancillary archipelagoes are the only surviving major emergent remnants of the now-submerged continent of Zealandia and many of the resident endemic taxa are relicts of Eocene and Miocene Zealandian and Antarctic malacofaunas (see the previous section on New Caledonia and the Norfolk Ridge). Some of the more important of these Neozealandic endemics include a large fauna of zidonine volute shells, comprising over 35 species in genera and subgenera such as *Alcithoe*, *Leporemax*, *Pachymelon*, *Palomelon*, and *Iredalina*, a large radiation of over 20 species of calliostomatids in the endemic genus *Maurea*, four species and subspecies in the genera *Struthiolaria* and *Pelicaria* of the relictual Miocene stromboidean family Struthiolariidae, at least nine species of the subantarctic buccinulid genus *Aeneator*, and the endemic abalone subgenus *Haliotis (Paua)* (some of these endemic genera are shown on Figure 6.14). With the exception of the northern tip of North Island, the entire provincial area is bathed in the cold water of the Tasman Current (Tasman Front) and the edge of the Subantarctic West Wind Drift, and the molluscan fauna reflects these cold-temperate oceanographic conditions by lacking all of the key tropical-paratropical index gastropod families. For this reason, the Neozealandic Province falls outside the scope of this book and is not included here.

The only exception to the temperate and cold-temperate marine climate of the Zealandian Province is found along the northern half of North Island, New Zealand, from the Bay of Plenty and Hauraki Gulf to the Three Kings Islands off Cape Reinga. Here, the sea temperature remains warm enough, year round, to allow some tropically-derived gastropods to become established within the local ecosystems. These atypical oceanic conditions are the result of the southward-flowing, warm-temperate East and West Auckland Currents, which transport warmer water southward from the area of Lord Howe Island and the Kermadec Islands. This warmer water area and its distinctive molluscan fauna was named the "Aupourian Province" by Powell (1979:7), but is here referred to as the **Aupourian Subprovince** and is recognized as a localized paratropical subprovincial area within the Neozealandic Province. Containing mostly classic New Zealand subantarctic genera and species, the subprovince does house a few tropical species, such as the cypraeids *Naria cernica kermadecensis* and *Lyncina vitellus orcina* (also found on Lord Howe Island and New South Wales, Australia), the strombid *Euprotomus kiwi*, and the cone shell *Calamiconus kermadecensis* (most shown here on Figure 6.24). These species also occur on the Kermadec Islands northeast of New Zealand and this shows that the Aupourian malacofauna has a slight faunal affinity to the Indo-Malaysian Molluscan Province to the north. Since the Aupourian Subprovince lacks representatives of many of the key tropical index subfamilies and families, such as the Modulidae, Melongenidae, Harpidae, Lyriinae, and Olivinae, the area is considered to be only marginally paratropical and represents the extreme southern fringe of the Western Pacific Tropical Region. See Daughenbaugh (2018: 36-54) for a detailed review of the Aupourian cowrie fauna and an overview of the geological history and oceanography of the Aupourian area.

ICONOGRAPHY OF GASTROPODS OF THE WESTERN PACIFIC TROPICAL REGION

(Principal Index Gastropods are shown on Figures 6.3 to 6.24)

Figure 6.3 Widespread Indo-Malaysian Provincial Index Gastropods.
A= *Callistocypraea aurantium* (Gmelin, 1791), length 102 mm, 50 m depth off Panglao, Bohol Island, Philippines; **B**= *Callistocypraea leucodon* (Broderip, 1828), length 88 mm, 100 m depth off Panglao, Bohol Island, Philippines; **C**= *Erronea fernandoi* Cate, 1969, length 25 mm, 20 m depth off Samal Island, Davao Gulf, Mindanao Island, Philippines; **D**= *Leporicypraea valentia* (Perry, 1811), length 90 mm, 100 m depth off Camotes Island, Philippines; **E**= *Amonovula piriei* (Petuch, 1973), length 16 mm, 50 m depth off Balut Island, Sarangani Strait, Philippines; **F**= *Morum (Herculea) ponderosum* (Hanley, 1858), length 30 mm, under coral slab in 25 m depth, off Horseshoe Reef, Okinawa, Ryukyu Islands, Japan; **G**= *Musteloliva xenos* (Petuch and Sargent, 1986), length 14 mm, 200 m depth off Samarai Island, China Straits, eastern Papua-New Guinea; **H**= *Cylinder gloriamaris* (Chemnitz, 1777), length 120 mm, 150 m depth off Panglao, Bohol Island, Philippines; **I**= *Turriconus excelsus* (Sowerby, 1908), length 88 mm, 150 m depth off Panglao, Bohol Island, Philippines; **J**= *Yeddoconus boholensis* (Petuch, 1979), length 30 mm, 150 m depth off Panglao, Bohol Island, Philippines; **K**=*Yeddoconus aphrodite* (Petuch, 1979), length 19 mm, 150 m depth off Panglao, Bohol Island, Philippines; **L**= *Ransoniella martini* (Schepman, 1907), length 16 mm, 30 m depth off Balicasag, Bohol Island, Philippines.

Figure 6.4 Index Gastropods of the Indonesian Subprovince and Malukuan Infraprovince, Indo-Malaysian Province. A= *Murex spinostreptos* Houart, 2010, length 50 mm, 20 m depth off Pangandaran, southern Java, Indonesia; **B=** *Babylonia semipicta* (Sowerby II, 1866), length 45 mm, 100 m depth off Pulau Belitung Island, Java Sea, Indonesia; **C=** *Cymbiola innexa* (Reeve, 1849), length 104.3 mm, 20 m depth off Pernang, West Nusa Tenggara, Sumbawa Island, Indonesia; **D=** *Cymbiola octogonalis* (Senders and Senders, 1995), length 124 mm, 60 m depth off Masalembu Island, Java Sea, Indonesia; **E=** *Galeola vicweei* (Recourt, 1989), holotype length 30 mm, Madura Island, Java, Indonesia; **F=** *Hexaplex bundharmai* (Houart, 1996), length 42 mm, 10 m depth off Togian Island, Sulawesi, Indonesia; **G=** *Pionoconus frauenfeldi* (Crosse, 1865), length 40 mm, 20 m depth off Popalo, Sulawesi Island, Celebes Sea, Indonesia; **H=** *Textilia cervus* (Lamarck, 1822), length 96 mm, 100 m depth, south of Balabac Island, southern Sulu Sea, Philippines (found in Indonesia and the Sulu Sea, Philippines); **I=** *Conus equestris* Röding, 1798, length 47 mm, 20 m depth off Popalo, Sulawesi Island, Celebes Sea, Indonesia; **Malukuan Infraprovince J=** *Conus nocturnus deburghiae* Sowerby, 1857, length 40 mm, 10 m depth off Prajas, western Waigeo Island, Raja Ampat Islands, Indonesia; **K=** *Conus nocturnus* Lightfoot, 1786, length 42 mm, on coral reefs off Yaba, Bacan Island, Maluku Islands, Indonesia.

Figure 6.5 Index Gastropods of the Nusa Tenggaran Infraprovince, Indo-Malaysian Province.
A= *Erronea vredenbergi* Schilder, 1927, length 25 mm, low tide, Kuta Selatan, Bali, Indonesia; **B**= *Cymbiola chrysostoma* (Swainson, 1824), length 47 mm, low tide, Maumere, Flores Island, East Nusa Tenggara, Indonesia; **C**= *Cymbiola cymbiola* (Gmelin, 1791), length 67.5 mm, 100 m depth off Enu Island, Aru Islands, Arafura Sea, Indonesia; **D**= *Agaronia lutraria fikasherinae* Dharma, 2009, length 52 mm, 10 m depth off Pantai Batu Bengkung, southeastern Java, Indonesia; **E**= *Cylinder abbas johnabbasi* Petuch and Berschauer, 2018, length 73 mm, 20 m depth off Pangandaran, southern Java, Indonesia; **F**= *Eugeniconus nobilis nobilis* (Linnaeus, 1758), length 36 mm, 10 m depth off Maumere, Flores Island, East Nusa Tenggara, Indonesia; **G**= *Eugeniconus nobilis abbai* Poppe and Tagaro, 2011, length 41 mm, 1 m depth, Pasar Wulandoni, Lamanuna, southern Lembata Island, East Nusa Tenggara, Indonesia; **H**= *Eugeniconus nobilis skinneri* (da Motta, 1982), length 45 mm, 5 m depth off Balian Beach, Bali, West Nusa Tenggara, Indonesia; **I**= *Eugeniconus nobilis victor* (Broderip, 1842), length 45 mm, 2 m depth off Cape Bahagia, Nereng, northern Lembata Island, East Nusa Tenggara, Indonesia; **J**= *Graphiconus wittigi* (Walls, 1977), length 31 mm, 10 m depth off Ritaebang, Solor Island, Flores, East Nusa Tenggara, Indonesia; **K**= *Phasmoconus giorossii* (Bozzetti, 2006), length 32 mm, 5 m depth off Kubu, Bali, West Nusa Tenggara, Indonesia.

Figure 6.6 Index Gastropods of the Melanesian Subprovince, Indo-Malaysian Province.
A= *Bistolida stolida crossei* (Marie, 1869), length 32 mm, 15 m depth, Prony Bay, New Caledonia; **B**= *Contradusta bregeriana* (Crosse, 1868), length 22 mm, 1 m depth along the western shore of Nimoa Island, Calvados Chain, Louisiade Archipelago, Papua-New Guinea; **C**= *Cribrarula catholicorum* Schilder and Schilder, 1938, length 18 mm, 3 m depth, in the main lagoon of Rossel Island, Louisiade Archipelago, Solomon Sea, Papua-New Guinea; **D**= *Erronea (Adusta) melanesiae* Schilder, 1937, length 36 mm, 1 m depth in sea grass, northern shore of Nimoa Island, Calvados Chain, Louisiade Achipelago, Papua-New Guinea; **E**= *Naria eburnea* (Barnes, 1824), length 48 mm, 1 m depth along the western shore of Nimoa Island, Calvados Chain, Louisiade Archipelago, Papua-New Guinea; **F**= *Talostolida pseudoteres* Lorenz and Barbier, 1992, length 17 mm, 15 m depth in Prony Bay, New Caledonia; **G**= *Phasmoconus santinii* Monnier and Limpalaër, 2014, length 18 mm, in coral rubble, 10 m depth off Nananu-i-Kake Island, Fijis; **H**= *Cymbiola rutila* (Broderip, 1826), length 77 mm, 2 m depth off Sideia Island, Milne Bay District, Papua-New Guinea; **I**= *Viduoliva rubrolabiata* (Fischer, 1903), length 45 mm, on black volcanic sand, 2 m depth, off Epao, Efate Island, Vanuatu; **J**= *Bathyconus fijiensis* (Moolenbeek, Rockel, and Bouchet, 2008), length 28 mm, 250 m depth off Tavua Bay, Viti Levu Island, Fiji Islands; **K**= *Phasmoconus sutanorcum* (Moolenbeek, Rockel, and Bouchet, 2008), length 38 mm, 200 m depth off Nggela Pile Island, Iron Bottom Sound, Solomon Islands.

Figure 6.7 Index Gastropods of the Solomonian Infraprovince, Indo-Malaysian Province.
A= *Contradusta bregeriana pervelata* (Lorenz, 2002), length 20 mm, 2 m depth, Marau Sound, Guadalcanal Island, Solomon Islands; **B=** *Eclogavena coxeni* (Cox, 1873), length 25 mm, low tide, western shore of Nimoa Island, Calvados Chain, Louisiade Archipelago, Papua-New Guinea; **C=** *Eclogavena hesperina* (Schilder and Summers, 1963), length 21 mm, low tide off Rabaul, New Britain Island, Papua-New Guinea; **D=** *Ecolgavena hesperina insolita* Bergonzoni, 2016, length 19 mm, 3 m depth, Marau Sound, Guadalcanal Island, Solomon Islands; **E=** *Eclogavena pseudohesperina* Bergonzoni, 2016, length 21 mm, low tide, western shore of Nimoa Island, Calvados Chain, Louisiade Archipelago, Papua-New Guinea; **F=** *Laevistrombus guidoi* Man in't Veld and de Turck, 1998, length 67 mm, 1 m depth, Naunonga, Santa Cruz Islands, Solomon Islands; **G=** *Acutoliva buelowi stoneorum* (Petuch and Sargent, 1986), length 20 mm, 50 m depth in the Marau Sound, Guadalcanal Island, Solomon Islands; **H=** *Graphiconus gabryae* Korn and Rockel, 1992, length 90 mm, 180 m depth in Marau Sound, Guadalacanal Island, Solomon Islands; **I=** *Phasmoconus solomonensis* (Delsaerdt, 1992), length 32 mm, 100 m depth in Marau Sound, Guadalcanal Island, Solomon Islands; **J=** *Pionoconus epistomioides* (Weinkauff, 1875), length 43 mm, 2 m depth in the main lagoon of Rossel Island, Louisiade Archipelago, Solomon Sea, Papua-New Guinea; **K=** *Phasmoconus nahoniaraensis* (da Motta, 1986), length 31 mm, 20 m depth off the Lungga River mouth, Guadalcanal Island, Solomon Islands; **L=** *Phasmoconus zebra* (Lamarck, 1810), length 36 mm, 5 m depth off Makira Island, Makira-Ulawa, Solomon Islands.

Figure 6.8 The *Cymbiola* Species Complex of the Solomonian Infraprovince, Indo-Malaysian Province.
A, B= *Cymbiola ceraunia* (Crosse, 1880), length 76 mm, 3 m depth off Poroporo, Choiseul Island, Solomon Islands (endemic to the northern Solomon Islands); **C=** *Cymbiola macgillvrayi* (Crosse, 1873), length 123 mm, 3 m depth in the main lagoon of Rossel Island, Louisiade Archipelago, Solomon Sea, Papua-New Guinea (endemic to Rossel Island and the western Louisiades); **D=** *Cymbiola norrisii* (Gray, 1838), length 95 mm, 5 m depth off Kitava Island, Trobriand Islands, Solomon Sea, Papua-New Guinea (endemic to the Trobriand and northern D'Entrecasteaux Islands); **E, F=** *Cymbiola piperita* (Sowerby II, 1842), length 64 mm, 2 m depth off Nuli Sapi, Rogeia Island, Milne Bay District, Papua-New Guinea (endemic to eastern Papua and the eastern Louisiades); **G=** *Cymbiola rueckeri* (Crosse, 1867), length 94 mm, 3 m depth off Kia, Santa Isabel Island, Solomon Islands (endemic to the southern Solomon Islands).

Figure 6.9 Index Gastropods of the Neocaledonian Subprovince, Indo-Malaysian Province.
A= *Gourmya gourmyi* (Crosse, 1861), length 41 mm, low tide on mud flats, off the mouth of the Thio River, Thio, New Caledonia; **B**= *Austrasiatica langfordi cavatoensis* (Lorenz, 2002), length 48 mm, 450 m depth off Banc Jumeau, New Caledonia; **C**= *Cymbiolacca* (*Magnavictoria*) *thatcheri* (McCoy, 1868), length 70 mm, 5 m depth on Bellona Reef, Chesterfield Islands, New Caledonia; **D**= *Leporicypraea mappa kanakinus* Batt, 2004, length 59 mm, 2 m depth on reef off Bay Goro, Grande Terre, New Caledonia; **E**= *Leporicypraea mappa montrouzieri* (Dautzenberg, 1903), length 70 mm, 5 m depth, Japan Bank, New Caledonia; **F**= *Fustifusus pinicola* Darragh, 1991, length 29 mm, 300 m depth south of the Ile des Pins, New Caledonia; **G**= *Mauritia eglantina* form *nigricans* (Montrouzier, 1875), length 43 mm, 5 m depth in Nickel Bay, Noumea, New Caledonia; **H**= *Cymbiola deshayesi* (Reeve, 1855), length 82 mm, 2 m depth off Gadji, Ile des Pins, New Caledonia; **I**= *Ransoniella martini superstes* (Schilder, 1930), length 17 mm, 20 m depth in Prony Bay, Grande Terre, New Caledonia; **J**= *Cymbiola rossiniana* (Bernardi, 1859), length 124 mm, 10 m depth off Gadji, Ile des Pins, New Caledonia; **K**= *Ipsa childreni novaecaledoniae* (Schilder and Schilder, 1952), length 27 mm, 10 m depth in Nickel Bay, Noumea, New Caledonia.

Figure 6.10 Conidae Species Radiations of the Neocaledonian Subprovince, Indo-Malaysian Province.
A= *Thoraconus exiguus* (Lamarck, 1810), length 20 mm, 5 m depth, Ile Tenia, New Caledonia; **B=** *Thoraconus bougei* (Sowerby III, 1907), length 29 mm, 2 m depth in Prony Bay, Grande Terre, New Caledonia; **C=** *Thoraconus cabritii* (Bernardi, 1858), length 22 mm, 2 m depth in Nickel Bay, Noumea, Grande Terre, New Caledonia; **D=** *Thoraconus plumbeus* (Reeve, 1844), length 21 mm, 2 m depth, Ile Ducos, Vincer Bay, Grande Terre, New Caledonia; **E=** *Thoraconus vayssetianus* (Crosse, 1872), length 20 mm, 5 m depth off Ile Ouen, New Caledonia; **F=** *Thoraconus optimus* (Sowerby III, 1913), length 50 mm, 5 m depth in Bay Goro, Grande Terre, New Caledonia; **G=** *Graphiconus lienardi* (Bernardi and Crosse, 1861), length 45 mm, 3 m depth off Brick Island, New Caledonia; **H=** *Graphiconus richeri* (Richard and Moolenbeek, 1988), length 49 mm, dredged from 250 m depth by the *Tui II*, Grand Passage, North Belep Island, New Caledonia; **I=** *Vituliconus swainsoni* (Estival and von Cosel, 1986), length 45 mm, 5 m depth on Bellona Reef, Chesterfield Islands, New Caledonia; **J=** *Conus suffusus* Sowerby III, 1870, length 43 mm, 2 m depth in Bourail Bay, New Caledonia; **K=** *Conus pseudomarmoreus* Crosse, 1875, length 49 mm, 3 m depth in Nickel Bay, Noumea, New Caledonia (note the characteristic strong spiral sculpture on the body whorl); **L=** *Conus crosseanus* Bernardi, 1861, length 51 mm, 2 m depth on the main reef off Gadji, Ile des Pins, New Caledonia.

Figure 6.11 Index Gastropods of the Norfolkian Infraprovince, Neocaledonian Subprovince.
A= *Serratifusus lineatus* Harasewych, 1991, length 40 mm, 500 m depth on the Norfolk Ridge, 100 km south of the Ile des Pins, New Caledonia; **B=** *Amalda aureomarginata* Kilburn and Bouchet, 1988, length 25 mm, 500 m depth on the Norfolk Ridge, 100 km south of the Ile des Pins, New Caledonia; **C=** *Afonsoconus bruuni* (Powell, 1958), length 34 mm, 400 m depth on the Norfolk Ridge, 100 km south of the Ile des Pins, New Caledonia; **D=** *Boucheticonus alisi* (Moolenbeek, Rockel, and Richard, 1995), length 20 mm, 400 m depth on the Norfolk Ridge, 100 km south of the Ile des Pins, New Caledonia; **E=** *Granulifusus bacciballus* Hadorn and Fraussen, 2005, length 22 mm, dredged from 250 m depth by the *Tui II*, Grand Passage, North Belep Island, New Caledonia; **F=** *Kioconus gondwanensis* (Rockel and Moolenbeek, 1995), length 25 mm, 300 m depth southeast of Belep Island, New Caledonia; **G=** *Kioconus plinthus* (Richard and Moolenbeek, 1988), length 28 mm, 400 m depth on the Norfolk Ridge, 100 km south of Ile des Pins, New Caledonia; **H=** *Yeddoconus boucheti* (Richard, 1983), length 15 mm, dredged from 250 m depth by the *Tui II*, Grand Passage, North Belep Island, New Caledonia; **I=** *Kurodaconus luciae* (Moolenbeek, 1986), length 30 mm, 500 m depth on the Norfolk Ridge, 100 km south of Ile des Pins, New Caledonia; **J=** *Chicoreus boucheti* (Houart, 1983), length 20 mm, trawled from 320 m depth on the Norfolk Ridge north of Burnt Pine, Norfolk Island; **K=** *Plicolyria poppei* Bail, 2002, length 31 mm, trawled from 400 m depth on the Norfolk Ridge, 100 km south of Ile des Pins, New Caledonia; **L=** *Babelomurex neocaledonicus* Kosuge and Oliviero, 2001, length 33 mm, 400 m depth on the Norfolk Ridge, 100 km south of Ile des Pins, New Caledonia.

Figure 6.12 Index Gastropods of the Philippinian Subprovince, Indo-Malaysian Province.
A= *Pustularia chiapponii* Lorenz, 1999, length 20 mm, 3 m depth off Surigao City, Mindanao Island, Philippines; **B=** *Turriconus beatrix* Tenorio, Poppe, and Tagaro, 2007, length 23 mm, 180 m depth off Aliguay Island, Philippines; **C=** *Acutoliva bathyalis* (Petuch and Sargent, 1986), length 20 mm, 250 m depth, off Balicasag, Bohol Island, Philippines; **D=** *Isoconus richardsae* (Korn and Rockel, 1992), length 28 mm, 10 m depth in Sogod Bay, Leyte Island, Philippines; **E=** *Calamiconus escondidai* Poppe and Tagaro, 2005, length 28 mm, 250 m depth off Balicasag, Bohol Island, Philippines; **F=** *Pionoconus leobottonii* (Lorenz, 2006), length 42 mm, 100 m depth in Coron Bay, Coron Island, Palawan, Philippines; **G=** *Strategoconus thalassiarchus* (Sowerby I, 1834), length 63 mm, 2 m depth off Culion Island, Palawan, Philippines; **H=** *Turriconus rizali* (Biggs, Watkins, Showers, *et al*., 2010), length 31 mm, 200 m depth, off Balicasag, Bohol Island, Philippines; **I=** *Yeddoconus gattegnoi* Poppe and Tagaro, 2017, length 28 mm, 200 m depth, in the Olango Channel off Olango Island, Philippines; **J=** *Yeddoconus olangoensis* Poppe and Tagaro, 2017, length 23 mm, 250 m depth off Olango Island, Philippines; **K=** *Yeddoconus nereis* (Petuch, 1979), length 24 mm, 200 m depth, off Balicasag, Bohol Island, Philippines; **L=** *Homalocantha dondani* d'Attilio and Kosuge, 1989, length 25 mm, 20 m depth off Balicasag, Bohol Island, Philippines.

Figure 6.13 The *Calibanus* Species Radiation of the Philippinian Subprovince, Indo-Malaysian Province.
A= *Calibanus aegrotus* (Reeve, 1849), length 43 mm, 2 m depth, Laminusa Island, Siasi, Sulu Sea, Philippines; **B**= *Calibanus albicans* (Sowerby II, 1857), length 37 mm, 1 m depth off southern Culion Island, Sulu Sea, Philippines; **C**= *Calibanus albus* (Sowerby, III, 1887), length 42 mm, Coron Island, Palawan, Sulu Sea, Philippines; **D**= *Calibanus cecilei* (Reeve, 1845), length 30 mm, in sand near sea grass bed, 5 m depth in Maimbung Bay, Jolo Island, Sulu Archipelago, Sulu Sea, Philippines; **E**= *Calibanus crepusculum* (Reeve, 1844), length 39 mm, 10 m depth off Balabac Island, Palawan, Philippines; **F**= *Calibanus furvus* (Reeve, 1843), length 38 mm, low tide, Cuyo Island, Palawan, Philippines; **G**= *Calibanus granifer* (Reeve, 1849), length 37 mm, 5 m depth off Zamboanga, Mindanao Island, Philippines; **H**= *Calibanus neobuxeus* (da Motta, 1991), length 48 mm, 2 m depth off Busuanga Island, Philippines; **I**= *Calibanus nivalis* (da Motta, 1985), length 45 mm, 1 m depth off Banggi Island, Sabah, southern Sulu Sea; **J**= *Calibanus polygrammus* (Tomlin, 1937), length 64 mm, low tide, Lapu-Lapu City, Cebu Island, Philippines; **K**= *Calibanus turritinus* (da Motta, 1985), length 39 mm, 20 m depth off Cuyo Island, Palawan, Philippines.

Figure 6.14 The *Cymbiola* Volute Species Radiation of the Philippinian Subprovince, Indo-Malaysian Province.
A= *Cymbiola alexisallaryi* Cossignani, 2018, length 117.4 mm, 20 m depth off Pangutaran Island, Sulu Archipelago, Philippines; B= *Cymbiola aulica* (Sowerby I, 1825), length 123.3 mm, 15 m depth off Cagayan de Oro, Mindanao Island, Philippines; C= *Cymbiola cathcartiae* (Reeve, 1856), length 94.8 mm, 3 m depth in San Antonio Bay, southern Palawan Island, Philippines; D= *Cymbiola laminusa* Poppe, Tagaro, and Bail, 2011, length 122.6 mm, 3 m depth, Siasi Island, Sulu Archipelago, Philippines; E= *Cymbiola malayensis* (Doute and Bail, 1999), length 122 mm, 20 m depth off Banggi Island, Sabah, Malaysia, southern Sulu Sea; F= *Cymbiola palawanica* Doute and Bail, 2000, length 94.9 mm, 10 m depth in Honda Bay, off Puerto Princessa, Palawan Island, Philippines; G= *Cymbiola vespertilio augustinensis* Bail and Doute, 1995, length 85 mm, 2 m depth in the Davao Gulf, off Cape San Augustin, Mindanao Island, Philippines; H= *Cymbiola vespertilio matiensis* Doute, 1985, length 81 mm, 10 m depth off Tictabon Island, Zamboanga, Mindanao Island, Philippines.

Figure 6.15 Index Gastropods of the Suluan Infraprovince, Philippinian Subprovince, Indo-Malaysian Province.
A= *Austrasiatica langfordi poppeorum* Lorenz and Chiapponi, 2017, length 55 mm, 200 m depth, off Balut Island, Mindanao, Philippines; **B=** *Lambis adami* Bozzetti and Cossignani, 2003, length 60 mm, 100 m depth, Aliguay Island, Mindanao, Philippines; **C=** *Homalocantha granpoderi* Merle and Garrigues, 2011, length 40 mm, under coral slabs, 1 m depth off Bantayan Island, Philippines; **D=** *Homalocantha ninae* Merle and Garrigues, 2011, length 39 mm, on dead coral branches, 10 m depth off Surigao, Mindanao Island, Philippines; **E=** *Pterynotus miyokoae* (Kosuge, 1979), length 55 mm, on dead coconut palm log, from 200 m depth off Balicasag, Bohol island, Philippines; **F=** *Oniscidia kurzi* (Petuch, 1979), length 23 mm, on pink hydrocoral beds, 200 m depth off Balicasag, Bohol Island, Philippines; **G=** *Recourtoliva poppei* (Sargent and Petuch, 2008), length 40 mm, 200 m depth off Aliguay Island, Mindanao, Philippines; **H=** *Boucheticonus pseudokimioi* (da Motta and Martin, 1982), length 23 mm, 250 m depth off Aliguay Island, Mindanao, Philippines; **I=** *Cylinder barbieri* Raybaudi Massilia, 1995, length 32 mm, 5 m depth on reef off Samar Island, Philippines; **J=** *Cylinder scottjordani* Poppe, Monnier, and Tagaro, 2012, length 51 mm, 3 m depth off Cuyo Island, Sulu Sea, Philippines; **K=** *Fulgiconus cebuensis* (Wils, 1990), length 32 mm, 10 m depth off Moalboal, Cebu Island, Philippines; **L=** *Homalocantha pisori* d'Attilio and Kosuge, 1989, length 30 mm, 20 m depth off Bantayan Island, Visayan Sea, Philippines.

Figure 6.16 Index Gastropods of the Suluan Infraprovince, Philippinian Subprovince, Indo-Malaysian Province.
A= *Kioconus (Ongoconus) vanvilstereni* (Moolenbeek and Zandbergen, 2013), length 26 mm, 10 m depth off Balabac, Palawan, Philippines; **B=** *Homalocantha vicdani* Attilio and Kosuge, 1989, length 26 mm, 20 m depth off Panglao, Bohol Island, Philippines; **C=** *Homalocantha anomaliae* Kosuge, 1979, length 49 mm, 200 m depth off Balut Island, Mindanao, Philippines; **D=** *Ficadusta pulchella aliguayensis* Van Heesvelde and Deprez, 2002, length 33 mm, 200 m depth off Aliguay Island, Mindanao, Philippines; **E=** *Musteloliva boholensis* (Petuch and Sargent, 1986), length 30 mm, 150 m depth off Zamboanga, Mindanao Island, Philippines; **F=** *Rolaniconus dedonderi* (Goethaels and Monsecour, 2013), length 22 mm, trawled from 150 m depth off Aliguay Island, Mindanao, Philippines; **G=** *Viduoliva zamboangensis* (Petuch and Sargent, 1986), length 36 mm, trawled from 100 m depth off Coron, Palawan Island, Philippines; **H=** *Darioconus viperinus* (Lauer, 1986), length 48 mm, in coral, 20 m depth off Zamboanga, Mindanao Island, Philippines; **I=** *Bistolida stolida aureliae* Cossignani, 2017, length 33 mm, 10 m depth off Balabac, Palawan, Philippines; **J=** *Pionoconus robini* Limpalaër and Monnier, 2012, length 22 mm, 10 m depth off Balabac, Palawan, Philippines; **K=** *Phasmoconus zapatosensis* (Rockel, 1987), length mm, 20 m depth off Mompong Island, Marinduque, Philippines; **L=** *Strategiconus thomae* (Gmelin, 1791), length 77 mm, 150 m depth off Balabac Island, Palawan, Philippines (also found in northern Indonesia and the Maluku Islands).

Figure 6.17 Index Gastropods of the Suluan Infraprovince, Philippinian Subprovince, Indo-Malaysian Province.
A= *Conus vidua* Reeve, 1843, length 55 mm, 10 m depth off Cuyo Island, Palawan, Philippines; **B**= *Eclogavena dani* (Beals, 2002), length 21 mm, 30 m depth off Balabac Island, Palawan, Philippines; **C**= *Eclogavena dayritiana mandejarorum* Petuch and Myers, 2014, length 17 mm, inside sponges, 20 m depth along the eastern coast of Nangalao Island, Palawan, Philippines; **D**= *Phasmoconus balabacensis* (Filmer, 2012), length 28 mm, 10 m depth in San Carlos Bay, Cuyo Island, Philippines; **E**= *Viduoliva mindanaoensis* (Petuch and Sargent, 1986), length 57 mm, 3 m depth off Isabela City, Basilan Island, Philippines; **F**= *Calamiconus tethys* (Petuch and Sargent, 2011), length 108 mm, 20 m depth, Smith Coral Garden, Coron Passage, Coron Island, Palawan, Philippines; **G**= *Cylinder glorioceanus* Poppe and Tagaro, 2009, length 46 mm, 30 m depth off Jolo Island, Sulu Archipelago, Sulu Sea, Philippines; **H**= *Cylinder tagaroae* Limpalaër and Monnier, 2013, length 51 mm, low tide, Calachuchi, Coron Island, Palawan, Philippines; **I**= *Cylinder telatus* (Reeve, 1848), length 50 mm, 12 m depth off Balabac Island, Palawan, Philippines; **J**= *Eugeniconus bitleri* (da Motta, 1984), length 33 mm, 1 m depth off Panglima Sugala, Tawitawi Island, southern Sulu Archipelago, Philippines; **K**= *Eugeniconus cordigera* (Sowerby II, 1866), length 45 mm, 15 m depth off Balabac Island, Palawan, Philippines; **L**= *Chicoreus jessicae* Houart, 2008, length 44 mm, in tangle nets, 100 m depth off Little Liguid Island, Samal, Davao Gulf, Mindianao, Philippines.

Figure 6.18 Widespread Index Gastropods of the Japonic Province.
A= *Naria cernica ogasawarensis* (Schilder, 1944), length 31 mm, 25 m depth, off Izu Oshima, Izu Islands, Honshu Island, Japan; **B=** *Perisserosa guttata azumai* Schilder, 1960, length 55 mm, 200 m depth south of Makurazaki, Kyushu Island, Japan; **C=** *Purpuradusta japonica* (Gaskoin, 1849), length 19 mm, 3 m depth off Tosashimizu, Kochi, Shikoku Island, Japan; **D=** *Charonia sauliae* (Reeve, 1844), length 155 mm, 2 m depth off Tosashimizu, Kochi, Shikoku Island, Japan; **E=** *Chicoreus asianus* Kuroda, 1942, length 81 mm, 100 m depth in Shibushi Bay, off Kushima, Kyushu Island, Japan; **F=** *Profundiconus profundorum* (Kuroda, 1956), length 71 mm, 300 m depth off Tosashimizu, Shikoku Island, Japan; **G=** *Paradusta hungerfordi* (Sowerby III, 1888), length 38 mm, 200 m depth in the East China Sea, north of Keelung, Taiwan; **H=** *Babylonia japonica* (Reeve, 1843), length 75 mm, 50 m depth off Atsumi Aichi-ken, Honshu, Japan; **I=** *Miliariconus fulgetrum* (Sowerby II, 1834), length 34 mm, 2 m depth in Tean Bay, Amami Oshima Island, Amami Islands, Ryukyu Archipelago, Japan; **J=** *Pionoconus fulmen* (Reeve, 1848), length 57 mm, trawled by fishermen from 25 m depth off Tosashimizu, Kochi, Shikoku Island, Japan; **K=** *Endemoconus sieboldi* (Reeve, 1848), length 75 mm, 100 m depth off Shirahama, Wakayama, Honshu Island, Japan.

Figure 6.19 Index Gastropods of the Shikokuan Subprovince, Japonic Province.
A= *Pomaulax japonicus* (Dunker, 1844), diameter 78 mm, 5 m depth, Tateyama, Boso Peninsula, Chiba, Honshu Island, Japan; **B**= *Japelion adelphicus* (Dall, 1907), length 91 mm, 500 m depth off Saikai, Kyushu Island, Japan; **C**= *Palmadusta artufelli* (Jousseaume, 1867), length 16 mm, low tide, Enoshima Island, Kanagawa, Sagami Bay, Honshu Island, Japan; **D**= *Doxander japonicus* (Reeve, 1851), length 64 mm, 30 m depth off Enshu-nada, Atsumi Bay, Aichi, Honshu Island, Japan; **E**= *Pteropurpura esycha* (Dall, 1825), length 30 mm, 100 m depth off Shirahama, Wakayama, Honshu Island, Japan; **F**= *Pteropurpura stimpsoni* (Adams, 1863), length 38 mm, 100 m depth off Shirahama, Wakayama, Honshu Island, Japan; **G**= *Pteropurpura falcata* (Sowerby II, 1834), length 55 mm, 100 m depth off Yaizu, Suruga Bay, Honshu Island, Japan; **H**= *Filifusus glaber* (Dunker, 1882), length 73 mm, 3 m depth off Tosashimizu, Kochi, Shikoku Island, Japan; **I**= *Metajapelion pericochlion* (Schrenck, 1863), length 110 mm, 100 m depth off Shirahama, Wakayama, Honshu Island, Japan; **J**= *Bathybembix aeola* Watson, 1879, height 45 mm, 200 m depth off Enshu-nada, Atsumi Bay, Aichi, Honshu Island, Japan; **K**= *Nitidiconus pauperculus* (Sowerby II, 1834), length 26 mm, under rocks, low tide, Enoshima Island, Kanagawa, Sagami Bay, Japan.

Figure 6.20 The Volutid Species Radiation of the Shikokuan Subprovince, Japonic Province.
A= *Fulgoraria tosaensis* Shikama 1967, length 95 mm, 250 m depth in the East China Sea, due west of Amami Oshima Island, Amami Islands, Ryukyu Archipelago, Japan; **B=** *Fulgoraria rupestris* (Gmelin, 1791), length 86 mm, 100 m depth in the Sumo-nada Sea, Nagasaki, Kyushu Island, Japan; **C=** *Musashia hirasei* (Sowerby III, 1912), length 128 mm, trawled from 100 m depth off Mie, Honshu Island, Japan; **D=** *Nipponomelon magna* (Kuroda and Habe, 1950), length 100 mm, 100 m depth off Hidaka, Tomikawa, Hokkaido Island, Japan; **E=** *Nipponomelon prevostiana* (Crosse, 1878), length 92 mm, 50 m depth off Mombetsu, Hokkaido Island, Japan; **F=** *Psephaea concinna* (Broderip, 1836), length 125 mm, trawled from 200 m depth off Enshu-nada, Shizuoka, Honshu Island, Japan; **G=** *Psephaea daviesi* (Fulton, 1838), length 124 mm, 200 m depth in the Sumo-nada Sea, Nagasaki, Kyushu Island, Japan; **H=** *Psephaea hayashii* (Habe and Ito, 1965), length 132 mm, 105 m depth off Hyogo, Honshu Island, Japan; **I=** *Psephaea kaneko* (Hirase, 1922), length 100 mm, 110 m depth off Hyogo, Honshu Island, Japan; **J=** *Saotomea delicata* (Fulton, 1940), length 53 mm, 200 m depth off Ensu-nada, Shizuoka, Honshu Island, Japan; **K=** *Fulgoraria hamillei* (Crosse, 1869) length 145 mm, 100 m depth, Yaizu, Suruga Bay, Honshu Island, Japan.

Figure 6.21 Index Gastropods of the Ryukyuan Subprovince, Japonic Province.
A= *Austrasiatica langfordi* (Kuroda, 1938), length 57 mm, 250 m depth, East China Sea north of Uotsuri Jima, Senkaku Islands, Japan; **B**= *Austrasiatica sakurai* (Habe, 1970), length 45 mm, 250 m depth off Kuba Jima, Senkaku Islands, Japan; **C**= *Eclogavena luchuana* (Kuroda, 1960), length 22 mm, 2 m depth in Tean Bay, Amami Oshima Island, Amami Islands, Ryukyu Archipelago; **D**= *Lyncina kuroharai* (Kuroda and Habe, 1961), length 45 mm, 50 m depth north of Ishigaki Island, Ryukyu Archipelago, Japan; **E**= *Nesiocypraea teremachii* (Kuroda, 1938), length 69 mm, 250 m depth, East China Sea off Uotsuri Jima, Senkaku Islands, Japan; **F**= *Raybaudia joyceae* (Clover, 1970), length 58 mm, 50 m depth north of Ishigaki Island, Ryukyu Islands, Japan; **G**= *Chicoreus ryukyuensis* Shikama, 1978, length 33 mm, 20 m depth in Oura Bay, off Kanucha, Okinawa, Ryukyu Archipelago, Japan; **H**= *Babylonia magnifica* Fraussen and Stratmann, 2005, length 65 mm, 50 m depth west of Toshima Island, northern Ryukyu Archipelago, Japan; **I**= *Viduoliva elegans hemiltona* (Duclos, 1835), length 41 mm, 3 m depth, Tean Bay, Amami Oshima Island, Amami Islands, Ryukyu Archipelago, Japan; **J**= *Leptoconus kawamurai* (Habe, 1962), length 80 mm, 200 m depth off Amami Ohama, Amami Oshima Island, Amami Islands, Ryukyu Archipelago, Japan; **K**= *Babylonia kirana* Habe, 1965, length 58 mm, trawled from 30 m depth off Hugaka, Okinawa, Ryukyu Archipelago, Japan.

Figure 6.22 Index Gastropods of the South China Subprovince, Japonic Province.
A= *Ancistrolepis vietnamensis* Sirenko and Goryachev, 1990, length 100 mm, 250 m depth, South China Sea east of DaNang, Vietnam; **B=** *Turriconus takahashii* Petuch and Berschauer, 2019, length 58 mm, 250 m depth south of the Pratas Islands, South China Sea; **C=** *Babylonia pieroangelai* Cossignani, 2008, length 80 mm, 100 m depth, South China Sea east of DaNang, Vietnam; **D=** *Canalilyria kurodai* (Kawamura, 1964), length 89 mm, 200 m depth, South China Sea off the Paracel Islands; **E=** *Fulgoraria bailorum* Thach, 2014, length 118 mm, 300 m depth, South China Sea between DaNang, Vietnam and the Paracel Islands; **F=** *Fulgoraria ericarum* Doute, 1997, length 130 mm, 300 m depth, South China Sea between Hong Kong, China and the Pratas Islands; **G=** *Fulgoraria thachi* Bail and Chino, 2010, length 86 mm, 250 m depth east of DaNang, Vietnam; **H=** *Musashia allaryi* (Bail, 2005), length 170 mm, 300 m depth, South China Sea between DaNang, Vietnam and the Paracel Islands; **I=** *Saotomea pratasensis* (Lan, 1997), length 42 mm, 250 m depth, South China Sea between Hong Kong, China and the Pratas Islands; **J=** *Marginellona gigas* (von Martens, 1904), length 100 mm, 600 m depth, South China Sea off the Paracel Islands; **K=** *Chicoreus exuberans* Cossignani, 2004, length 52 mm, 50 m depth, South China Sea off the Pratas Islands; **L=** *Bondarevia minima* (Bondarev, 1994), length 28 mm, 250 m depth south of the Pratas Islands, South China Sea.

Figure 6.23 Panbathyal Japonic Molluscan Province gastropods that are shared with the Indo-Malaysian Molluscan Province.

A= *Austrasiatica hirasei* (Roberts, 1913), length 51 mm, 100 m depth south of Makurazaki, Kagoshima, Kyushu Island, Japan; **B**= *Palmulacypraea musumea* (Kuroda and Habe, 1961), length 19 mm, 200 m depth south of Kushimoto, Kii Peninsula, Wakayama, Japan; **C**= *Kurodaconus stupa* (Kuroda, 1956), length 86.5 mm, 300 m depth off Orchid Island, Bashi Channel, Taiwan; **D**= *Afonsoconus kinoshitai* (Kuroda, 1956), length 67 mm, 150 m depth south of Kaohsiung, Taiwan; **E**= *Graphiconus armadillo* (Shikama, 1971), length 72 mm, 200 m depth south of Kushimoto, Kii Peninsula, Wakayama, Japan; **F**= *Graphiconus kuroharai* (Habe, 1965), length 59 mm, 200 m depth off Tosashimizu, Kochi, Shikoku Island, Japan; **G**= *Kioconus hirasei* (Kuroda, 1956), length 41.8 mm, 100 m depth of Balicasag, Bohol Island, Philippines; **H**=*Profundiconus lani* (Crandall, 1979), length 45 mm, 200 m depth in the East China Sea, north of Keelung, Taiwan; **I**= *Kurodaconus stupella* (Kuroda, 1956), length 75 mm, 350 m depth off Balut Island, Mindanao, Philippines; **J**= *Textilia dusaveli* (A. Adams, 1872), length 82 mm, 30 m depth off Onna Point, Okinawa, Ryukyu Archipelago, Japan; **K**= *Turriconus miniexcelsus* (Olivera and Biggs, 2010), length 30 mm, 250 m in the East China Sea, due north of Uotsumi Jima, Senkaku Islands, Japan.

Figure 6.24 Index Gastropods of the Aupourian Subprovince, Neozealandic Molluscan Province.
A= *Naria cernica kermadecensis* (Powell, 1958), length 26 mm, 5 m depth under rocks, Cascade Bay, Great King Island, Three Kings Islands, New Zealand; **B, C**= *Euprotomus kiwi* (Bozzetti and Sargent, 2011), paratype #2, length 89.4 mm, 15 m depth off Raoul Island, Kermadec Islands, New Zealand (also found at Berghan Point, mainland North Island, New Zealand); **D**= *Calamiconus kermadecensis* (Iredale, 1912), length 37 mm, Parengarenga Harbour, Northland, North Island, New Zealand; **E**= *Alcithoe larochei* Marwick, 1926, length 115 mm, trawled from 450 m depth off the Bay of Plenty, North Island, New Zealand; **F**= *Pachymelon (Palomelon) fissurata* Dell, 1963, length 177 mm, trawled from 300 m depth off Aldermen Islands, North Island, New Zealand; **G**= *Alcithoe (Leporemax) davegibbsi* Hart, 1999, length 48 mm, trawled from 100 m depth in Spirits Bay, northernmost North Island, New Zealand; **H**= *Alcithoe (Leporemax) haurakiensis* Dell, 1956, length 51 mm, 50 m depth off Kawau Island, Hauraki Gulf, North Island, New Zealand; **I**= *Struthiolaria papulosa* (Martyn, 1784), length 73 mm, 2 m depth on mud, Whangarei Harbour, North Island, New Zealand; **J**= *Struthiolaria (Pelicaria) vermis* (Martyn, 1784), length 41 mm, 10 m depth off Taukikura, North Island, New Zealand; **K**= *Struthiolaria (Pelicaria) vermis flemingi* Neef, 1970, length 37 mm, 30 m depth in scallop nets, off Leigh, Northland, North Island, New Zealand.

CHAPTER 7. Indian Tropical Region

The Indian Tropical Region, named for the Indian Subcontinent, spans the northern half of the Indian Ocean, from southern Mozambique to southern Somalia along East Africa, the Red Sea and Gulf of Aden, the Persian Gulf, the Arabian Sea coast from Pakistan to southern India, the entire Bay of Bengal, the western coast of Thailand to northern Sumatra, and the Andaman Sea, but does not include the west coast of Australia. This vast region also encompasses a myriad of tropical islands, including Madagascar and its adjacent groups, the Dahlak Archipelago, Socotra Island, the Mascarene Archipelago, the Cocos Keeling Islands, the Chagos Archipelago, the Seychelles Islands, the Laccadive and Maldives Atolls, the Andaman and Nicobar Archipelgoes, and Sri Lanka. Within this extended eutropical area, two distinct molluscan faunal provinces are now known to exist: the **Lemurian Province** and its subprovinces (the **Andamanian**, the **Bengalian**, the **Malabaran**, the **Mascarenean**, the **Madagascan**, and the **Mozambican**) and the **Eritrean Province** and its subprovinces (the **Aqaban**, the **Dahlakian**, the **Omanian**, and the **Somalian**). A number of wide-ranging gastropods demarcate the boundaries of the Indian Tropical Region, some of which include the cypraeids *Naria ocellata* and *Naria lamarcki*, the muricid *Indothais lacera*, the olivid *Carmione bulbosa*, and the conids *Rhizoconus sumatrensis* and *Harmoniconus parvatus*. The evolutionary patterns and geologic history of the Indian Ocean corals were outlined by Obura (2015) and these parallel the biogeographical patterns seen in the Lemurian molluscan faunas.

Lemurian Molluscan Province

Named for Lemuria, the mythical "lost continent" of the central Indian Ocean, the Lemurian Province is the largest biogeographical unit within the Indian Tropical Region, extending from Mozambique to Somalia along East Africa, and including the island of Madagascar, the entire northern Indian Ocean coastline from Pakistan to western Thailand, the southern and eastern Arabian Sea, the Bay of Bengal, the Andaman Sea, and encompassing all of the islands in the central Indian Ocean (shown here on Figure 7.1). Numerous very large rivers empty into the Indian Ocean basin, increasing the nutrient levels along much of the adjacent coastlines and supporting species-rich, high-productivity ecosystems. Some of these high-effluent rivers include the Irrawaddy of Myanmar, the Ganges, Godavari, and Indus of India and Pakistan, and the Zambezi, Ruvuma, and Tana of East Africa. Like the Western Pacific, the Indian Ocean contains a wide variety of eutropical marine environments, including coastal mangrove jungles and brackish lagoons, extensive mud flats along river deltas, and coral reef-lined volcanic islands and coral atolls in the open ocean areas. These different habitats support rich and distinctive molluscan faunas, including many widespread Lemurian index taxa such as the cypraeids *Leporicypraea rosea*, *Palmadusta diliculum*, and *Palmadusta asellus armadillo*, the cassid *Phalium fimbria*, the olivid *Miniaceoliva tremulina*, and the conids *Cylinder archiepiscopus*, *Cylinder paulucciae*, *Kioconus caillaudi*, *Pionoconus gubernator*, *Pionoconus barthelemyi*, *Rhombiconus zonatus*, and *Virgiconus berdulinus* (most shown here on Figure 7.3). Recent biochemical and genetic research has shown that many of the widespread

Indo-Pacific taxa, which originally were thought to be single wide-ranging species, are now known to actually constitute a pair of subspecies, with Pacific and Indian Ocean siblings. In the case of the common widespread cowrie, *Cypraea tigris*, the Indian Ocean populations were found to be genetically-distinct from those found in the Pacific Ocean; with the Indian Ocean subspecies being the true *C. tigris tigris* while the Pacific Ocean subspecies is the sibling *C. tigris pardalis* (see Lorenz, 2017; Meyer, 1998, 2003, 2004). More biochemical analyses may show that many, if not all, of the wide-ranging Indo-Pacific gastropods actually represent a pair of subspecies, with Indian and Pacific components, and not single genetic entities.

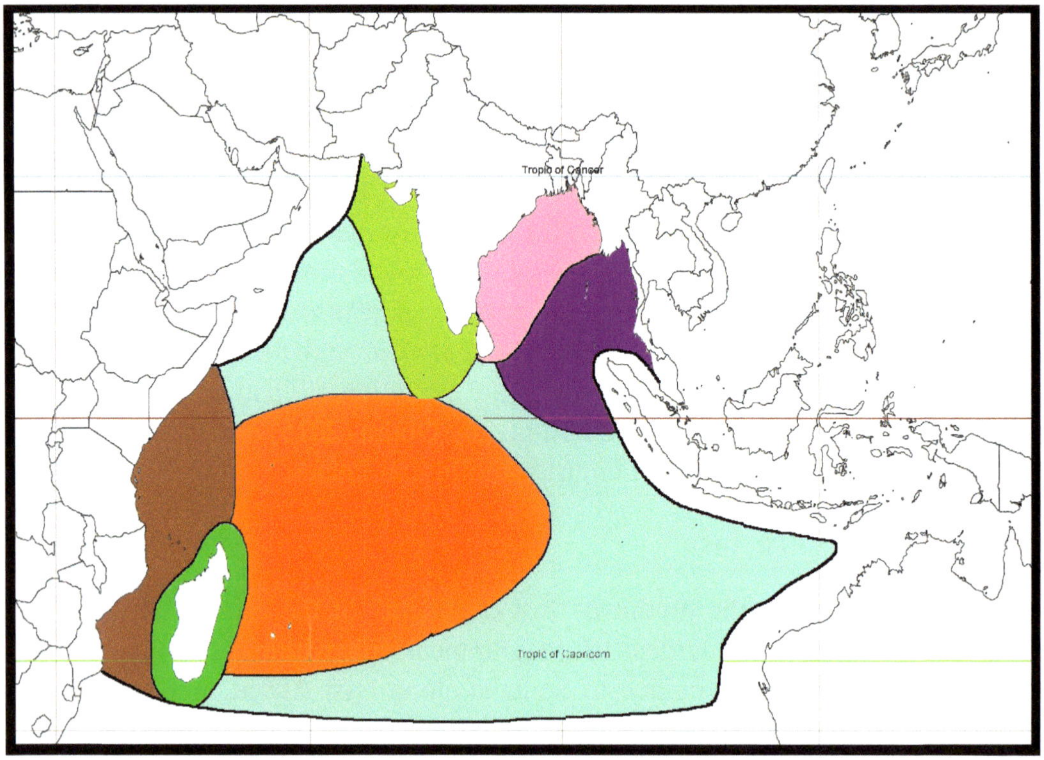

Figure 7.1 Map of the Lemurian Molluscan Province, showing the areal extents of its subprovinces: the **Andamanian Subprovince** (purple), the **Bengalian Subprovince** (light rose), the **Deccan Subprovince** (lime green), the **Mascarenean Subprovince** (orange), the **Madagascan Subprovince** (green), and the **Mozambican Subprovince** (brown). The light blue color demarcates deep water areas such as the Mid-Indian Basin, the Wharton Basin, and the Crozet Basin in the east and the Mascarene Basin and Arabian Basin in the northwest, all of which contain abyssal plain environments and faunas.

Andamanian Subprovince

Named for the Andaman Sea, the Andamanian Subprovince extends from the Delta of the Irrawaddy River in the north, to the west coasts of Myanmar and Thailand and the Mergui Archipelago in the east, to the Malacca Strait in the south, and to the Andaman and Nicobar Islands on the west. The semi-enclosed basin of the Andaman Sea has evolved its own characteristic molluscan fauna with a large enough endemic component to have reached subprovincial status. Some of these classic Andamanian index taxa, some with very restricted geographical ranges, include the cypraeids *Callistocypraea nivosa* and *Perisserosa guttata surinensis*, the strombid *Mirabilistrombus listeri*, the olivid *Arctoliva pacifica*, and the cone shells *Darioconus thailandis*, *Graphiconus ranonganus*, *Kioconus simanoki*, *Stellaconus sukhadwalai*, *Phasmoconus andamanensis*, and *Textilia vicweei* (all shown here on Figure 7.4). Of special interest within the Andamanian Subprovince is the small olivid, *Galeola andamanensis*, which is endemic to South Andaman Island and the adjacent John Lawrence Island. This important Andaman Islands endemic has been confused with the common, widespread *Galeola carneola adspersa* for decades now by many shell dealers, researchers, and collectors. To rectify this misidentification, we here illustrate (Figure 7.4 E) a typical specimen of *Galeola andamanensis*, showing the classic yellow and brown oblique banding and purple aperture. Other important Andamanian index species include the cowries *Mauritia arabica merguina*, *Erronea caurica chrismeyeri*, and *Naria lamarckii phuketensis*, the olivids *Oliva februaryana* and *Acutoliva buelowi phuketensis*, and the cone shells *Graphiconus maculospira*, *Eugeniconus renatae*, and *Pionoconus floridus* (= *chusaki*).

Bengalian Subprovince

Named for the Bay of Bengal, the Bengalian Subprovince encompasses the entire embayment area west of the Andaman Islands, from the Irrawaddy River Delta to the Ganges River Delta, and the entire eastern coast of India and Sri Lanka. Because of the immense concentrations of nutrients amd organic matter present in the effluent from these rivers, and also from the Mahanadi, Godavari, and Krishna Rivers, the Bay of Bengal is dominated by muddy, organic-rich sea floors and high-productivity, eutrophic water masses. The southern end of the subprovince, along Tamil Nadu and Sri Lanka, is far enough away from the eutrophic conditions of the northern Bay that coral reefs can flourish in shallow coastal waters. All of these Bengalian habitats support species-rich molluscan faunas which exhibit high levels of endemism, particularly within the muddy sea floor ecosystems south of the Ganges River. Some of these Bengalian index taxa include the cypraeid *Erronea (Ipserronea) angioyorum*, the olivids *Miniaceoliva kremerorum* and *Viduoliva macleaya*, the volutes *Harpularia lapponica* and *Harpularia aurisiaca*, and the cone shells *Cylinder bengalensis* and *Rhizoconus rawiensis* (all shown here on Figure 7.5). On the isolated coral reef complexes and carbonate environments of the Tamil Nadu and Sri Lankan coasts, a large number of endemic, regionally-restricted taxa have also evolved, some of which include the cowrie *Cribrarula abaliena ganteri* and the cone shells *Cylinder abbas*, *Eugeniconus friedae*, and *Pionoconus subfloridus* (illustrated on Figure 7.5). The muricid *Chicoreus palmarosae* is especially common on Sri Lanka, where it grows to a larger size, and is much more colorful, than specimens from elsewhere in the Indo-Pacific Super-Region.

Malabaran Subprovince

Named for the Malabar Coast of southwestern India (Goa south to Kanyakumari), the Malabaran Subprovince encompasses the entire eastern half of the Arabian Sea and also the Laccadive and Maldive Atolls. The entire western coast of India is fringed by a broad continental shelf, often more than 150 km wide, which drops precipitously into the Arabian Basin. This broad shelf area receives a large sediment budget from the effluent of major river systems such as the Narmada, Mahi, Tapi, Vasai, Savitri, Vaitarna, Vashishti, Bav, and many others, all of which flow from the Deccan Plateau and Western Ghats mountain range. These rivers enrich the Malabar shelf with organic material and abundant nutrients, and the entire offshore area supports extremely rich marine ecosystems and a diverse molluscan fauna with a high degree of endemism. Some of these Malabaran endemic taxa include the strombid *Amabiliplicatus sibbaldi*, the large olivid *Miniaceoliva flammeacolor*, the zebra-striped volutid *Harpularia loroisi*, and the conids *Graphiconus cuneiformis*, *Graphiconus indomaris*, *Fusiconus dictator*, *Leptoconus milneedwardsi clytospira*, *Quasiconus tuticorinensis*, *Stellaconus bayani*, and *Virgiconus malabaricus*. These are all illustrated here on Figure 7.6. The Laccadive and Maldive Island component of the Malabaran Subprovince has a malacofauna composed primarily of widespread Lemurian coral reef-dwelling taxa. These atoll molluscan faunas have a more impoverished biodiversity than do the faunas of the Malabar Coast, with only a few endemic taxa being known. One of the more prominent endemic Laccadive reef-dwelling species is the large cone shell, *Pionoconus leehmani* (Figure 7.6 G).

Mascarenean Subprovince

Named for the Mascarene Archipelago, which consists of Mauritius, Reunion, and Rodrigues Islands and their adjacent smaller islands and shoals, the Mascarenean Subprovince encompasses the hundreds of volcanic islands and coral atolls found within the central Indian Ocean between the Maldives and Madagascar. This widely-distributed insular subprovince also includes the Seychelles Islands, the British Indian Ocean Territories of Diego Garcia and the Chagos Archipelago, the Agalega Islands, Desroches, Tromelin, and Coetivy Islands, and the Salha de Malha and Nazareth Banks. Because of their remoteness and great distance from the African and Eurasian mainlands, the islands and banks contained within the Mascarenean Subprovince exhibit high levels of localized endemism. The large volcanic island of Mauritius, with its surrounding reef complexes, houses the largest number of Mascarenean endemic taxa, including species such as the cowries *Bistolida oweni menkeana*, *Cribrarula esontropia*, and *Cribrarula cribellum*, the olivid *Miniaceoliva trenulina* form *hayesi*, and the cone shells *Darioconus rubiginosus*, *Darioconus magoides*, and *Darioconus episcopus* (all shown here on Figure 7.7). The island of Reunion also houses a number of endemic species, with the cone shells *Darioconus rubropennatus* (Figure 7.7J) and *Pionoconus mascarenensis* being the most noteworthy. These localized endemic taxa occur together with a large fauna of widespread Mascarenean index species, some of which include the cypraeid *Leporicypraea rosea subsignata*, the strombid *Ophioglossolambis violacea*, the olivids *Miniaceoliva pica* (=*M. mascarena*) and *Annulatoliva maculata*, the harpid *Harpa costata*, and the cone shells *Textilia julii*, *Cylinder corbula*, *Cylinder loman*, and *Rolaniconus moussoni* (shown here on Figures 7.7 and 7.8). The Salha de Malha Bank north of Mauritius also houses numerous endemic taxa, such as the rare conid *Phasmoconus primus* and the volutes *Indolyria surinamensis* and *Indolyria doutei*. The adjacent Nazareth Bank is known to house yet another endemic volute in the same genus, *Indolyria bondarevi*.

Of special interest within the Mascarenean Subprovince is the atoll of Diego Garcia in the remote Chagos Archipelago. Here, the large cowrie *Erronea (Adusta) nymphae* (Figure 7.8I) has evolved in isolation and is only known from the patch reefs within the confines of the atoll lagoon. This highly-restricted endemic cowrie is now known to be endangered, primarily due to military construction on the atoll islands and by infilling of portions of the atoll lagoon. Another area of special interest within the Mascarenean Subprovince is St. Brandon's Shoal northeast of Mauritius. This large coral atoll, also known as the Cargados Carajos Shoals, contains over 40 small, low islands and hundreds of separate smaller reef systems, and recently has been discovered to be an evolutionary hot spot, referred to here as the **Brandonian Infraprovince**. A number of highly-restricted endemic gastropods have evolved here, some of which include the cowries *Bistolida nanostraca* and *Bistolida piae*, the ficid *Ficus dandrimonti*, the harpid *Morum lorenzi* (all shown here on Figure 7.8), and the conid *Darioconus brandonensis*. Felix Lorenz (*personal communication*) has conjectured that St. Brandon Shoals, and the surrounding smaller banks and islands, represent an area of geographical heterochrony and have acted as a refugium for Miocene-Pliocene mollusks. The entire infraprovince can be considered to be a secondary relict pocket.

Madagascan Subprovince

Named for the island of Madagascar off the southeastern coast of Africa, the Madagascan Subprovince encompasses the entire island and several adjacent banks, such as the Sakalaves Seamounts and the Davie Ridge. As the largest continental island in the Indian Ocean, Madagascar houses a wide variety of tropical marine environments, including coral reef complexes, rocky shorelines, muddy estuaries, coastal lagoons, and extensive sand beaches. The island also contains numerous large rivers that flow down from the central massifs of the Ankaratra, Tsaratanana, and Ivakoany Mountains, and their effluent enriches the adjacent marine environments with organic matter and dissolved nutrients. Some of these rivers, such as the Maevarono, Mahajamba, Betsiboka, Mahavavy, Maningoza, Mangoky, and Onilahy, have created large coastal estuaries and deltaic environments which support high-biodiversity ecosystems. Originally part of Africa, what was to become Madagascar was torn off the main continent when India broke away and moved eastward during the early Cretaceous, around 135 million years ago. The island remained attached to the Indian Subcontinent until the late Cretaceous, around 88 million years ago, when it separated from that rapidly-moving landmass and became a separate island all to itself. Madagascar has remained an isolated island off the East African coast ever since and this genetic isolation has allowed for the evolution of highly endemic terrestrial and marine faunas.

Oceanographically, the island is divided into two general areas; a northern half that is dominated by the South Equatorial Current component of the Indian Ocean Gyre and a southern half that is dominated by cooler water upwelling systems that are driven by the confluences of the Mozambique and Agulhas Currents. The northern area exhibits eutropical water conditions, with extensive coral reef systems and carbonate environments, and contains a standard Lemurian Province fauna along with a number of widespread Madagascan endemics such as the cypraeids *Cribrarula pellisserpentis*, *Palmadusta androyensis*, and *Naria citrina dauphinensis*, the volute *Indolyria delessertiana*, the olivid *Acutoliva leonardhilli*, and the conids *Darioconus gracianus* and *Darioconus convolutus* (shown here on Figure 7.9). The southern half of the island, on the other hand, is a cooler water area which houses a species-rich malacofauna with an extremely high level of endemism. This localized evolutionary hot spot is here referred to as the **Toliaran Infraprovince**. Named for the city of Toliara, at the southwestern end of the island, this evolutionary hot spot was originally described as a full subprovince by Felix Lorenz (2004). The infraprovincial area is now known to contain a mixed fauna of classic western Lemurian and East African taxa, widespread Madagascan index species, and a number of localized endemic taxa such as the cone shells *Textilia chiapponorum*, *Textilia solangeae*, *Rolaniconus olgiatii*, *Cylinder vezzarochristophei*, and *Dendroconus medoci*, along with a species radiation of the volute genus *Indolyria*, including *I. pauljohnsoni*, *I. tulearensis*, *I. patbaili*, and *I. solangeae* (all shown here on Figure 7.10). Of special interest within the Toliaran Infraprovince is a large incipient species radiation taking place among sequestered populations of the textile cone, *Cylinder archiepiscopus*, found on remote reefs and within isolated bays. This radiation, which is unique within the Lemurian Province, constitutes a swarm of at least 18 distinct taxa, all of which are here considered to be subspecies of the widespread Lemurian Province *Cylinder archiepiscopus*. Some of these remarkable endemic Toliaran textile cones include *Cylinder archiepiscopus albospiratus*, *C. archiepiscopus duofasciatus*, *C. archiepiscopus auriger*, *C. archiepiscopus concatenatus* (all shown here on Figures 7.9 and 7.10), *C. archiepiscopus sirventi*, *C. archiepiscopus tenuiornata*, *C. archiepiscopus aquata*, *C. archiepiscopus tenebrosus*, and at least ten others. A similar, but smaller, radiation of the conid genus *Darioconus* is also evolving within the Toliaran Infraprovince, including species such as *Darioconus rosiae*, *D. behelokensis*, *D. corbieri*, *D. vezoi* (all shown here on Figures 7.9 and 7.10), *D. pseudoracemosus*, *D. pseudoecho*, and several others. The high level of endemism seen in the cone shells of the Toliaran Infraprovince (the Madagascar "Deep South") was recently described by Tenorio *et al.* (2018).

Mozambican Subprovince

Named for the southeastern African country of Mozambique, the Mozambican Subprovince extends from southern Somalia to Ponta do Ouro on the South Africa-Mozambique border, and also encompasses the area of the Mozambique Channel between mainland Africa and Madagascar. The molluscan fauna encompassed within this large biogeographical area is one of the richest found anywhere within the Lemurian Province, rivalling those of India and Madagascar. The southern part of the subprovince, in the area nearest to KwaZulu-Natal, is under the influence of cooler water and contains a malacofauna that is transitional with the Natalean Subprovince of the South African Province (see Chapter 10). In this area, numerous Natalean-South African taxa, such as the strepsidurid *Melapium elatum*, the cowrie *Paradusta barclayi*, the volutes *Fusivoluta clarkei* and *Fusivoluta barnardi*, and the cone shells *Bathyconus elokismenos* and *Kioconus (Ongoconus) typhon* (all shown here on Figures 7.11 and 7.12), occur together with eutropical western Lemurian taxa such as the cowrie *Leporicypraea geographica africola* and the cone shell *Leptoconus milneedwardsi* (shown here on Figures 7.11 and 7.12). The molluscan fauna of the Mozambican Subprovince exhibits a very high level of endemism, with many taxa being restricted to small geographical areas or offshore islands. Like the neighboring Madagascan Subprovince, the Mozambican Subprovince has evolved large species radiations of the conid genera *Cylinder* and *Darioconus*, including species complexes such as *Cylinder eumitis*, *C. cholmondeleyi*, and *C. suzannae* and *Darioconus bazarutensis*, *D. lohri* (all shown on Figures 7.11 and 7.12), *D. colubrinus*, and the true *D. pennaceus.* These remarkable endemic cones are sympatric with a large fauna of other endemic taxa, including the cowrie *Barycypraea fultoni massieri,* the turbinellids *Tudivasum zanzibaricum* and *Coluzea juliae*, a swarm of volute species including *Lyria lyraeformis*, *Athleta semirugata*, *Athleta epigona, Similyria kosibayensis*, and *Similyria aikeni*, and the conids *Dendroconus zulu* and *Bathyconus ramalhoi* (all shown here on Figures 7.11 and 7.12).

Eritrean Molluscan Province

Named after the Ancient Greek term for the modern-day Red Sea ("Erythra Thalassa"), the Eritrean Molluscan Province houses one of the richest eutropical malacofaunas found anywhere on Earth and encompasses the entire Red Sea, the northern coast of Somalia, the coasts of Yemen and Oman, and the entire Persian Gulf. Although small when compared to the neighboring Lemurian Province, the Eritrean is faunistically-complex, with four distinct and definable subprovinces; the **Aqaban**, **Dahlakian**, **Omanian**, and **Somalian**. The evolution of the Eritrean molluscan fauna is tied directly to environmental instability, the result of the catastrophic aspects of plate tectonic movement and oceanographic eustatic fluctuations. Having formed as a rift zone during the late Eocene, the Red Sea basin began to widen during the Oligocene and was flooded with sea water during the late Oligocene and early Miocene. Throughout the Miocene and Pliocene, the basin continued to widen and the first large coral reef systems became established. These reef systems supported rich carbonate environments that were colonized by mollusks that had evolved in the adjacent Gulf of Aden and the East African coast of the Indian Ocean. During the 100+ m sea level drops at the beginning of the Pleistocene (Gelasian Age), the narrow Strait of Bab al-Mandab, at the southern end of the Red Sea, became emergent and formed a land bridge that connected the African mainland to the Arabian Peninsula (modern Djibouti with Yemen). A very shallow and narrow ephemeral salt water river may have flowed across the land bridge, but this intermittent feature had little or no impact on the oceanography of the Red Sea. The narrow terrestrial barrier across Bab al-Mandab effectively isolated the Red Sea from the Indian Ocean, creating what was essentially a giant salt water lake, referred to here as the **Eritrean Paleosea**.

Although completely land-locked, the salinity within the Eritrean Paleosea retained open-oceanic levels throughout the Pleistocene, primarily because of the increased rainfall and wetter climatic conditions in northeastern Africa and the Arabian Peninsula at that time. This large, narrow inland sea remained isolated from the Indian Ocean for over 2 million years and did not reconnect with the Gulf of Aden and the main oceanic basin until the latest Pleistocene (Tarantian Age). During this extended period of genetic isolation, the Eritrean Paleosea mollusks experienced enhanced genetic drift and quickly evolved entire suites of new endemic species. With a lowered sea level during the Pleistocene, the Eritrean Paleosea was also spatially smaller than the Red Sea is today, with the entire Gulf of Suez, and most of its southern third, being emergent and covered by desert environments. Only the northern two-thirds of the Eritrean Paleosea and the Gulf of Aqaba retained deep oceanic conditions and these areas acted as refugia for many of the present-day Red Sea endemic mollusks. After the Bab al-Mandab Strait re-opened in the Late Pleistocene, the endemic species that had evolved in the isolated Eritrean Paleosea spread southward into the Gulf of Aden, down the northeastern coast of Africa, across the southern coast of the Arabian Peninsula, and into the Gulf of Oman. Around 15,000 years ago, the flood plain of the Tigris and Euphrates Rivers in present-day Iraq began to fill with sea water and these marine conditions spread northward up the valley to form the Persian Gulf. The hardiest of the Gulf of Oman mollusks then migrated into this newly-formed shallow sea and gave rise to the impoverished molluscan fauna that is seen in the Persian Gulf today. DiBattista, *et al.* (2016) gave a detailed synopsis of the historical geology, oceanography, and paleoclimatology of the Red Sea and Eritrean Paleosea. To date, this is the single best work on the paleoceanography of that ancient body of water.

The Eritrean Province can be defined and demarcated by the presence of numerous wide-ranging provincial index taxa, including the cerithiid *Cerithium erythraeonense*, the strombids *Conomurex fasciatus* and *Tricornis tricornis*, the ranellid *Septa marerubrum*, the muricids *Homalocantha digitata*, *Chicoreus corrugatus*, and *Hexaplex kusterianus*, and the fasciolariids *Marmarofusus polygonoides* and *Pleuroploca gracilior* (all illustrated here on Figure 7.13). A very large and highly-endemic cowrie fauna has also evolved within the Red Sea, primarily as a result of having been genetically isolated from the Indian Ocean faunas during the Pleistocene sea level drops. Some of these classic Eritrean Province cypraeids include *Cypraea pantherina*, *Bistolida erythraeensis*, *Luria pulchra*, *Lyncina camelopardalis*, *Naria nebrites*, *Talparia exusta*, *Nucleolaria sturanyi*, *Pustularia marerubrum*, *Purpuradusta notata*, *Cribrarula cribraria perstata* (all shown here on Figure 7.14), and many other species and subspecies. Likewise, a very large and highly-endemic fauna of cone shells has also evolved within the Eritrean Province, including species such as *Phasmoconus erythraeensis*, *Gastridium cuvieri*, *Leptoconus locumtenens*, *Darioconus quasimagnificus*, *Miliariconus taeniatus*, *Phasmoconus jickelii*, *Rhizoconus semivelatus*, *Virgiconus thomasi*, *Rhizoconus fumigatus*, *Pionoconus nigropunctatus*. *Puncticulis aequipunctatus*, *Cylinder neovicarius* (all shown here on Figure 7.15), and numerous other species and subspecies.

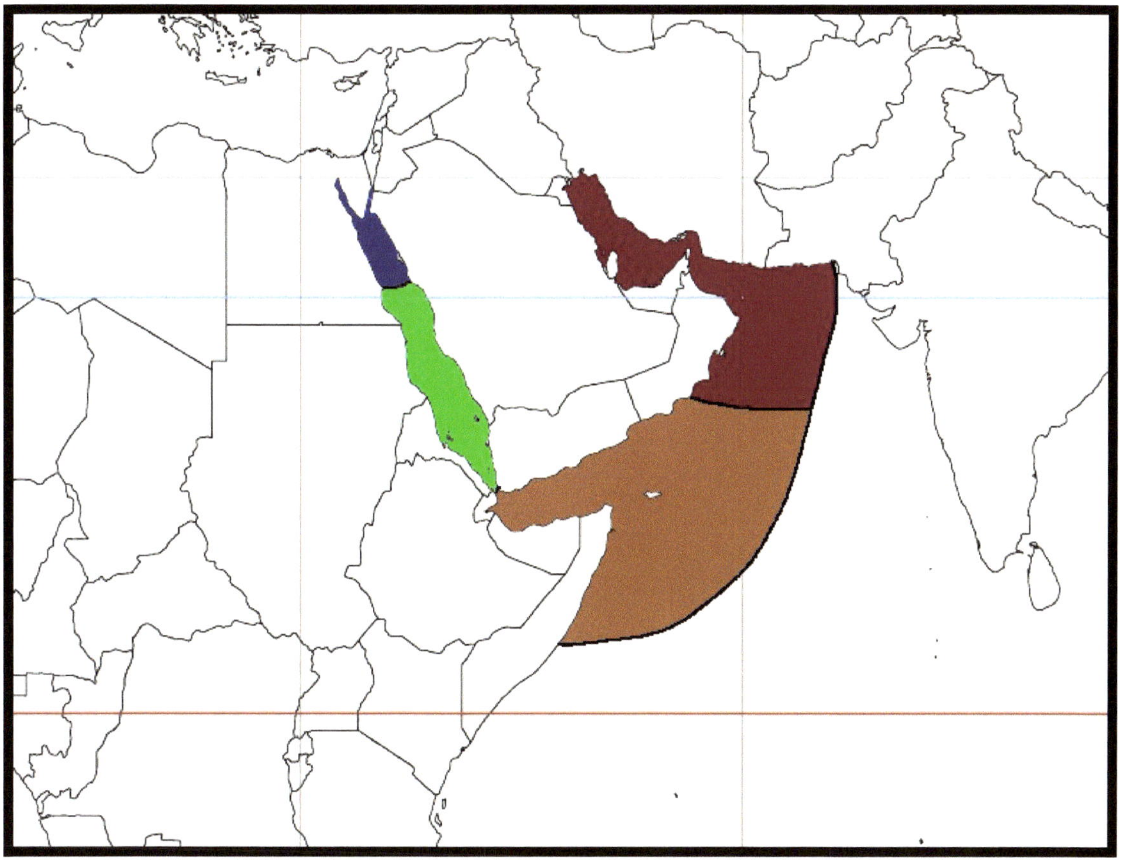

Figure 7.2 Map of the Eritrean Molluscan Province, showing the areal extents of its four subprovinces: the **Aqaban Subprovince** (blue), the **Dahlakian Subprovince** (green), the **Omanian Subprovince** (dark rose), and the **Somalian Subprovince** (brown).

Aqaban Subprovince

Named for the Gulf of Aqaba in the northern Red Sea, the Aqaban Subprovince encompasses the entire Gulf of Aqaba as well as the adjacent Egyptian coastal areas from Ras Ghareb to Safaga and also the southern part of the Gulf of Suez. The Gulf formed as a tectonic rift zone during the Pliocene and the area has been gradually widening and expanding northward into the present-day Dead Sea region. During the mid-Pleistocene, approximately 1 million years ago and at the end of the Calabrian Age, the shallow Strait of Tiran along the southern end of the Gulf closed off due to the low sea level and the entire body of water became an isolated salt water lake, referred to here as the **Sinai Paleosea**. Although having formed in an arid region with little rainfall, the Sinai Paleosea did not evaporate to become a hypersaline brine lake, but retained its open ocean salinity levels because of the inflow of fresh water from the Jordan River. Even today, the Jordan River effluent allows the semi-enclosed Gulf of Aqaba to maintain normal salinities and to retain large fringing coral reef systems along its steep, cliff-like edges and submarine canyons.

During the times of genetic isolation caused by the mid-Pleistocene low sea levels, the enclosed Sinai Paleosea evolved an entirely new fauna of gastropods, derived from the faunas of the main Red Sea basin. When the sea returned to normal levels during the late Pleistocene and the Strait of Tiran re-opened, these Aqaban endemic species and subspecies migrated out into the northern end of the Red Sea basin and into the Gulf of Suez. Some of the more important Aqaban Subprovince endemics include the cowries *Luria pulchra sinaiensis*, *Cypraea pantherina rasnasraniensis*, *Lyncina camelopardalis sharmiensis*, *Bistolida rubiginosa aegyptica*, *Erronea caurica nabeqensis*, *Naria turdus pardalina*, and *Talostolida teres elatensis*, the muricids *Homalocantha dovpeledi*, *Homalocantha elatensis*, *Chicoreus peledi*, and *Naquetia fosteri*, the mitrid *Pseudonebularia dovpeledi*, and the conids *Calamiconus akabensis*, *Harmoniconus sharmiensis*, *Miliariconus sinaiensis*, *Darioconus bratcherae*, and *Pionoconus elatensis*, all of which evolved within the confines of the Pleistocene Sinai Paleosea. These are illustrated here on Figures 7.16 and 7.17.

Dahlakian Subprovince

In the southern Red Sea, off the coast of Eritrea, the large Dahlak Archipelago of over 120 desert islands harbors a species-rich gastropod fauna with a high percentage of endemism. This island group, the namesake of the Dahlakian Subprovince, is now known to contain the same type of malacofauna that is found throughout the entire southern one-third of the Red Sea. The island-dominated subprovince also encompasses the neighboring Hanish and Halib Islands of Eritrea, the Farasan Islands of Saudi Arabia, the Kamaran and Zubair Archipelagoes of Yemen (with active volcanoes), and the adjacent coastlines of Eritrea, southwestern Saudi Arabia, and southwestern Yemen. The areal extent of the Dahlakian Subprovince corresponds directly to the bathymetry of the basin at the southern end of the Pleistocene Red Sea, where these isolated coral reef ecosystems evolved new taxa within the enclosed inland sea. The endemic Dahlakian gastropods appear to be relictual species, left-overs from the time when the Red Sea was, essentially, an extended salt water lake. Some of the more important endemic index gastropods include the muricid *Chicoreus ethiopius*, and the cone shells *Cylinder neovicarius dahlakensis*, *Phasmoconus nigromaculatus*, and *Gastridium fragilissimum* (all shown here on Figure 7.17).

Omanian Subprovince

Named for the country of Oman, on the southeastern end of the Arabian Peninsula, the Omanian Subprovince contains one of the richest molluscan faunas found anywhere in the northern Indian Tropical Region. Of the four Eritrean Subprovinces, the Omanian has the highest level of endemism, approaching full provincial status. Extending from near Nishtun in southeasternmost Yemen, across the Oman coast to Kuwait, and encompassing the entire Gulf of Oman, the Persian Gulf, and the adjacent coast of Iran as far east as Chabahar Bay, the Omanian Subprovince is also the largest single biotic subdivision found within the Eritrean area. During the Pleistocene sea level drops, what is now the shallow Persian Gulf was dry land and was bisected by the parallel flow channels of the Tigris and Euphrates Rivers. These rivers, which now empty into the northern end of the Persian

Gulf, then terminated at what is now the narrow Strait of Hormuz and formed waterfalls as they tumbled into the deep Gulf of Oman. As sea level began to rise during the latest Pleistocene, the ocean overtopped the Strait of Hormuz cascades and flowed northward, gradually drowning the flood plain of the Tigris and Euphrates Rivers and creating the modern Persian Gulf. At this time, the mollusks from the Gulf of Oman moved northward into the newly-formed Persian Gulf and formed the impoverished fauna that is found there today. Since the water mass of the Persian Gulf is so geologically young, no endemic species are present within the confines of the basin and only a few of the hardiest Omanian taxa can live there.

During the late Pleistocene sea level lows, the Gulf of Oman and the Oman coast acted as a refugium for older Pliocene and early Pleistocene faunas and many of these have survived into recent times as relictual taxa. One of the most important of these Omanian prehistoric left-overs is the primitive cowrie, *Barycypraea teulerei* (Figure 7.18 A), which represents a survivor of the large *Barycypraea* species radiation of Miocene southern Asia and Indonesia. The Omanian Subprovince also houses many distinctive endemic index taxa, some of which include the turbinid *Turbo (Aspilaturbo) jonathani*, the cowries *Naria turdus winkworthi* and *Monetaria annulus dilatissima*, the ranellid *Ranularia boschi*, the muricids *Murex echinodes*, *Hexaplex rileyi*, *Hexaplex kuesterianus blazeki*, and *Pygmaepterys paulbochi*, the pisaniid *Cantharus vezzarochristofei*, the marginellids *Persicula masirana* and *Volvarina arabica*, the olivid *Ancilla boschi*, and the conids *Fusiconus stocki*, *Quasiconus melvilli*, *Pionoconus koukae*, *Darioconus laueri*, and *Rhizoconus ardisiaceus* (all shown here on Figures 7.18 and 7.19). Of special interest within the Omanian Subprovince is the **Masiran Infraprovince**, a localized evolutionary hot spot centered on Al Masirah Island off the central coast of Oman. The protected bays along the shoreline of this large desert island house extensive Turtle Grass beds and closely resemble the coastal environments seen within the Gulf of Venezuela (see Chapter 2, Venezuelan Subprovince). Some of the more important Masiran endemic gastropods include the strombid *Conomurex coniformis masirensis*, the cypraeids *Erronea (Adusta) persica dilatata* and *Erronea caurica masirensis*, the left-handed fasciolariid *Sinistralia gallagheri*, and the extremely beautiful acteonid *Punctacteon eloisae* (all shown here on Figure 7.19). The very rare lyriine volute, *Indolyria leslieboschae*, is also only known from the Al Masirah Island area.

Somalian Subprovince

Named for the East African country of Somalia, the Somalian Subprovince extends from Djibouti and the entire Gulf of Aden, south around the "Horn of Africa" to near Merca, Somalia, and also includes the islands of Socotra and Kilmia and the adjacent coast of southern Yemen. The southern half of the subprovince, from Xaafuun south to Merca, is under the influence of massive deep sea upwelling systems and the coastal waters are seasonally cold and nutrient-rich. These fertile water conditions support immense plankton blooms, which, in turn, support a rich and diverse benthic fauna, including numerous endemic gastropods. The northern half of the subprovince, encompassing the entire Gulf of Aden and Socotra Island, has extremely warm water temperatures and supports extensive and very rich coral reef complexes, with some of the best-developed occurring around Djibouti, Berbera, and Socotra Island. These two main types of Somalian environments house numerous distinctive endemic gastropods, with some of the most prominent and characteristic being the cowries *Callistocypraea broderipi somalica*, *Cypraea pantherina catulus*, *Naria marginalis pseudocellata*, *Naria marginalis quadriangula*, and *Talparia exusta meridiana*, the strombid *Tricornis oldi*, the muricid *Hexaplex bozzadamii*, the left-handed fasciolariid *Sinistralia somaliensis*, the marginellids *Glabella mirabilis* and *Glabella obtusa*, the volute *Festilyria festiva deceptrix*, and the cone shells *Strategiconus splendidulus*, *Phasmoconus salzmanni*, *Conasprella bozzettii*, *Calamiconus subroseus*, *Phasmoconus angioiorum*, *Darioconus echo*, and *Profundiconus neotorquatus*. Most of these are illustrated here on Figures 7.20 and 7.21.

ICONOGRAPHY OF GASTROPODS OF THE INDIAN TROPICAL REGION

(Principal Index Gastropods are shown on Figures 7.3 to 7.21)

Figure 7.3 Widespread Index Gastropods of the Lemurian Province.
A= *Leporicypraea rosea* (Gray, 1824), length 71 mm, 3 m depth off Nosy Mitsio Island, Madagascar; **B**= *Palmadusta diliculum* (Reeve, 1845), length 28 mm, under dead coral at low tide, Inhaca Island, Mozambique; **C**= *Phalium fimbria* (Gmelin, 1791), length 83 mm, 100 m depth off Saint Pierre, La Reunion Island, Mascarene Islands; **D**= *Miniaceoliva tremulina* (Lamarck, 1811), length 96 mm, 3 m depth on sand, Mamoudzou, Mayotte Island, Comoros Islands; **E**= *Cylinder archiepiscopus* (Hwass, 1792), length 83 mm, 5 m depth off Thoothukudi, Tamil Nadu, India; **F**= *Cylinder paulucciae* (Sowerby III, 1877), length 52 mm, 50 m depth off Tombant de Grand Ravine, La Reunion Island, Mascarene Islands; **G**= *Kioconus caillaudi* (Kiener, 1846), length 32 mm, 200 m depth off Morone, Grande Comore Island, Comoros Islands; **H**= *Pionoconus gubernator* (Hwass, 1792), length 78 mm, Fort Dauphin and Nante Nina, Madagascar; **I**= *Rhombiconus zonatus* (Hwass, 1792), length 53 mm, 2 m depth off Thoothukudi, Tamil Nadu, India; **J**= *Virgiconus berdulinus* (Veillard, 1972), length 87 mm, 100 m depth, off Park Rynie, Natal, South Africa; **K**= *Palmadusta asellus armadillo* Bergonzoni, 2015, length 18 mm, 5 m depth off Fort Dauphin, Madagascar (found along East Africa, Madagascar, and surrounding islands).

Figure 7.4 Index Gastropods of the Andamanian Subprovince, Lemurian Province.
A= *Callistocypraea nivosa* (Broderip, 1827), length 45 mm, 10 m depth off Letsok-aw Island, Mergui Archipelago, Myanmar; **B=** *Perisserosa guttata surinensis* Raybaudi Massilia, 1978, length 55 mm, 100 m depth south of Ko Chang Island, Ranong, Thailand; **C=** *Mirabilistrombus listeri* (Gray, 1852), length 136 mm, 100 m depth south of Ko Chang Island, Ranong, Thailand; **D=** *Arctoliva pacifica* (Marrat, 1870), length 49 mm, 50 m depth, Phuket Island, Thailand; **E=** *Galeola andamanensis* (Bridgman, 1909), length 24 mm, 3 m depth off Guptapara, South Andaman Island, Andaman Islands (this Andaman endemic has been incorrectly synonymized with the widespread *Galeola carneola adspersa*; it differs in being a more elongated shell, in having a deep purple aperture, and in having purple-brown oblique zebra stripes); **F=** *Darioconus thailandis* (da Motta, 1978), length 45 mm, 10 m depth, Ko Yao Yai Island, Thailand; **G=** *Graphiconus ranonganus* (da Motta, 1978), length 78 mm, 100 m depth off Ranong Province near Myanmar border, Thailand; **H=** *Kioconus simanoki* Tenorio, Poppe, and Tagaro, 2007, length 69 mm, 120 m depth off Ko Racha Yai Island, Thailand; **I=** *Stellaconus sukhadwalai* (Rockel and da Motta, 1983), length 57 mm, 120 m depth south of Ko Racha Yai Island, Thailand; **J=** *Textilia vicweei* (Old, 1973), length 76 mm, 50 m depth off Zadetkyi Island, Mergui Archipelago, Myanmar; **K=** *Phasmoconus andamanensis* (E.A. Smith, 1878), length 22 mm, 10 m depth off Phuket Island, Thailand.

Figure 7.5 Index Gastropods of the Bengalian Subprovince, Lemurian Province.
A= *Erronea (Ipserronea) angioyorum* (Biraghi, 1978), length 32 mm, 70 m depth on a muddy sea floor, off Kollam, Kerala, India; **B**= *Rhizoconus rawaiensis* (da Motta, 1978), length 28 mm, 20 m depth off Phuket Island, Thailand; **C**= *Harpularia aurisiaca* (Lightfoot, 1786) banded form *vexillum* (Gmelin, 1791), length 81 mm, 50 m depth due east of Trincomalee, Sri Lanka; **D**= *Harpularia lapponica* (Linnaeus, 1758), length 78 mm, 20 m depth off Rameswaram, Tamil Nadu, India; **E**= *Cribrarula abaliena ganteri* (Lorenz, 1998), length 20 mm, under rocks at low tide, Trincomalee, Sri Lanka; **F**= *Miniaceoliva kremerorum* (Petuch and Sargent, 1986), length 50 mm, 30 m depth off Kollam, Kerala, India; **G**= *Viduoliva macleaya* (Duclos, 1840), length 48 mm, 20 m depth off Rameswaram, Tamil Nadu, India; **H**= *Cylinder abbas* (Hwass, 1792), length 45 mm, 10 m depth off Trincomalee, Sri Lanka; **I**= *Cylinder bengalensis* (Okutani, 1968), length 87 mm, 50 m depth south of Sittwe, northeastern Bay of Bengal, Myanmar; **J**= *Eugeniconus friedae* (da Motta, 1991), length 44 mm, 2 m depth north of Mannar, Sri Lanka; **K**= *Pionoconus subfloridus* (da Motta, 1985) , length 68 mm, 5 m depth off Thoothukudi, Tamil Nadu, India.

Figure 7.6 Index Gastropods of the Malabaran Subprovince, Lemurian Province.
A= *Amabiliplicatus sibbaldi* (Sowerby I, 1842), length 33 mm, 10 m depth off Ratnagiri, Maharashtra, India; **B**= *Miniaceoliva flammeacolor* (Petuch and Sargent, 1986), length 85 mm, 25 m depth, due west of Mumbai, Maharashtra, India; **C**= *Fusiconus dictator* (Melvill, 1898), length 43 mm, 100 m depth due west of Harihareshwar, Maharashtra, India; **D**= *Graphiconus cuneiformis* (E.A. Smith, 1877), length 50 mm, 40 m depth due west of Kochi, Kerala, India (note the violet color within the aperture); **E**= *Graphiconus indomaris* Bozzetti, 2014, length 56 mm, 70 m depth off Kochi, Kerala, India; **F**= *Leptoconus milneedwardsi clytospira* (Melvill and Standen, 1899), length 137 mm, 50 m depth off Ratnagiri, Maharashtra, India; **G**= *Pionoconus leehmani* (da Motta and Rockel, 1979), length 64 mm, 5 m depth on a coral reef off Amini Island, Amindivi Group, Laccadive Islands, Lakshadweep, India; **H**= *Quasiconus tuticorinensis* (Rockel and Korn, 1990), length 25 mm, 50 m depth off Kochi, Kerala, India; **I**= *Harpularia loroisi* (Valenciennes, 1863), length 98 mm, 30 m depth, southwest of Kochi, Kerala, India; **J**= *Stellaconus bayani* (Jousseaume, 1872), length 60 mm, 50 m depth off Kochi, Kerala, India; **K**= *Virgiconus malabaricus* Monnier, Limpalaër, and Tenorio, 2017, length 66 mm, 50 m depth off Kochi, Kerala, India.

Figure 7.7 Index Gastropods of the Mascarenean Subprovince, Lemurian Province.
A= *Bistolida oweni menkeana* (Deshayes, 1863), length 12 mm, 5 m depth off Mahebourg, Mauritius Island, Mascarene Islands; **B=** *Cribrarula cribellum* (Duclos, 1833), length 14 mm, 5 m depth off Mahebourg, Mauritius Island, Mascarene Islands; **C=** *Cribrarula esontropia* (Duclos, 1833), length 23 mm, 3 m depth off Grand Baie, Mauritius Island, Mascarene Islands; **D=** *Miniaceoliva tremulina hayesi* Sargent and Petuch, 2012, length 44 mm, 3 m depth off Trou-aux-Biches, Mauritius Island, Mascarene Islands; **E=** *Miniaceoliva pica* (Lamarck, 1811), length 66 mm, 5 m depth off Le Morne, Mauritius Island, Mascarene Islands; **F=** *Harpa costata* (Linnaeus, 1758), length 69 mm, 5 m depth in Blue Bay, Mauritius Island, Mascarene Islands (also found in NE Madagascar); **G=** *Darioconus rubiginosus* (Hwass, 1792), length 40.3 mm, 5 m depth off Grande Baie, Mauritius Island, Mascarene Islands; **H=** *Darioconus magoides* (Melvill, 1900), length 55 mm, 5 m depth off Mahebourg, Mauritius, Mascarene Islands; **I=** *Darioconus episcopus* (Hwass, 1792) (*marmoricolor* Melvill, 1900 is a synonym), length 42.2 mm, 20 m depth off Le Morne, Mauritius Island, Mascarene Islands; **J=** *Darioconus rubropennatus* (da Motta, 1982), length 38.4 mm, 20 m depth off Saint-Gilles, La Reunion Island, Mascarene Islands; **K=** *Textilia julii* (Lienard, 1870), length 38.5 mm, 20 m depth off Poudre d'Or, Mauritius Island, Mascarene Islands.

Figure 7.8 Index Gastropods of the Mascarenean Subprovince and the Brandonian Infraprovince, Lemurian Province. (Mascarenean Subprovince) **A**=*Annulatoliva maculata* (Duclos, 1840), length 55 mm, 3 m depth off Ile Sapaille, Egmont Atoll, Chagos Archipelago; **B**= *Leporicypraea rosea subsignata* (Melville, 1888), length 73 mm, 3 m depth off Coco Island, St. Brandon Shoals, Cargados Carajos Islands (shell base intense purple); **C**= *Cylinder corbula* (Sowerby II, 1858), length 61 mm, 3 m depth on reef off Port St. James, North Island, Agalega Islands; **D**= *Rolaniconus moussoni* (Crosse, 1865), length 21 mm, 2 m depth off Coco Island, St. Brandon Shoals, Cargados Carajos Islands; **E**= *Ophioglossolambis violacea* (Swainson, 1821), length 111 mm, 10 m depth off Coco Island, St. Brandon Shoals, Cargados Carajos Islands; **F**= *Erronea (Adusta) nymphae* (Jay, 1850), length 44 mm, 3 m depth on Spur Reef, Middle Island, Diego Garcia Atoll, Chagos Archipelago (endemic to Diego Garcia Atoll); **G**= *Cylinder loman* (Dautzenberg, 1937), length 32 mm, 5 m depth off Port Louis, Mauritius; (**Brandonian Infraprovince**) **H**= *Bistolida nanostraca* Lorenz and Chiapponi, 2005, length 14 mm, 2 m depth off Coco Island, St. Brandon Shoals, Cargados Carajos Islands; **I**= *Bistolida piae* Lorenz and Chiapponi, 2005, length 16 mm, 2 m depth, Coco Island, St. Brandon Shoals, Cargados Carajos Islands; **J**= *Morum lorenzi* Monsecour, 2011, length 23 mm, 1 m depth off Coco Island, St. Brandon Shoals, Cargados Carajos Islands; **K**= *Ficus dandrimonti* Lorenz, 2012, length 38 mm, 2 m depth, Coco Island, St. Brandon Shoals, Cargados Carajos Islands.

Figure 7.9 Widespread Index Gastropods of the Madagascan Subprovince, Lemurian Province.
A= *Naria citrina dauphinensis* Lorenz, 2002, length 23 mm, 3 m depth on a reef off Fort Dauphin, Taolagnaro, Madagascar; **B=** *Cribrarula pellisserpentis* Lorenz, 1999, length 22 mm, 3 m depth off Ambaro, Nossi Be Island, Madagascar; **C=** *Palmadusta androyensis* Blocher and Lorenz, 1999, length 18 mm, 2 m depth on a reef off Fort Dauphin, Taolagnaro, Madagascar; **D=** *Indolyria delessertiana* (Petit de la Saussaye, 1842), length 55 mm, 1 m depth, Morumbe Bay, Madagascar; **E=** *Acutoliva leonardhilli* (Petuch and Sargent, 1986), length 25 mm, low tide on sand bars, off Toliara, Madagascar; **F=** *Cylinder vezzarochristophei* Bozzetti, 2018, length 60 mm, 10 m depth off Toliara, Madagascar; **G=** *Darioconus convolutus* (Sowerby II, 1858), length 73 mm, 2 m depth off Salary Bay, Madagascar; **H=** *Cylinder archiepiscopus auriger* (Röding, 1798), length 56 mm, 20 m depth off Anakao, Toliara, Madagascar; **I=** *Cylinder archiepiscopus albospiratus* Bozzetti, 2017, length 58 mm, 5 m depth on coral reef off Fort Dauphin, Madagascar; **J=** *Darioconus gracianus* (da Motta and Blocher, 1982), length 52 mm, 2 m depth on coral reef off Toliara, Madagascar; **K=** *Darioconus vezoi* Korn, Niederhofer, and Blocher, 2000, length 55 mm, 3 m depth off Lavanono, Madagascar; **L=** *Cylinder archiepiscopus concatenatus* (Kiener, 1850), length 49.4 mm, 10 m depth on reef off Itampolo, Madagascar.

Figure 7.10 Index Gastropods of the Toliaran Infraprovince, Lemurian Province.
A= *Textilia chiapponorum* Lorenz, 2004, length 32 mm, in coral rubble, 10 m depth off Lavanono, Madagascar; **B=** *Indolyria pauljohnsoni* (Poppe and Terryn, 2002), length 37 mm, 1 m depth in sea grass bed, off Lavanono, Madagascar; **C=** *Indolyria solangeae* (Bozzetti, 2008), length 55 mm, 1 m depth off Manantenina, north of Fort Dauphin, Madagascar; **D=** *Indolyria tulearensis* (von Cosel and Blocher, 1977), length 57 mm, low tide in sand, Toliara, Madagascar; **E=** *Indolyria patbaili* Bouchet, 1999, length 65 mm, 20 m depth off Lavanono, Madagascar; **F=** *Darioconus behelokensis* (Lauer, 1989), length 53 mm, 10 m depth under coral, Beheloka, Madagascar; **G=** *Darioconus corbieri* (Blocher, 1994), length 54 mm, 5 m depth under coral, Salary Bay, Madagascar; **H=** *Darioconus rosiae* Monnier, Batifoix, and Limpalaër, 2018, length 51 mm, 10 m depth off Anakao, Madagascar; **I=** *Dendroconus medoci* Lorenz, 2004, length 45 mm, found dead on the beach, Itampolo, Madagascar; **J=** *Rolaniconus olgiatii* (Bozzetti, 2004), length 33 mm, 5 m depth in Salary Bay, Madagascar; **K=** *Textilia solangeae* (Bozzetti, 2004), length 23 mm, 2 m depth, Lavanono, Madagascar; **L=** *Cylinder archiepiscopus duofasciatus* Bozzetti, 2016, length 48 mm, low tide on reef, Tulear, Madagascar.

Figure 7.11 Index Gastropods of the Mozambican Subprovince, Lemurian Province.
A= *Barycypraea fultoni massieri* Lorenz, 1991, length 75 mm, collected by a Russian trawler, 150 m depth off Quissico, Mozambique; **B**= *Melapium elatum* (Schubert and Wagner, 1829), length 47 mm, 100 m depth off Maputo, Mozambique; **C**= *Lyria lyraeformis* Swainson, 1821, length 87 mm, trawled from 40 m depth north of Pemba Island, Tanzania; **D**= *Athleta semirugata* (Rehder and Weaver, 1974), length 47 mm, trawled from 350 m depth off Inhambane, Mozambique; **E**= *Similyria aikeni* (Veldsman, 2012), length 66 mm, 250 m depth off Inhaca Island, Mozambique; **F**= *Cylinder eumitis* (Tomlin, 1926), length 75 mm, 20 m depth off Zavora Point, Inhambane, Mozambique; **G**= *Darioconus bazarutensis* (Fernandes and Monteiro, 1988), length 44 mm, 2 m depth off Asneira, Bazaruto Island, Mozambique; **H**= *Darioconus lohri* (Kilburn, 1972), length 42 mm, low tide, Ponta Dobela, Maputo, Mozambique; **I**= *Dendroconus zulu* (Petuch, 1979), length 77 mm, 20 m depth off Chinde, Zambezi River Delta, Mozambique ("*zulu*" from Madagascar is a different species); **J**= *Kioconus (Ongoconus) typhon* (Kilburn, 1975), length 42 mm, 150 m depth off Inhaca Island, Mozambique; **K**= *Leptoconus milneedwardsi* (Jousseaume, 1894), length 85 mm, 100 m depth off Nacala, Mozambique; **L**= *Tudivasum zanzibaricum* (Abbott, 1958), length 39 mm, trawled by fishermen from 100 m depth east of Pemba Island, Zanzibar Archipelago, Tanzania.

Figure 7.12 Index Gastropods of the Mozambican Subprovince, Lemurian Province.
A= *Paradusta barclayi* (Reeve, 1857), length 28 mm, 100 m depth off Nacal, Mozambique; **B=** *Leporicypraea geographica africola* Lorenz, 2017, length 67 mm, 10 m depth off Nungwi, Zanzibar, Tanzania; **C=** *Fusivoluta clarkei* Rehder, 1969, length 77 mm, 150 m depth off Inhaca Island, Mozambique; **D=** *Similyria kosibayensis* Veldsman, 2012, length 40 mm, 100 m depth off Ponta de Ouro, Mozambique; **E=** *Athleta epigona* (von Martens, 1904), length 32 mm, 360 m depth off Dar es Salaam, Tanzania; **F=** *Cylinder cholmondeleyi* (Melvill, 1900), length 67 mm, 2 m depth, Shimoni, Kenya; **G=** *Darioconus elisae* (Kiener, 1845), length 57 mm, 20 m depth off Cabaceira Pequena, Mozambique; **H=** *Bathyconus elokismenos* (Kilburn, 1975), length 62 mm, 250 m depth off Inhambane, Mozambique; **I=** *Fusivoluta barnardi* Rehder, 1969, length 110 mm, 250 m depth off Maputo, Mozambique; **J=** *Cylinder suzannae* (von Rossum, 1990), length 41 mm, Malindi, Island, Kenya; **K=** *Bathyconus ramalhoi* (Coomans, Moolenbeek, and Wils, 1986), length 25 mm, 12 m depth off Zanzibar, Tanzania; **L=** *Coluzea juliae* Harasewych, 1987, length 68 mm, 250 m depth off Inhambane, Mozambique.

Figure 7.13 Widespread Index Gastropods of the Eritrean Province.
A= *Conomurex fasciatus* (Born, 1778), length 40 mm, low tide, Muzahdorf Island, Dahlak Archipelago, Eritrea; **B**= *Tricornis tricornis* (Humphrey, 1786), length 120 mm, 1 m depth off Port Sudan, Sudan; **C**= *Chicoreus corrugatus* (Sowerby II, 1841), length 44 mm, 10 m depth off Eilat, Israel, Gulf of Aqaba; **D**= *Homalocantha digitata* (Sowerby II, 1841), length 47 mm, 5 m depth off Khor Angar, Djibouti; **E**= *Darioconus quasimagnificus* (da Motta, 1982), length 55 mm, 2 m depth off Dafiyat, Masirah Island, Oman; **F**= *Cerithium erythraeonense* Lamarck, 1822, length 51.8 mm, 2 m depth off Shaab Abu Hashish, Hurghada, Egypt; **G**= *Hexaplex kuesterianus* (Tapparone-Canefri, 1875), length 84 mm, 1 m depth off Dafiyat, Masirah Island, Oman; **H**= *Septa marerubrum* (Garcia-Talavera, 1985), length 39 mm, 3 m depth off Sharm El-Sheikh, Sinai Peninsula, Egypt; **I**= *Marmarofusus polygonoides* (Lamarck, 1822), length 92 mm, trawled from 20 m depth off Suakin, Sudan; **J**= *Pleuroploca gracilior* (Tapparone-Canefri, 1875), length 128 mm, 3 m off Dafiyat, Masirah Island, Oman; **K**= *Pionoconus striatus juliaallaryae* (Cossignani, 2013), length 91 mm, 10 m depth off Eilat, Israel, Gulf of Aqaba (note sloping shoulder area); **L**= *Luria isabella erythraea* Raybaudi, 1986, length 37 mm, under coral rubble, 1 m depth off Socotra Island, Gulf of Aden.

Figure 7.14 Cowrie Species Radiations of the Eritrean Province.
A= *Bistolida erythraeensis* (Sowerby II, 1837), length 18 mm, 3 m depth off Sheik Said Island, Eritrea; **B**= *Cypraea pantherina* Lightfoot, 1786, length 80 mm, 3 m depth off Obock, Djibouti; **C**= *Luria pulchra* (Gray, 1824), length 46 mm, 5 m depth off Khor Angar, Djibouti; **D**= *Lyncina camelopardalis* (Perry, 1811), length 63 mm, on coral reef, 5 m depth off Sheik Said Island, Eritrea; **E**= *Naria nebrites* (Melvill, 1888), length 32 mm, 1 m depth off northern Muzahdorf Island, Dahlak Archipelago, Eritrea; **F**= *Talparia exusta* (Sowerby II, 1832), length 69 mm, 10 m depth, off Port Sudan, Sudan; **G**= *Erronea caurica quinquefasciata* (Röding, 1798), length 49 mm, low tide, Dafiyat, Al Masirah Island, Oman; **H**= *Naria macandrewi* (Sowerby III, 1870), length 13 mm, 3 m depth off Eilat, Israel, Gulf of Aqaba; **I**= *Purpuradusta notata* (Gill, 1858), length 14 mm, 1 m depth off Hurghada, Egypt; **J**= *Nucleolaria sturanyi* Schilder and Schilder, 1938, length 28 mm, 20 m depth off Abu Ramada Island, Hurghada, Egypt; **K**= *Pustularia marerubra* Lorenz, 2009, length 13 mm, 20 m depth off Abu Ramada Island, Hurghada, Egypt; **L**= *Cribrarula cribraria perstata* Lorenz, 2017, length 30 mm, 2 m depth in sponges, off Bosaso, northern Somalia, Gulf of Aden.

Figure 7.15 Cone Shell Species Radiations of the Eritrean Province.
A= *Phasmoconus erythraeensis* (Reeve, 1843), length 25 mm, 3 m depth off Harmil Island, Dahlak Archipelago, Eritrea; **B**= *Cylinder neovicarius* (da Motta, 1982), length 94 mm, in sand and coral, 15 m depth off Eilat, Israel, Gulf of Aqaba; **C**= *Gastridium cuvieri* (Crosse, 1858), length 38 mm, on coral slabs, 3 m depth off Khor Angar, Djibouti; **D**= *Leptoconus locumtenens* (Blumenbach, 1791), length 42 mm, in sand and algae, 5 m depth off Tadjoura, Djibouti; **E**= *Miliariconus taeniatus* (Hwass, 1792), length 27 mm, on rocks at low tide, Hurghada, Egypt; **F**= *Phasmoconus jickelii* (Weinkauff, 1873), length 39 mm, in sand, 5 m depth off Tadjoura, Djibouti; **G**= *Pionoconus nigropunctatus* (Sowerby II, 1858), length 31.2 mm, low tide, Hirghigo, Eritrea; **H**= *Puncticulis aequipunctatus* (Dautzenberg, 1937), length 36 mm, 1 m depth off Dishet El Dhaba, Hurghada, Egypt; **I**= *Rhizoconus semivelatus* (Sowerby III, 1882), length 11 mm, under coral slabs, 1 m depth off Harmil Island, Dahlak Archipelago, Eritrea; **J**= *Rhizoconus fumigatus* (Hwass, 1792), length 35 mm, 2 m depth on a reef off Tadjoura, Djibouti; **K**= *Virgiconus thomasi* (Sowerby III, 1881), length 49 mm, 1 m depth off Port Sudan, Sudan; **L**= *Phasmoconus dillwynii* (Reeve, 1849), length 20 mm, 2 m depth off Farasan Island, Saudi Arabia.

Figure 7.16 Index Gastropods of the Aqaban Subprovince, Eritrean Province.
A= *Cypraea pantherina rasnasraniensis* Heiman and Meinis, 1999, length 86 mm, 5 m depth off Sharm El-Sheikh, Sinai Peninsula, Egypt; **B=** *Luria pulchra sinaiensis* Heiman and Meinis, 2000, length 68 mm, 25 m depth off Nebk, Sinai Peninsula, Egypt; **C=** *Lyncina camelopardalis sharmiensis* Heiman and Meinis, 2002, length 76 mm, 8 m depth off Nebk, Egypt; **D=** *Naria turdus pardalina* (Dunker, 1852), length 40 mm, on exposed rock platforms at low tide, Eilat, Israel, Gulf of Aqaba; **E=** *Talostolida teres elatensis* (Heiman and Meinis, 2002), length 37 mm, 10 m depth off Eilat, Israel, Gulf of Aqaba; **F=** *Naria nebrites oblonga* (Heiman, 2002), length 29 mm, 1 m depth off Dahab, Sinai Peninsula, Egypt; **G=** *Erronea caurica nabeqensis* (Heiman and Meinis, 1999), length 41 mm, low tide, Dishet El Dhaba, Hurghada, Egypt; **H=** *Darioconus bratcherae* Petuch and Berschauer, 2019, length 46 mm, 5 m depth in coral rubble, off Eilat, Israel, Gulf of Aqaba; **I=** *Bistolida rubiginosa aegyptica* Lorenz, 2017, length 27.5 mm, in coral rubble, 20 m depth off Abu Ramada Island, Hurghada, Egypt; **J=** *Homalocantha dovpeledi* Houart, 1982, length 49 mm, 40 m depth on a rock cliff, off Taba, Sinai Governorate, Egypt, northern Gulf of Aqaba; **K=** *Homalocantha elatensis* Heiman and Mienis, 2009, length 52 mm, on sand and coral, 25 m depth off Dahab, Sinai Governorate, Egypt, Gulf of Aqaba.

Figure 7.17 Index Gastropods of the Aqaban Subprovince and Dahlakian Infraprovince, Eritrean Province. **(Aqaban Subprovince) A=** *Calamiconus akabensis* (Sowerby III, 1887), length 62 mm, 5 m depth off Eilat, Israel, Gulf of Aqaba; **B=** *Pionoconus elatensis* (Wils, 1971), length 35 mm, 2 m depth in coral rubble, off Eilat, Israel, Gulf of Aqaba; **C=** *Harmoniconus sharmiensis* (Wils, 1986), length 13 mm, low tide, Ras Mohammed, Sinai Peninsula, Egypt; **D=** *Miliariconus sinaiensis* Petuch, Berschauer, and Poremski, 2017, length 30 mm, low tide, Eilat, Israel, Gulf of Aqaba; **E=** *Chicoreus peledi* E. Vokes, 1978, length 85 mm, on a rock cliff in 45 m depth, off Eilat, Israel, Gulf of Aqaba; **F=** *Pseudonebularia dovpeledi* (Turner, 1997), length 24 mm, 20 m depth off Taba, Sinai Governorate, Egypt, northern Gulf of Aqaba; **G=** *Naquetia fosteri* D'Attilio and Herz, 1987, length 74 mm, on an exposed cliff face in 46 m depth, off Dahab, Sinai Governorate, Egypt, Gulf of Aqaba; **(Dahlakian Subprovince) H=** *Chicoreus ethiopius* (E. Vokes, 1978), length 29 mm, on dead coral, 3 m depth off Harmil Island, Dahlak Archipelago, Eritrea; **I=** *Phasmoconus nigromaculatus* (Rockel and Moolenbeek, 1992), length 20 mm, in sand between coral, 5 m depth off Harmil Island, Dahlak Archipelago, Eritrea; **J=** *Cylinder neovicarius dahlakensis* (da Motta, 1982), length 66 mm, 3 m depth off Harmil Island, Dahlak Archipelago, Eritrea; **K=** *Gastridium fragilissimum* (Petuch, 1979), length 30 mm, 2 m depth off Harmil Island, Dahlak Archipelago, Eritrea; **L=** *Naria turdus dilatata* (Dunker, 1852), length 43 mm, 1 m depth on reef off Harmil Island, Dahlak Archipelago, Eritrea.

Figure 7.18 Index Gastropods of the Omanian Subprovince, Eritrean Province.
A= *Barycypraea teulerei* (Cazenavette, 1846), length 50 mm, low tide, Sur Masirah, Al Masirah Island, Oman; **B**= *Persicula masirana* Roth and Petit, 1972, length 10 mm, 3 m depth off Mirbat, Al Masirah Island, Oman; **C**= *Ranularia boschi* (Abbott and Lewis, 1970), length 70 mm, 20 m depth off Al Ashkharah, Oman; **D**= *Murex echinodes* Houart, 2011, length 123 mm, 20 m depth off Dafiyat, Al Masirah Island, Oman; **E**= *Pygmaepterys paulboschi* (Smythe and Houart, 1984), length 14 mm, found dead in shell rubble, Dafiyat, Al Masirah Island, Oman; **F**= *Fusiconus stocki* (Coomans and Moolenbeek, 1990), length 30 mm, trawled by fishermen from 20 m depth off Al Ashkharah, Oman; **G**= *Quasiconus melvilli* (Sowerby III, 1879), length 23 mm, found at 1 m depth off Sur Masirah, Al Masirah Island, Oman; **H**= *Rhizoconus ardisiaceus* (Kiener, 1850), length 31 mm, low tide, Dafiyat, Al Masirah Island, Oman; **I**= *Pionoconus koukae* Monnier, Limpalaër, and Robin, 2013, length 50 mm, 1 m depth off Dafiyat, Al Masirah Island, Oman; **J**= *Hexaplex rileyi* (Attilio and Myers, 1984), length 56 mm, on rocks at low tide, Al Jubail Island, Abu Dhabi, United Arab Emirates; **K**= *Monetaria annulus dilatissima* Lorenz, 2017, length 30 mm, on rocks at low tide, Sur Masirah, Al Masirah Island, Oman; **L**= *Darioconus laueri* Monnier and Limpalaër, 2013, length 42 mm, dark color form, 2 m depth off Az Zafaram, Sohar, Oman, Gulf of Oman (endemic to the Persian Gulf and Gulf of Oman).

Figure 7.19 Index Gastropods of the Omanian Subprovince and Masiran Infraprovince, Eritrean Province.
A= *Naria turdus winkworthi* (Schilder and Schilder, 1938), length 30 mm, low tide, Sur Masirah, Al Masirah Island, Oman; **B=** *Hexaplex kuesterianus blazeki* Cossignani, 2017, length 57 mm, on rocks at low tide, Khorfakkan, Fujairah, United Arab Emirates; **C=** *Cantharus vezzarochristofei* Cossignani, 2017, length 32 mm, low tide, Al Khaluf, Oman; **D=** *Volvarina arabica* Boyer, 2012, length 19 mm, low tide in sand, Sur Masirah, Al Masirah Island, Oman; **E=** *Ancilla boschi* Kilburn, 1980, length 22 mm, Al Khaluf, Al Masirah Island, Oman; **F=** *Turbo (Aspilaturbo) jonathani* Dekker, Moolenbeek, and Dance, 1992, width 25 mm, under rocks at low tide, Sur Masirah, Al Masirah Island, Oman; (**Masiran Infraprovince) G=** *Conomurex coniformis masirensis* Moolenbeek and Dekkers, 1993, length 48 mm, low tide, Sur Masirah, Al Masirah Island, Oman; **H=** *Sinistralia gallagheri* (Smythe and Chatfield, 1981), length 18 mm, 20 m depth off Al Ashkharah, Oman; **I=** *Punctacteon eloisae* Abbott, 1973, length 29 mm, low tide in sand, Sur Masirah, Al Masirah Island, Oman; **J=** *Erronea (Adusta) persica dilatata* Bergonzoni, 2014, length 42 mm, 2 m depth off Mirbat, Al Masirah Island, Oman; **K=** *Erronea caurica masirensis* Bergonzoni, 2018, length 44 mm, low tide, Sur Masirah, Al Masirah Island, Oman.

Figure 7.20 Index Gastropods of the Somalian Subprovince, Eritrean Province.
A, B= *Tricornis oldi* (Emerson, 1965), length 111 mm, trawled from 25 m depth off Berbera, Somalia; **C=** *Hexaplex bozzadamii* (Franchi, 1990), length 58 mm, trawled from 20 m depth off El Hur, Somalia; **D=** *Sinistralia somaliensis* (Smythe and Chatfield, 1984), length 23 mm, 30 m depth, Adale, Somalia; **E=** *Phasmoconus salzmanni* (Raybaudi and Rolan, 1997), length 34 mm, trawled from 100 m depth off Berbera, Somalia; **F=** *Glabella mirabilis* (Adams, 1869), length 32 mm, trawled from 50 m depth off Ras Hafun, Somalia; **G=** *Conasprella bozzettii* (Lauer, 1991), length 52 mm, trawled from 200 m depth off Mogadishu, Somalia; **H=** *Callistocypraea broderipii somalica* (Raybaudi Massilia, 1981), length 78 mm, 150 m depth off Ras Hafun, Somalia; **I=** *Calamiconus subroseus* (Rockel and Korn, 1992), length 33 mm, trawled from 100 m depth off Mogadishu, Somalia; **J=** *Strategiconus splendidulus* form *anadema* (Tomlin, 1937), length 40 mm, in sand and coral rubble, 2 m depth off Xaafuun, northern Somalia; **K=** *Naria marginalis pseudocellata* (Schilder and Schilder, 1938), length 29 mm, under dead coral, 2 m depth off Merca, Somalia.

Figure 7.21 Index Gastropods of the Somalian Subprovince, Eritrean Province.
A= *Festilyria festiva deceptrix* (Palazzi, 1981), length 123 mm, trawled from 100 m depth off Berbera, Somalia (also found in the Omanian Subprovince); **B=** *Cypraea pantherina catulus* Schilder, 1924, length 74 mm, under coral, 10 m depth off Berbera, Somalia; **C, D=** *Naria marginalis quadriangula* Bozzetti, 2011, length 24 mm, 20 m depth off Merca, Somalia; **E=** *Darioconus echo* (Lauer, 1988), length 74 mm, trawled from 30 m depth south of Mogadishu, Somalia; **F=** *Glabella obtusa* (Sowerby II, 1846), length 19 mm, trawled from 100 m depth off Berbera, Somalia; **G=** *Phasmoconus angioiorum* (Rockel and Moolenbeek, 1992), length 37 mm, trawled from 50 m depth off Adale, Somalia; **H=** *Profundiconus neotorquatus* (da Motta, 1985), length 81 mm, 200 m depth off Mogadishu, Somalia; **I=** *Strategoconus splendidulus* (Sowerby I, 1833), length 57 mm, 10 m depth off Bir Firqua, Little Aden, Aden; **J, K=** *Graphiconus adenensis* (E.A. Smith, 1891), length 60 mm, 50 m depth, Berbera, Somalia.

Some interesting endemic gastropods of the Indian Tropical Region: *Graphiconus primus* (left), *Volegalea dirki* (top), *Turbinella pyrum* (right), and *Turbo lajonkairi* (bottom).

CHAPTER 8. Australian Super-Region and the North Australian Tropical Region

Of the two oceanic super-regions, the Australian Super-Region is the smaller, covering only around one-tenth the surface area of the surrounding Indo-Pacific Super-Region. Although comparatively small, the super-region is oceanographically complex, encompassing the entire coastline of the Australian continent and the island of Tasmania, along with the southern half of the Timor and Arafura Seas, the Gulf of Carpentaria, the western half of the Coral and Tasman Seas, the Great Australian Bight, and the far-eastern edge of the Indian Ocean. Within this circumcontinental biotic area, two regions occur; the **North Australian Tropical Region** and the **South Australian Paratropical Region**. These two regions, together, encompass five distinct provinces, including the eutropical **Solanderian** and **Dampierian Provinces**, and the paratropical **Peronian**, **Maugean**, and **Flindersian Provinces**. Although the Dampierian and Solanderian Provinces of the northern Australian coast share numerous gastropod taxa with the neighboring eutropical Indo-Malaysian Province, they also contain unique Australian taxa that occur all around the continent. Although no species-level taxa are ubiquitous to all five provinces, several widespread genera are found around the entire continent and these can be used to define the biogeographic limits of the Australian Super-Region. Some of these super-regional

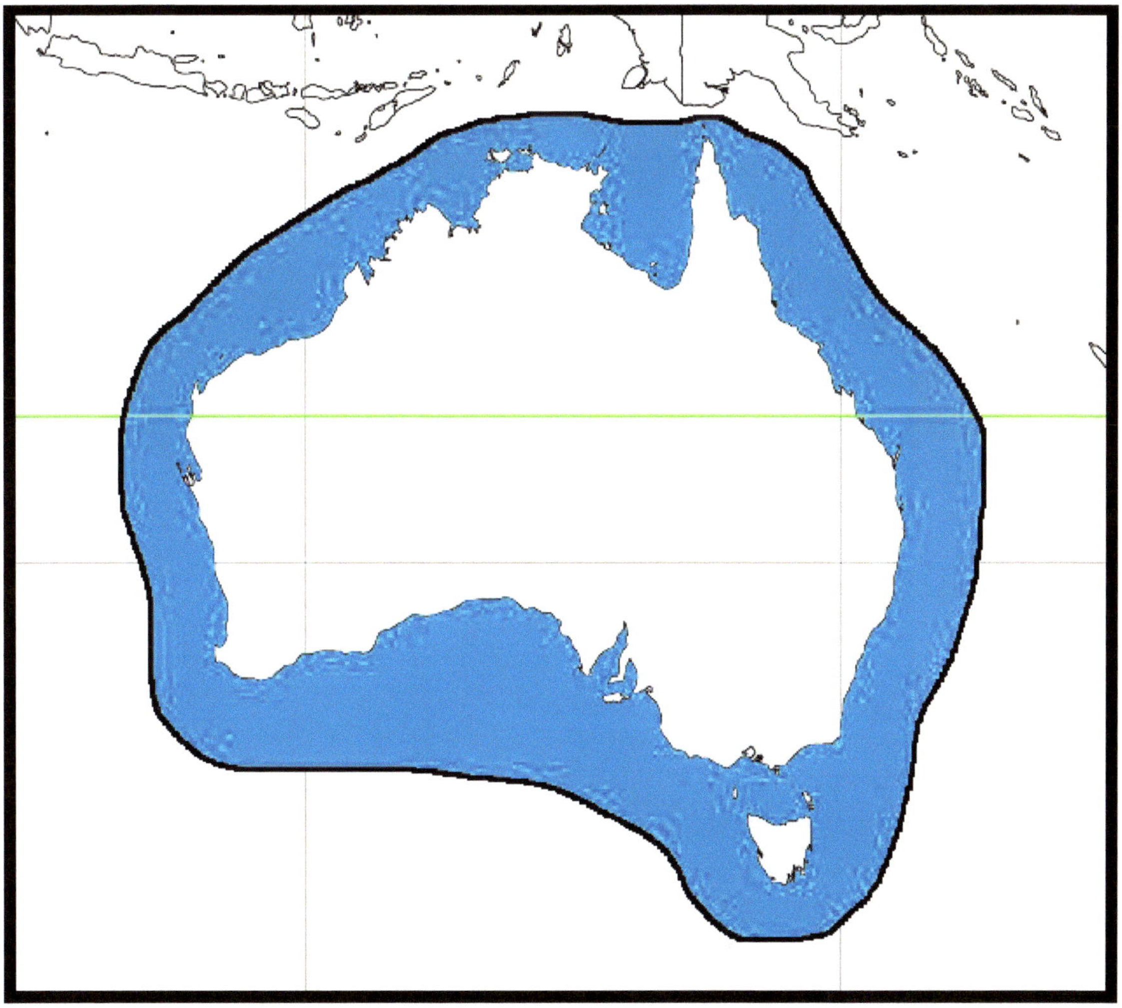

Figure 8.1. Map of the **Australian Super-Region** (blue) showing its areal extent.

index genera, with representative species in all five provinces, include the volutids *Amoria*, *Notovoluta*, and *Livonia*, the harpid *Austroharpa*, and the conid *Floraconus*. Many other endemic genera are found in only three or four of the five Australian provinces, such as the cypraeids *Umbilia*, *Zoila*, *Austrocypraea*, and *Notocypraea*, the volutid *Volutoconus*, and the cone shells *Eremiconus*, *Parviconus*, and *Austroconus*. Representative species of all of these wide-ranging Australian genera are illustrated here on Figures 8.5-8.16 and 9.5-9.11. As a tradition among Australian biogeographers, the molluscan provinces of Australia have been named after people and not geographical locations like the other provinces from around the world (see Waters *et al*., 2010 and Ebach, 2012 for reviews of the nomenclatural history).

North Australian Tropical Region

Extending from Shark Bay, Western Australia in the west to Cape Moreton, Queensland in the east, the Northern Australian Tropical Region includes all of the eutropical marine environments on the Australian continent. Included in this wide tropical region are the stromatolites of Shark Bay, the extensive mangrove jungles of Northern Territory, and the carbonate environments of the Great Barrier Reef. The North Australian Tropical Region contains only two molluscan provinces, the western

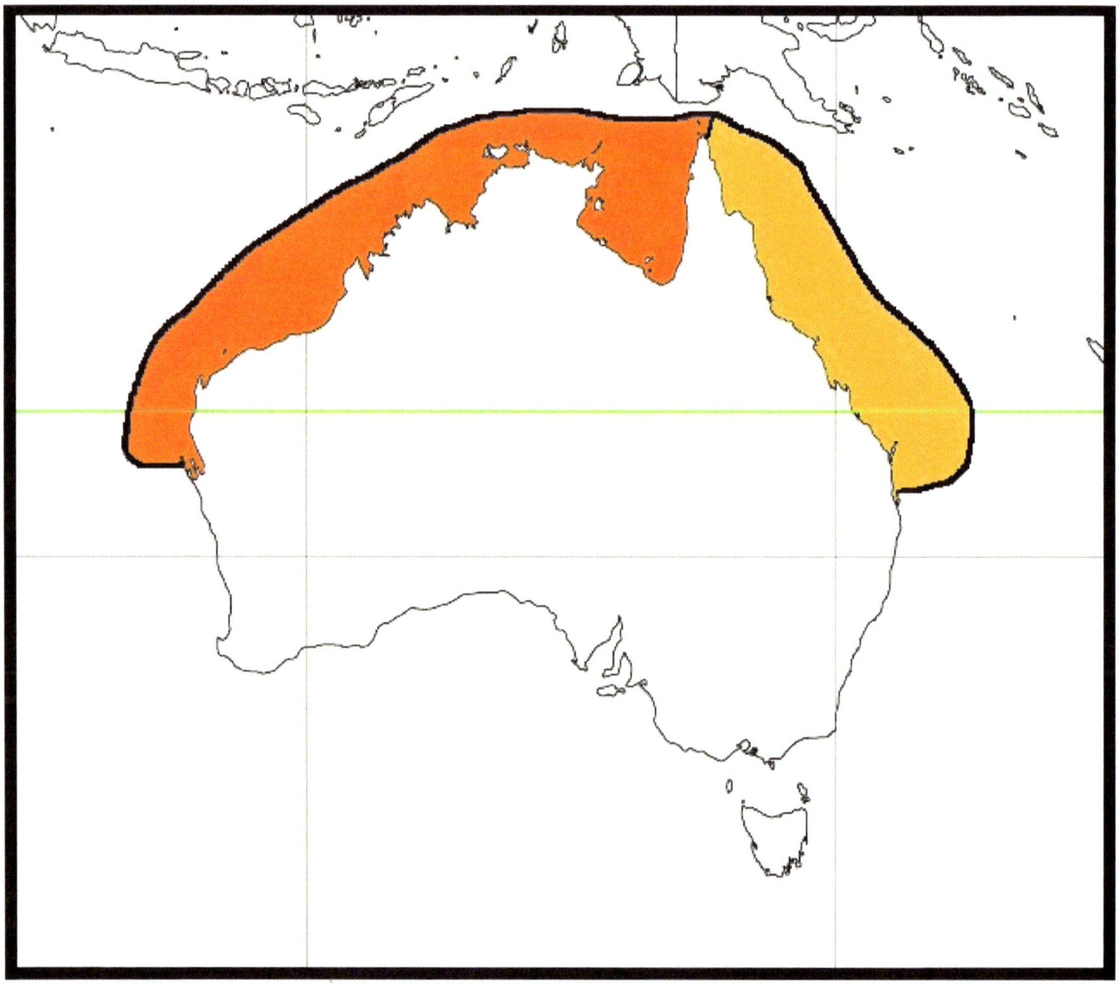

Figure 8.2 Map of the **North Australian Tropical Region**, showing the areal extent of its provinces: the **Solanderian Province** (gold) and the **Dampierian Province** (orange).

Dampierian Province of Western Australia State and the Northern Territory, and the eastern **Solanderian Province** of Queensland State. A large number of wide-ranging gastropods are found in both provinces and their ranges can be used to demarcate the boundaries of the region. Some of these North Australian Tropical Region index taxa include the turbinid *Turbo (Marmarostoma) perspeciosus*, the strombid *Doxander campbelli*, the cypraeids *Erronea (Solvadusta) subviridis dorsalis*, and *Purpuradusta macula*, the muricid *Chicoreus cervicornis*, the volutids *Amoria damoni*, *Amoria praetexta*, *Amoria turneri*, and *Cymbiola sofia*, the olivids *Proxoliva brettinghami* and *Viduoliva westralis*, and the conid *Plicaustraconus trigonus* (all shown here on Figure 8.5).

Solanderian Molluscan Province

Named for Daniel Solander, the student of Carolus Linnaeus who accompanied Captain James Cook on his explorations of Australia, the Solanderian Province extends from the central islands of the Torres Strait of northernmost Queensland, south to Tweed Heads, northernmost New South Wales. The entire province is dominated by the Great Barrier Reef and its associated island groups, but also

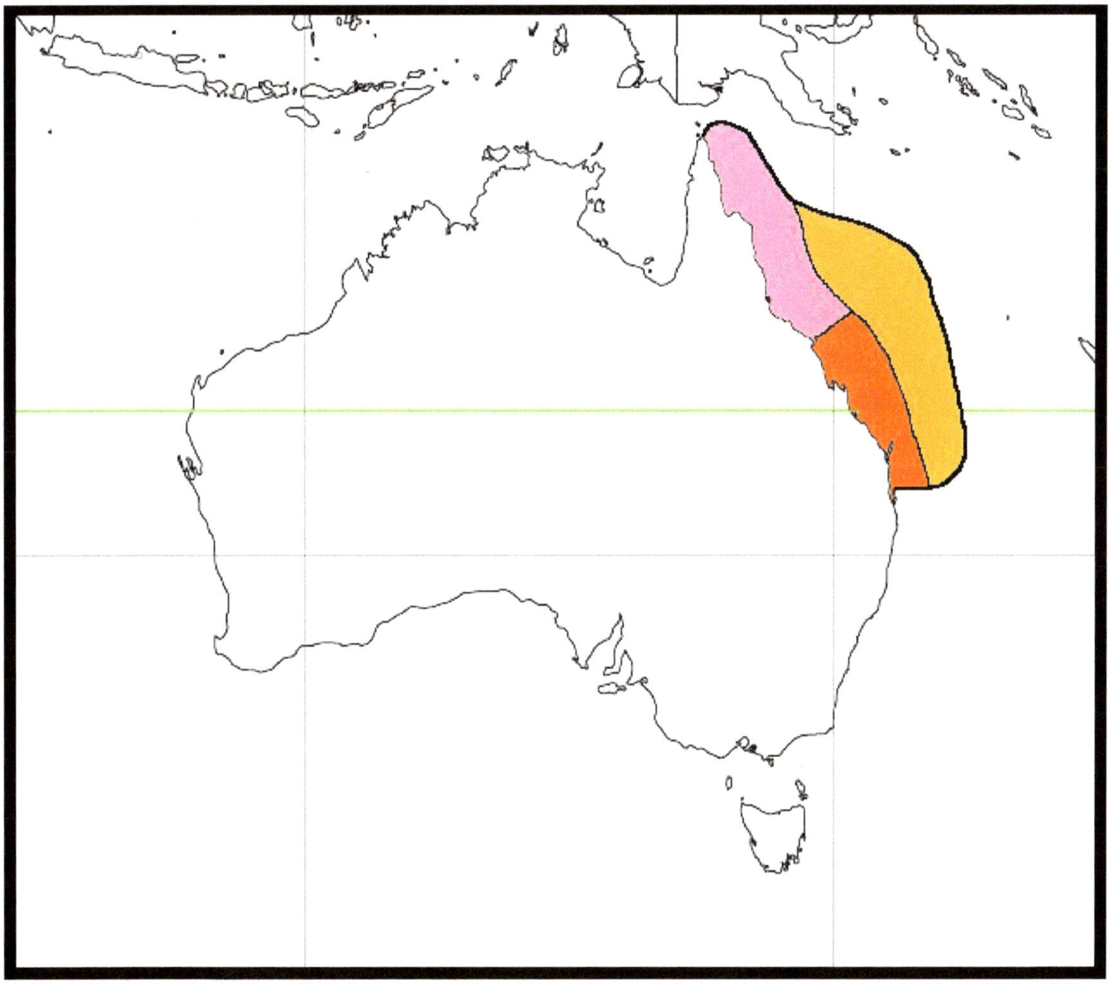

Figure 8.3 Map of the Solanderian Molluscan Province, showing the areal extents of its subprovinces: the **Moretonian Subprovince** (orange), the **Cairnsian Subprovince** (light rose), and the **Coralian Subprovince** (gold).

encompasses the myriad of isolated coral reefs and islands of the western Coral Sea east of the Great Barrier Reef. Because of its broad latitudinal distribution, from high eutropical areas to those that are only marginally-eutropical, the Solanderian can be divided into three distinct subprovinces, which include the **Moretonian Subprovince**, the **Cairnsian Subprovince**, and the **Coralian Subprovince.** A large number of widespread endemic taxa can be found throughout the entire province and these can be used to define the provincial boundaries. Some of the more important Solanderian index taxa include the cowries *Erronea xanthodon* and *Bistolida stolida lorrainae*, the volutes *Amoria guttata*, *Cymbiolacca pulchra*, *Cymbiolista hunteri*, and *Volutoconus grossi*, and the cone shells *Eremiconus limpusi*, *Kioconus (Ongoconus) nielsenae*, and *Leptoconus ammiralis temnes* (all shown here on Figure 8.6).

Moretonian Subprovince

The southern part of the Solanderian Province, extending from Yeppoon and the Keppel Islands, Queensland in the north to Tweed Heads, New South Wales in the south, is here referred to as the Moretonian Subprovince. Named for Moreton Island near Brisbane, the coastal waters of this subprovince are only marginally eutropical and are often cooled below tropical temperatures during the cold winter months. The area is also dominated by wide sandy beaches, scattered rocky headlands, and heavy surf conditions (as at Surfers Paradise on the Gold Coast), and lacks the extensive carbonate and coral reef environments seen farther north along the Queensland coast. Because of these different ecological parameters, the Moretonian Subprovince has evolved its own characteristic gastropod fauna, with a very high percentage of endemism, often exceeding 30%. Some of the more important Moretonian endemics include the cypraeids *Umbilia oriettae*, *Umbilia petilirostris*, *Paradusta coucomi*, and *Austrasiatica langfordi moretonensis*, the volutids *Cymbiolacca pulchra flindersi*, *Cymbiolacca complexa fraserensis*, *Cymbiolacca pulchra subelongata*, and *Cymbiolacca complexa moretonensis*, and the cone shells *Floraconus rufimaculosus*, *Tesselliconus devorsinei*, *Plicaustraconus wallangra*, and *Endemoconus sculletti* (all shown here on Figures 8.6 and 8.7). Many of these index species, especially the cowries, are confined to the deep water areas in Moreton Bay and off Moreton Island.

Cairnsian Subprovince

The northern half of the Solanderian Province, extending from the Torres Strait on the Papua-New Guinea-Queensland border south to Yeppoon, southern Queensland, is here referred to as the Cairnsian Subprovince (named for Cairns, Queensland, near the center of this biotic unit). Encompassing the entire Great Barrier Reef system of coral islands and reef complexes, this eutropical area houses the most species-rich molluscan communities found anywhere within the entire Australian Super-Region. With over 2,900 separate reefs and over 900 coral islands, and extending for over 2,300 km, the Great Barrier Reef is the world's largest coral reef system and houses a malacofauna with a very high level of endemicity, in some places approaching 40%. The reef tracts of the Great Barrier Reef are often widely-separated and have deep intervening channels, effectively creating barriers to dispersal for shallow water nonvagile groups with direct development. This allopatric pattern has resulted in the evolution of large species complexes of gastropods, with individual species being restricted to discrete island chains, reef systems, or within the deep channels themselves.

One of the most characteristic and impressive endemic species radiations from the Cairnsian Subprovince is seen in the volutid genus *Cymbiolacca*, where over 12 species and subspecies have been described from different parts of the Great Barrier Reef. Some of these include *Cymbiolacca cracenta* (Lady Musgrave Island), *C. houarti* (Fitzroy Reef), *C. persticta* (Sandy Cay, Swain Reefs) *C. wisemani* (Michaelmas Reef), *C. wisemani randalli* (Keeper Reef), *C. excelsior* (Departure Reef), *C. pulchra woolacottae* (Heron Island, Capricorn Group), and *C. coucomorum* (Swain Reefs) (all illustrated here on Figures 8.8 and 8.9). These endemic Cairnsian volutes occur along with a large number of other endemic taxa, some of which include the cowries *Bistolida brevidentata fluctuans*, *Erronea (Ipserronea) carnicolor*, *Bistolida brevidentata deceptor* (Torres Strait Islands), *Cribrarula melwardi*, and *Umbilia capricornica* (in deep channels), the volutes *Amoria volva*, *A. spenceriana*, *A. canaliculata*, *A. lineola*, *A. molleri capricornica*, and *A. molleri reducta* (both subspecies in deep channels), *Volutoconus multiformis*, *Nannamoria inflata*, and *Nannamoria gotoi*, the cone shells *Lividiconus biliosus imperator*, *Eremiconus colmani*, *Kioconus whiteheadae*, and *Kioconus queenslandicus* (all shown here on Figures 8.8, 8.9, and 8.10).

Coralian Subprovince

Named for the Coral Sea, the Coralian Subprovince encompasses the hundreds of cays and isolated coral reefs that are found in the open ocean east of the Great Barrier Reef. Extending from the latitude of Port Douglas to the latitude of Bundaberg, these Coral Sea reef systems parallel the Great Barrier Reef and can be as much as 400-600 km from the Queensland shoreline. Some of the largest and best-developed of these western Coral Sea reefs include Diamond Reef, Lihou Reef, Osprey Reef, Tregrosse Islets, Marion Reefs, Magdelaine Cays, Willis Cays, Coringa Islets, Moore Reefs, and Flinders Reefs. Although there is a large percentage of faunal overlap with the neighboring Cairnsian Subprovince, the Coralian Subprovince still contains a unique fauna with around 25% endemism. Some of the more important Coralian endemics include the harpid *Harpa queenslandica* and the volutids *Cymbiolacca* (*Magnavictoria*) *perplicata* and *Cymbiolacca intruderi* (shown here on Figure 8.10). Of special interest within the Coralian Subprovince is the isolated Lord Howe Island, which lies over 600 km due east of Port Macquarie, New South Wales, at the northern end of the Tasman Sea. Containing an impoverished gastropod fauna that has its closest affinities with the Coralian Subprovince, Lord Howe Island is here considered to be an extreme outlier of the Solanderian Province. Since it houses several endemic species, the island also can be considered to be an evolutionary hot spot, which is referred to here as the **Howeian Infraprovince**. Some of the Howeian endemic gastropods include the small abalone *Haliotis rubiginosa*, the turbinid *Turbo* (*Marmarostoma*) *cepoides*, and the volutid *Lyria howensis* (all illustrated here on

Figure 8.10). This last-mentioned small volute is closest to *Lyria deliciosa* from the Coral Sea Islands and New Caledonia, showing that the Howeian molluscan fauna is derived from the malacofaunas of those areas.

Dampierian Molluscan Province

Named for William Dampier, a 17th century English navigator and privateer who explored the Australian coast, the Dampierian Province extends from Shark Bay, Western Australia to the western side of the Torres Strait, Queensland. Within this immense expanse of coastline, the province encompasses the coral reefs complexes of Barrow, Cunningham, and Bedwell Islands and the Ashmore Reefs, the extensive mangrove jungles of the northern coast of Western Australia, the Arnhemland Peninsula of Northern Territory, and the entire Gulf of Carpentaria. Since this eutropical province spans four separate oceanographic systems, the Indian Ocean, the Timor Sea, the Arafura Sea, and the Gulf of Carpentaria, the Dampierian molluscan

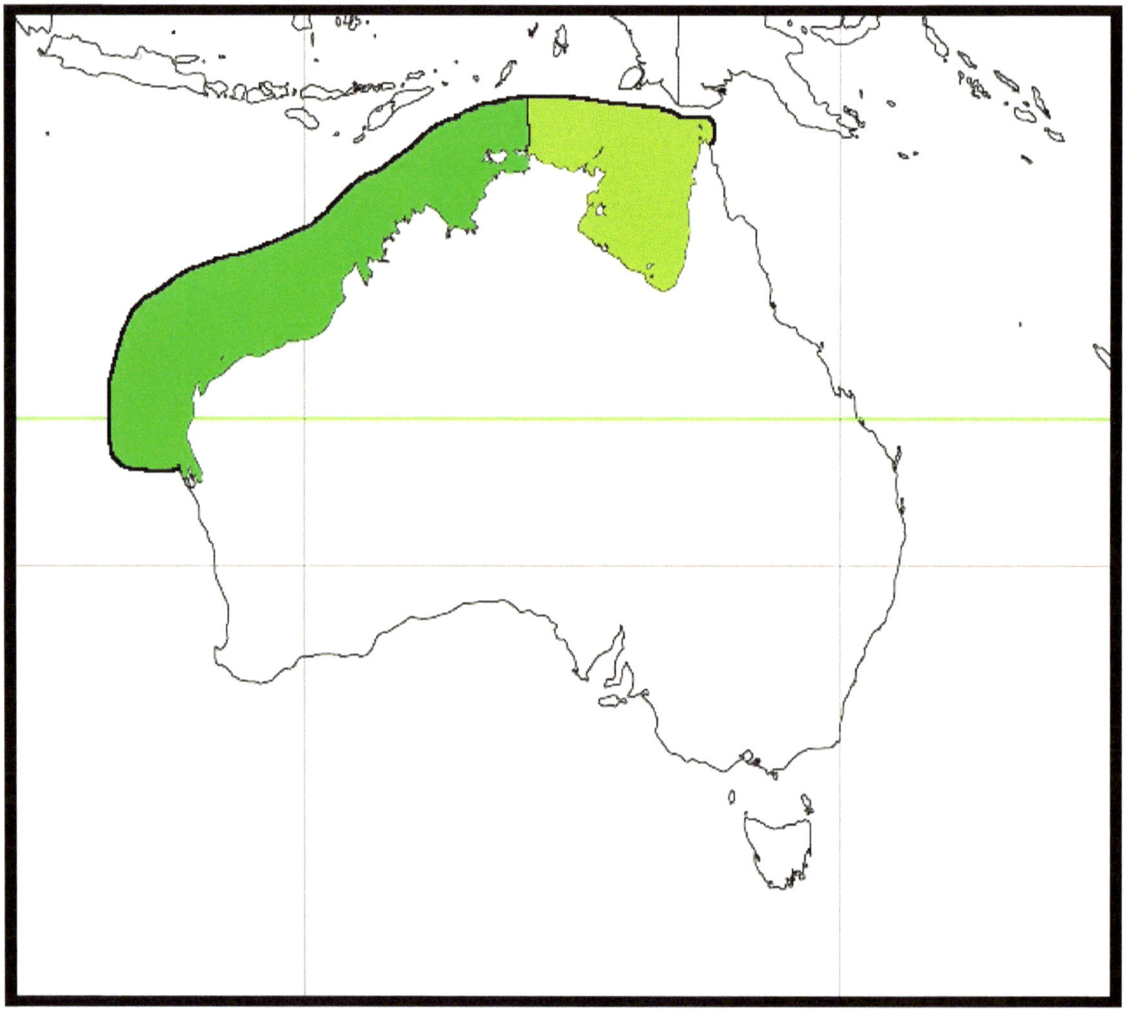

Figure 8.4 Map of the Dampierian Molluscan Province, showing the areal extent of its two subprovinces: the **Exmouthian Subprovince** (green), and the **Carpentarian Subprovince** (lime green).

fauna is very species-rich and can be divided into two large subprovinces; the **Exmouthian** in the west and the **Carpentarian** in the east. A large number of gastropods are found throughout the entire province and these widespread Dampierian endemics can be used to define the biogeographical boundaries. Some of the more important of these include the trochid *Austrocochlea zeus*, the cypraeid *Notocypraea pulicaria*, the fasciolariid *Marmarofusus tessellatus*, the muricid *Chicoreus stainforthi*, the volutes *Amoria ellioti*, *Amoria jamrachi*, *Amoria grayi*, *Cymbiola nivosa*, and *Cymbiola oblita*, the olivids *Miniaceoliva ornata* and *M. caerulea ponderi*, and the conids *Cylinder victoriae* and *Virroconus doreensis* (some shown here on Figure 8.11).

Exmouthian Subprovince

Named for the Exmouth Gulf of Western Australia, the Exmouthian Subprovince extends from Shark Bay north to the York Sound and Kimberley Region of northernmost Western Australia and comprises the open Indian Ocean component of the Dampierian Province. With its sandy beaches and extensive sand flats, offshore coral reefs, and mangrove jungles, the Exmouthian Subprovince supports a rich molluscan fauna that contains a high level of endemism. The geographical isolation of Western Australia from the other tropical areas of the continent has led to the evolution of many regional subspecies, many of which are confined to geographically-restricted localities. Some of these classic Exmouthian endemic gastropod species and subspecies include the cowries *Cribrarula exmouthensis*, *Cribrarula exmouthensis magnifica*, *Cribrarula abaliena australiensis*, *Erronea (Ipserronea) garyi*, *Erronea (Ipserronea) smithi*, and *Leporicypraea geographica rovae*, the strombid *Euprotomus iredalei*, the muricid *Timbellus bednalli*, the olivid *Acutoliva kurzi*, the volutids *Amoria dampieria*, *Amoria macandrewi*, *Volutoconus hargreavesi* and *Volutoconus coniformis*, and the cone shells *Eremiconus dampierensis*, *Floraconus novaehollandiae*, *Pionoconus barbara*, and *Calamiconus garywilsoni* (all shown here on Figures 8.12 and 8.13). Of special interest within the Exmouthian Subprovince is a species radiation of cowries in the genus *Zoila*, representing the northernmost members of this normally cold water cowrie group (see Chapter 9 for illustrations of the majority of the species from southwestern Australia). Some of these, which range only as far as the Kimberley Region and its offshore banks, include *Zoila perlae*, *Z. jeaniana aurata*, *Z. jeaniana inscripta*, *Z. eludens*, *Z. decipiens*, *Z. decipiens suprasinum*, and *Z. marginata ketyana* (some shown here on Figure 8.14). At least 12 species and subspecies of *Zoila* are known from the Exmouthian area. Also of interest within the subprovince is Ashmore Reef, an isolated coral bank located between southern Timor, Indonesia and the Kimberley Region. Recent studies have shown that this small coral bank houses an evolutionary hot spot, which is here referred to as the **Ashmorean Infraprovince**. Several distinctive endemic taxa have evolved there, with the best known being the giant volutid *Melo ashmorensis* and the cone shell *Pionoconus morrisoni*.

Carpentarian Subprovince

Named for the Gulf of Carpentaria of northeastern Northern Territory and northwestern Queensland, the Carpentarian Subprovince contains the least-studied molluscan fauna in all of Australia and is also the most unexplored area in the Australian Super-Region. Extending from the York Sound, Western Australia to the Torres Strait of Queensland, the subprovince also encompasses the Joseph Bonaparte Gulf, Melville Island, Groote Eylandt, Arnhemland, and the Pellew and Wellesley Islands of the southern Gulf. Besides the city of Darwin, the entire coastline is essentially an Aboriginal Preserve and is either sparsely populated or uninhabited. The small amount of data that has been gleaned on the malacology of this vast area has come from collections taken by shrimp boats out of Thursday Island, Torres Strait, or from local collectors around Darwin and the Van Dieman and Beagle Gulfs. As currently understood, the Carpentarian fauna is quite different from the neighboring Exmouthian and Cairnsian Subprovinces and exhibits a high level of endemism. Some of the more important subprovincial index taxa include the cypraeids *Bistolida brevidentata*, the ancillariid *Amalda herlaari vernedei*, the volutes *Volutoconus bednalli*, *Amoria keatsiana*, *Amoria newmanae*, *Cymbiola flavicans*, and *Calliotectum dupreyae arafurensis*, and the conids *Isoconus austroviola*, *Pionoconus arafurensis*, *Pionoconus vinctus*, and *Plicaustraconus adami* (all illustrated here on Figure 8.13).

ICONOGRAPHY OF GASTROPODS OF THE AUSTRALIAN SUPER-REGION AND THE NORTH AUSTRALIAN TROPICAL REGION

(Principal Index Gastropods are shown on Figures 8.5 to 8.14)

Figure 8.5 Widespread Gastropods of the North Australian Tropical Region (Australian Super-Region), occurring in the Solanderian and Dampierian Provinces.
A= *Erronea (Solvadusta) subviridis dorsalis* Schilder and Schilder, 1938, length 33 mm, 15 m depth off the Wellesley Islands, Queensland, Gulf of Carpentaria, Australia; **B**= *Purpuradusta macula* (Angas, 1867), length 11 mm, low tide off Moa (Banks) Island, Queensland, Torres Straits, Australia; **C**= *Doxander campbelli* (Griffith and Pidgeon, 1834), length 57 mm, low tide, Prince of Wales Island, Queensland, Torres Straits, Australia; **D**= *Chicoreus cervicornis* (Lamarck, 1822), length 45 mm, 15 m depth off the Wellesley Islands, Queensland, Gulf of Carpentaria, Australia; **E**= *Amoria damoni* Gray, 1864, length 83 mm, low tide, Broome, Western Australia; **F**= *Amoria praetexta* (Reeve, 1849), length 38 mm, low tide, Rosemary Island, Dampier Archipelago, Western Australia; **G**= *Amoria turneri* (Gray, 1834), length 63 mm, 25 m depth off Port Darwin, Northern Territory, Australia; **H**= *Cymbiola sofia* (Gray, 1846), length 68 mm, 25 m depth off Groote Eylandt, Northern Territory, Gulf of Carpentaria, Australia; **I**= *Proxoliva brettinghami* (Bridgman, 1909), length 10 mm, low tide, Broome, Western Australia; **J**= *Viduoliva westralis* (Petuch and Sargent, 1986), length 50 mm, low tide, Broome, Western Australia; **K**= *Plicaustraconus trigonus* (Reeve, 1848), length 49 mm, low tide off the Maitland River mouth, Karratha, Western Australia, Australia (also found in northern Queensland); **L**= *Turbo (Marmarostoma) perspeciosus* (Iredale, 1929), width 32 mm, on coral at low tide, Lamont Reef, Capricorn Group, Great Barrier Reef, Queensland, Australia.

Figure 8.6 Widespread Index Gastropods of the Solanderian Province and the Moretonian Subprovince.
A= *Erronea xanthodon* (Sowerby II, 1832), length 27 mm, low tide, Shoal Point, Mackay, Queensland, Australia (also northern New South Wales); **B=** *Bistolida stolida lorrainae* Lorenz, 2017, length 28.5 mm, 5 m depth off Thursday Island, Queensland, Australia; **C=** *Amoria guttata* McMichael, 1964, length 52 mm, 100 m depth off Cairns, Queensland, Australia; **D=** *Cymbiolacca pulchra* (Sowerby I, 1825), length 70 mm, 50 m depth off Bustard Head, Queensland, Australia; **E=** *Cymbiolista hunteri* (Iredale, 1931), length 116 mm, 70 m depth off Brunswick Heads, northern New South Wales, Australia; **F=** *Volutoconus grossi* (Iredale, 1927), length 87 mm, 50 m depth off Tin Can Bay, Queensland, Australia; **G=** *Eremiconus limpusi* (Rockel and Korn, 1990), length 35 mm, 100 m depth in the Capricorn Channel, Queensland, Australia; **H=** *Kioconus (Ongoconus) nielsenae* (Marsh, 1962), length 45 mm, 50 m depth in Tin Can Bay, Queensland, Australia; **I=** *Leptoconus ammiralis temnes* (Iredale, 1930), length 75 mm, 100 m depth off Cape Moreton, Queensland, Australia; **(Moretonian Subprovince) J=** *Plicaustraconus advertex* (Garrard, 1961), length 32 mm, 200 m depth off Moreton Island, Cape Moreton, Queensland, Australia; **K=** *Endemoconus sculletti* (Marsh, 1962), length 42 mm, 100 m depth in the Capricorn Channel, Queensland, Australia.

Figure 8.7 Index Gastropods of the Moretonian Subprovince, Solanderian Province.
A= *Austrasiatica langfordi moretonensis* (Schilder, 1965), length 63 mm, 250 m depth off Cape Moreton, Queensland, Australia; **B**= *Umbilia oriettae* Lorenz and Massiglia, 2005, length 86 mm, 150 m depth off Cape Moreton, Queensland, Australia; **C**= *Umbilia petilirostris* Darragh, 2002, length 76 mm, 250 m depth off the Capricorn Islands, Great Barrier Reef, Queensland, Australia; **D**= *Cymbiolacca pulchra flindersi* Bail and Limpus, 2017, length 70 mm, 30 m depth in the Capricorn Channel, Great Barrier Reef, Queensland, Australia; **E**= *Cymbiolacca complexa fraserensis* Bail and Limpus, 1998, length 67 mm, 50 m depth off Fraser Island, Queensland, Australia; **F**= *Cymbiolacca pulchra subelongata* Bail and Limpus, 2013, length 81 mm, 70 m depth off Lady Musgrave Island, Queensland, Australia; **G**= *Cymbiolacca complexa moretonensis* (Bail and Limpus, 1998), length 57 mm, 150 m depth off Cape Moreton, Queensland, Australia; **H**= *Paradusta coucomi* (Schilder, 1964), length 37 mm, 200 m depth off Moreton Island, Cape Moreton, Queensland, Australia; **I**= *Floraconus rufimaculosus* (Macpherson, 1959), length 43 mm, 100 m depth off Foster, New South Wales, Australia; **J**= *Plicaustraconus wallangra* Garrard, 1961), length 35 mm, 100 m depth off Moreton Island, Queensland, Australia; **K**= *Tesselliconus devorsinei* Petuch, Berschauer, and Poremski, 2015, length 35 mm, 10 m depth, Mooloolaba, Queensland, Australia.

Figure 8.8 Index gastropods of the Cairnsian Subprovince, Solanderian Province.
A= *Bistolida brevidentata fluctuans* (Iredale, 1935), length 25 mm, Swain Reefs, Great Barrier Reef, Queensland, Australia; **B=** *Cribrarula melwardi* (Iredale, 1930), length 19 mm, Heron Island, Capricorn Group, Great Barrier Reef, Queensland, Australia; **C=** *Bistolida brevidentata deceptor* (Iredale, 1935), length 28 mm, under algae and rocks at low tide, Thursday Island, Torres Strait, Queensland, Australia; **D=** *Amoria volva* (Gmelin, 1791), length 59 mm, Rudder Reef, Great Barrier Reef, Queensland, Australia; **E=** *Amoria canaliculata* (McCoy, 1869), length 41 mm, 40 m depth off Lady Musgrave Island, Queensland, Australia; **F=** *Amoria lineola* Bail and Limpus, 2009, length 52 mm, 100 m depth off Swain Reefs, Great Barrier Reef, Queensland, Australia; **G=** *Cymbiolacca cracenta* McMichael, 1963, length 64 mm, 20 m depth off Lady Musgrave Island, Queensland, Australia; **H=** *Cymbiolacca houarti* (Poppe, 1985), length 73 mm, 100 m depth off Fitzroy Reef, Capricorn Group, Great Barrier Reef, Queensland, Australia; **I=** *Cymbiolacca peristicta* McMichael, 1963, length 56 mm, 5 m depth off Sandy Cay, Swain Reefs, Great Barrier Reef, Queensland, Australia; **J=** *Amoria spenceriana* (Gatliff, 1908), length 75 mm, 50 m depth east of Ashmore Banks, Great Barrier Reef, Queensland, Australia; **K=** *Cymbiolacca wisemani* (Brazier, 1870), length 65 mm, 20 m depth off Michaelmas Reef, Great Barrier Reef, Queensland, Australia.

Figure 8.9 Index Gastropods of the Cairnsian Subprovince, Solanderian Province.
A= *Umbilia capricornica* Lorenz, 1989, length 65 mm, 90 m depth off Swain Reefs, Great Barrier Reef, Queensland, Australia; **B**= *Nannamoria inflata* Bail and Limpus, 2008, length 51 mm, 250 m depth between Bowen and Townsville, Queensland, Australia; **C**= *Amoria molleri capricornica* Bail and Limpus, 2001, length 86 mm, 200 m depth off Lady Musgrave Island, Capricorn Group, Great Barrier Reef, Queensland, Australia; **D**= *Cymbiolacca wisemani randalli* (Stokes, 1961), length 62 mm, 5 m depth off Keeper Reef, Great Barrier Reef, Queensland, Australia; **E**= *Cymbiolacca excelsior* Limpus and Bail, 1998, length 45 mm, in sand and rubble, 12 m depth off Departure Reef, Swain Reefs, Great Barrier Reef, Queensland, Australia; **F**= *Volutoconus multiformis* Bail and Limpus, 2013, length 60 mm, from 80 m depth in the Capricorn Channel, Queensland, Australia; **G**= *Cymbiolacca pulchra woolacottae* (McMichael, 1958), length 70 mm, low tide, Heron Island, Capricorn Group, Great Barrier Reef, Queensland, Australia; **H**= *Cymbiolista jansae* (van Pel and Moolenbeek, 2010), length 111 mm, 200 m depth off Myrmidon Reef, Great Barrier Reef, Queensland, Australia; **I**= *Amoria molleri reducta* Bail and Limpus, 2001, length 50 mm, 150 m depth in the Capricorn Channel, Queensland, Australia; **J**= *Cymbiolacca coucomorum* Bail and Limpus, 2000, length 66 mm, 18 m depth off Swain Reefs, Great Barrier Reef, Queensland, Australia; **K**= *Lividiconus biliosus imperator* (Woolacott, 1956), length 47 mm, low tide, Dingo Beach, Cairns, Queensland, Australia.

Figure 8.10 Index Gastropods of the Cairnsian and Coralian Subprovinces and the Howeian Infraprovince, Solanderian Province.

(Cairnsian Subprovince) A= *Nannamoria gotoi* Poppe, 1992, length 28 mm, 180 m depth, 100 km east of Swain Reefs, Great Barrier Reef, Queensland, Australia; **B=** *Eremiconus colmani* (Rockel and Korn, 1990), length 38 mm, 250 m depth due east of Fitzroy Reef, Capricorn Group, Queensland, Australia; **C=** *Kioconus queenslandicus* (da Motta, 1984), length 92 mm, 200 m depth due east of Fitzroy Reef, Capricorn Group, Queensland, Australia; **D=** *Kioconus whiteheadae* (da Motta, 1985), length 105 mm, 250 m depth due east of Fitzroy Reef, Capricorn Group, Queensland, Australia; **E=** *Erronea (Ipserronea) carnicolor* (Preston, 1909), length 35 mm, 40 m depth off Tanum Sands, Queensland, Australia; **(Coralian Subprovince) F=** *Cymbiolacca intruderi* (Poppe, 1985), length 56 mm, 250 m east of Diamond Reef, Coral Sea, Queensland, Australia; **G=** *Cymbiolacca (Magnavictoria) perplicata* (Hedley, 1902), length 87 mm, 15 m depth off Diamond Reef, Coral Sea, Queensland, Australia; **H=** *Harpa queenslandica* Berschauer and Petuch, 2016, length 36 mm, 15 m depth off Diamond Reef, Coral Sea, Queensland, Australia; **(Howeian Infraprovince) I=** *Haliotis rubiginosa* Reeve, 1846, length 29.1 mm, 1 m depth off Dawsons Point, Lord Howe Island, Tasman Sea; **J=** *Lyria howensis* Iredale, 1937, length 25 mm, 2 m depth on Front Reef, Lord Howe Island, Tasman Sea; **K=** *Turbo (Marmarostoma) cepoides* E.A. Smith, 1880, height 78.7 mm, 8 m depth off Front Reef, Lord Howe Island, Tasman Sea.

Figure 8.11 Widespread Index Gastropods of the Dampierian Province.
A= *Miniaceoliva ornata* (Marrat, 1867), length 44 mm, low tide on sand flats, Broome, Western Australia, Australia; **B**= *Marmarofusus tessellatus* (Sowerby III, 1880), length 38 mm, 25 m depth off Freemantle, Western Australia, Australia; **C**= *Hexaplex stainforthi* (Reeve, 1843), length 50 mm, low tide, off Broome, Western Australia, Australia; **D**= *Amoria ellioti* (Sowerby II, 1864), length 63 mm, low tide, Port Hedland, Western Australia; **E**= *Amoria jamrachi* (Gray, 1864), length 55 mm, low tide, Broome, Western Australia, Australia; **F**= *Cymbiola nivosa* (Lamarck, 1804), length 61 mm, low tide, Karratha, Roebourne, Western Australia, Australia; **G**= *Cymbiola oblita* (E.A. Smith, 1909), length 74 mm, low tide, off Broome, Western Australia, Australia; **H**= *Austrocochlea zeus* (Fischer, 1874), width 20 mm, 2 m depth off Dirk Hartog Island, Shark Bay, Western Australia, Australia; **I**= *Cylinder victoriae* (Reeve, 1843), length 57 mm, low tide, Broome, Western Australia; **J**= *Virroconus dorreensis* (Peron, 1807), length 34 mm, low tide, Dirk Hartog Island, Shark Bay, Western Australia; **K**= *Amoria grayi* Ludbrook, 1953, length 69 mm, low tide off Broome, Western Australia, Australia.

Figure 8.12 Index Gastropods of the Exmouthian Subprovince, Dampierian Province.
A= *Volutoconus coniformis* Cox, 1871, length 59 mm, on sand, 15 m depth off Onslow, Western Australia, Australia; **B**= *Euprotomus iredalei* (Abbott, 1860), length 65 mm, 5 m depth off Port Hedland, Western Australia, Australia; **C**= *Timbellus bednalli* (Brazier, 1878), length 45 mm, low tide, Broome, Western Australia, Australia; **D**= *Amoria dampieria* Weaver, 1960, length 32 mm, 60 m depth off Weld Island, Western Australia, Australia; **E**= *Amoria macandrewi* (Sowerby III, 1887), length 40 mm, 1 m depth off Onslow, Western Australia; **F**= *Volutoconus hargreavesi* (Angas, 1872), length 92 mm, 50 m depth off Thevenard Island, Western Australia, Australia; **G**= *Acutoliva kurzi* (Petuch and Sargent, 1986), length 23 mm, on sand bars at low tide, off Broome, Western Australia, Australia; **H**= *Eremiconus dampierensis* (Filmer and Coomans, 1985), length 24 mm, under rock rubble, 10 m depth off Burrup, Dampier Peninsula, Western Australia, Australia; **I**= *Floraconus novaehollandiae* (A. Adams, 1854), length 33 mm, low tide off Port Hedland, Western Australia, Australia; **J**= *Pionoconus barbara* (Brazier, 1898), length 50 mm, 6 m depth off Burrup, Dampier Peninsula, Western Australia, Australia; **K**= *Calamiconus garywilsoni* (Lorenz and Morrison, 2004), length 18 mm, 20 m depth off Victor Island, Exmouth Gulf, Western Australia; **L**= *Cribrarula exmouthensis* (Melvill, 1888), length 24 mm, 15 m depth off North West Cape, Exmouth Gulf, Western Australia, Australia.

Figure 8.13 Index Cowries of the Exmouthian Subprovince, Dampierian Province.
A= *Zoila jeaniana inscripta* Lorenz, 2017, golden color form, length 70 mm, 80 m depth NW of Bernier Island, Western Australia, Australia; **B=** *Cribrarula exmouthensis magnifica* (Lorenz, 2002), length 36 mm, 8 m depth off Broome, Western Australia, Australia; **C=** *Zoila decipiens* (E.A. Smith, 1880), length 56 mm, 10 m depth off Mangrove Point, Broome, Western Australia, Australia; **D=** *Zoila eludens* Raybaudi, 1991, length 63 mm, 24 m depth off Murion Island, Exmouth Gulf, Western Australia, Australia; **E=** *Zoila aurata* Raybaudi, 1979, length 74 mm, 180 m depth off Dirk Hartog Island, Shark Bay, Western Australia; **F=** *Leporicypraea geographica rovae* Lorenz, 2017, length 59 mm, 50 m depth off Coulomb Point, Dampier Peninsula, Western Australia; **G=** *Zoila marginata ketyana* (Raybaudi, 1978) form *hypermarginata* Raybaudi, 1993, length 49 mm, 120 m depth off northwestern Dirk Hartog Island, Shark Bay, Western Australia, Australia (apertural view; dorsum is pure white); **H=** *Zoila perlae* Lopez and Chiang, 1975, length 43 mm, 100 m depth, 50 km due north of Dampier, Western Australia, Australia; **I=** *Zoila rosselli edingeri* Raybaudi, 1990, length 55 mm, 33 m depth, off Point Quobba, Carnarvon, Western Australia, Australia; **J=** *Cribrarula abaliena australiensis* Lorenz, 2002, length 23 mm, in sponges, 2 m depth off Broome, Western Australia, Australia; **K=** *Erronea (Ipserronea) smithi* (Sowerby III, 1881), length 24 mm, 2 m depth off Mangrove Point, Broome, Western Australia; **L=** *Erronea (Ipserronea) garyi* Petuch, Berschauer, and Waller, 2019, length 18 mm, inside dead oysters, low tide on mudflats off Marv Island, King Sound, Western Australia.

Figure 8.14 Index Gastropods of the Carpentarian Subprovince, Dampierian Province.
A= *Bistolida brevidentata* (Sowerby II, 1870), length 28 mm, low tide, Groote Eylandt, Northern Territory, Australia; **B=** *Calliotectum dupreyae arafurensis* (Doute, 1988), length 175 mm, trawled from 300 m depth, 100 km south of Enu Island, Aru Islands, Arafura Sea; **C=** *Amoria keatsiana* Ludbrook, 1953, length 80 mm, low tide off Groote Eylandt, Northern Territory, Gulf of Carpentaria, Australia; **D=** *Amoria newmanae* Cotton, 1949, length 68 mm, trawled from 35 m depth, due west of Thursday Island, Torres Strait, Queensland, Australia; **E=** *Cymbiola flavicans* (Gmelin, 1791), length 71 mm, low tide, Numbulwar, Northern Territory, Gulf of Carpentaria, Australia; **F=** *Volutoconus bednalli* (Brazier, 1878), length 124 mm, 25 m depth north of the Wellesley Islands, Queensland, Gulf of Carpentaria, Australia; **G=** *Amalda vernedei herlaari* van Pel, 1989, length 75 mm, 100 m depth north of Melville Island, Northern Territory, Arafura Sea, Australia; **H=** *Isoconus austroviola* (Rockel and Korn, 1992), length 56 mm, low tide off Gunn Point, Darwin, Northern Territory, Australia; **I=** *Pionoconus arafurensis* Monnier, Limpalaër, and Robin, 2013, length 47 mm, low tide, South West Vernon Island, Vernon Islands, Van Dieman Gulf, Northern Territory, Australia; **J=** *Pionoconus vinctus* (A. Adams, 1855), length 57 mm, low tide off Lee Point, Darwin, Northern Territory, Australia; **K=** *Plicaustraconus adami* Wils, 1988, length 57 mm, 100 m depth north of Melville Island, Northern Territory, Arafura Sea, Australia.

CHAPTER 9. South Australian Paratropical Region

Extending from Shark Bay, Western Australia in the west to Cape Moreton, Queensland in the east, and encompassing the Great Australian Bight, Tasmania, and the New South Wales coast, the South Australian Paratropical Region includes all of the temperate and warm temperate marine environments found along the Australian continent. The region contains a true paratropical molluscan fauna, having large radiations of tropically-derived gastropod groups such as the Cypraeidae, Conidae, and Volutidae, but completely lacking the classic tropical index families Potamididae, Strombidae, and Modulidae and index genera such as *Turbinella* and *Harpa* (s.s.). This large temperate water area encompasses three separate molluscan provinces, the **Peronian Province**, the **Maugean Province**, and the **Flindersian Province** and their seven subprovinces. Although being in close proximity to the cold Antarctic Circumpolar Current, the three subprovincial areas have retained temperate-to-warm temperate oceanographic conditions due to a complex of warmer coastal currents and counter-currents that flow across southern Australia. The Peronian Province marine climate is moderated by the southward-flowing East Australian Current, which brings in a large amount of warm temperate water from the Tasman Sea. The oceanic temperatures of the Maugean and Flindersian Provinces are also moderated by warm

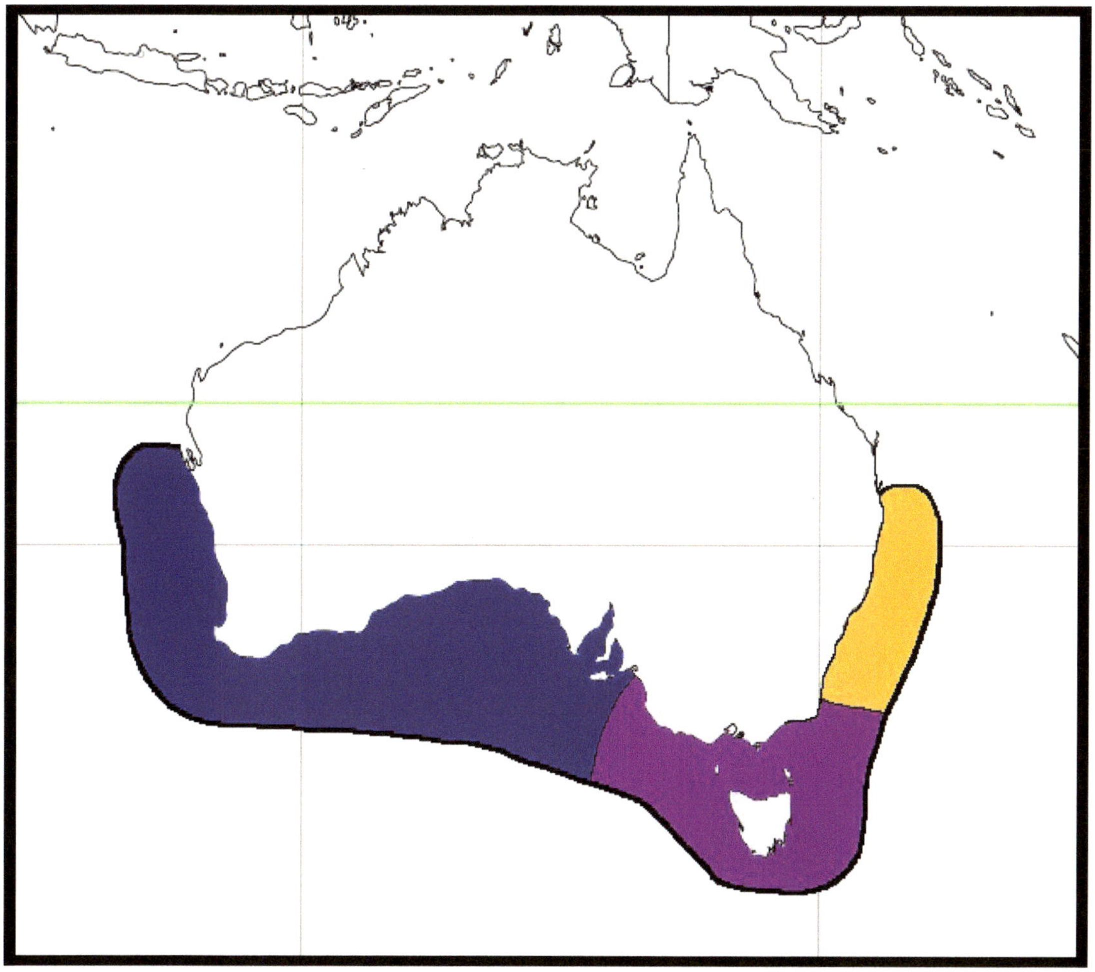

Figure 9.1 Map of the **South Australian Paratropical Region**, showing the areal extent of its provinces: the **Peronian Province** (gold), the **Maugean Province** (magenta), and the **Flindersian Province** (blue).

temperate currents, in this case by a complex of eastward-flowing currents that start along Western Australia and then flow across the Great Australian Bight and around Tasmania. The western end of this large current is referred to as the Leeuwin Current, the central section is referred to as the South Australian Current, and the eastern end, which flows around Tasmania, is referred to as the Zeehan Current. A cooler water counter-current, the Flinders Counter-Current, flows from east to west in the area between the Antarctic Circumpolar Current and the Leeuwin-South Australian-Zeehan current complex and causes many small gyres to form along the current interface. Based on late Neogene fossil mollusks from the Roe Calcarenite of the Great Australian Bight area of Western Australia, these oceanic conditions have existed, unchanged, since the late Pliocene, producing a stable marine climate that has allowed for the evolution of a distinctive molluscan fauna unlike any other seen on Earth. Some of these Pliocene relicts are now widespread regional index species and are found throughout all three provinces, including the cassids *Hypocassis fimbriata* and *Semicassis semigranosum*, the ranellid *Austrotriton subdistorta*, the muricid *Pterochelus triformis*, the volutes *Amoria undulata* (and its subspecies), *Mitraelyria mitraeformis*, and *Livonia roadknightae*, the harpid *Austroharpa exquisita*, the olivid *Acutoliva australis*, and the conids *Floraconus anemone* and *Floraconus peronianus* (most shown here on Figure 9.5).

Peronian Molluscan Province

Named for Francois Peron, an early 19th century French naturalist who explored the Australian coastline, the Peronian Province extends from Tweed Heads, northern New South Wales south to Jervis Bay and Point Perpendicular, New South Wales. The province breaks up into two broad oceanographic areas: a warmer northern section that extends from Tweed Heads to Seal Rocks, near Newcastle, New South Wales (referred to here as the **Macquarian Subprovince**; described and discussed in the following section) and a cooler southern section that extends from Newcastle south to Jervis Bay, New South Wales (the typical Peronian Subprovince, which contains an impoverished and less diverse fauna composed only of widespread species). This split marine climate is the result of the confluence of two major current systems: the southward-flowing East Australian Current, which brings subtropical water down from the Coral Sea, sometimes as far south as Sydney; and the Tasman Front, which is composed of a series of complex gyres and eddies that bring in cooler, temperate water from the adjacent Tasman Sea and which often abuts the southern New South Wales coast from Newcastle south to Jervis Bay. The Peronian Province has evolved a number of characteristic index gastropods which can be used to define the provincial boundaries. Some of these include the struthiolariid *Tylospira scutulata* (with the genus being endemic to the Peronian Province), the cowrie *Umbilia hesitata beddomei*, and the cone shells *Floraconus papilliferus* and *Austroconus sydneyensis* (all shown here on Figure 9.6).

Macquarian Subprovince

Named for Lachlan Macquarie, an early 19th century Governor of Australia who was known for his kindness to the convicts in the New South Wales penal colony, the Macquarian Subprovince extends from Tweed Heads to Seal Rocks near

Newcastle, New South Wales. This subprovincial area represents a broad area of faunal intergradation between the tropical Indo-Pacific and Solanderian malacofaunas to the north and the temperate southern Peronian and Maugean malacofaunas to the south. Besides containing all of the classic Peronian index taxa, the Macquarian Subprovince also contains a number of Indo-Pacific species that are represented there by endemic New South Wales subspecies. Some of these include the cypraeids *Lyncina vitellus orcina* (also on Lord Howe Island and North Island, New Zealand), and *Ovatipsa chinensis sydneyensis*. These endemic subspecific taxa occur along with a large fauna of Macquarian index species, including the rare cypraeid *Cribrarula gravida*, the volutes *Amoria benthalis*, *Amoria molleri*, *Lyreneta laseroni*, and *Cymbiolacca complexa*, and the conids *Floraconus aplustre*, *Plicaustraconus angasi*, and *Eremiconus minnamurra* (all shown here on Figure 9.6). Occasional specimens of strombids, such as *Canarium* and *Conomurex* are rarely found as far south as Sydney. These records reflect occasional individuals whose long lived planktonic larvae were introduced by rare seasonal warm currents and do not represent established breeding populations.

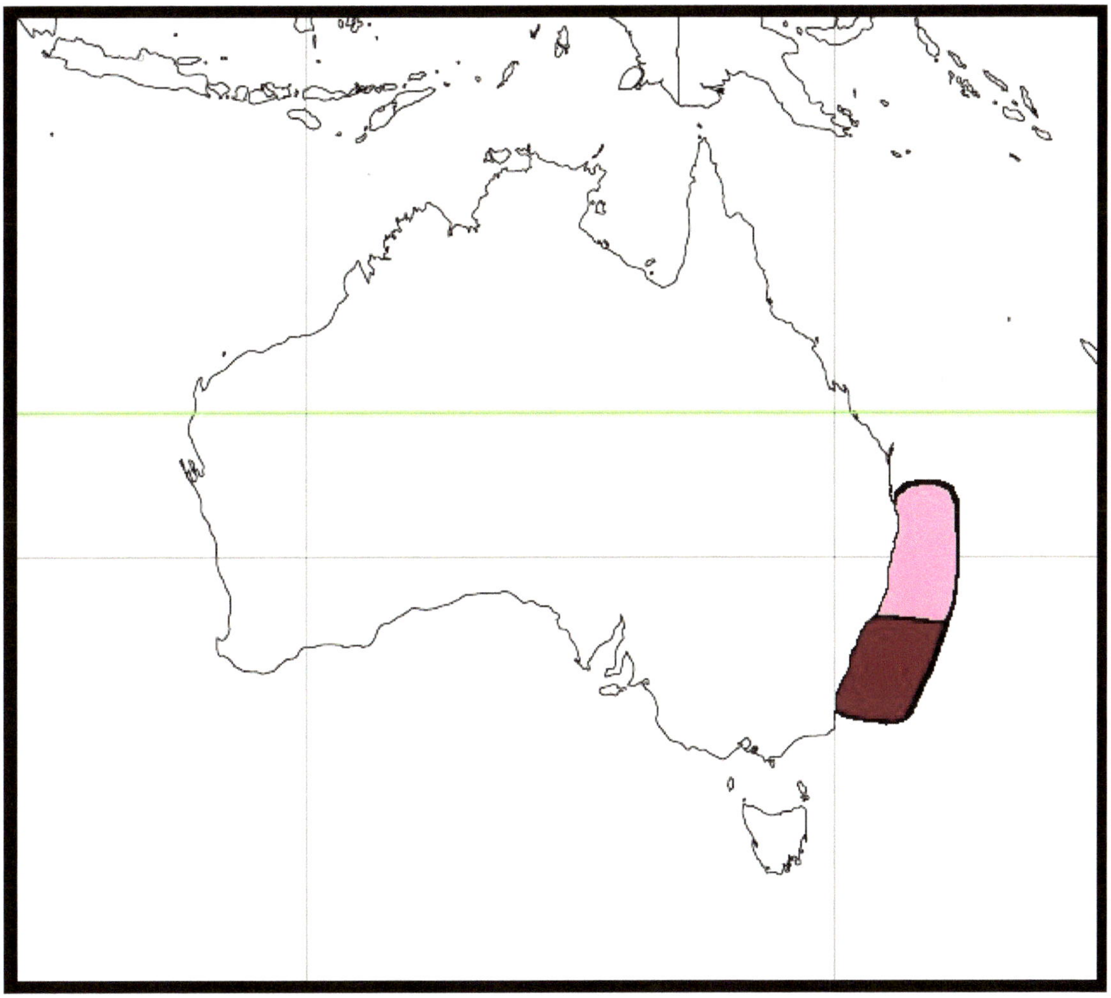

Figure 9.2 Map of the Peronian Molluscan Province, showing the areal extent of its single subprovince, the **Macquarian Subprovince** (light rose). A typical impoverished Peronian fauna, composed only of widespread index species, extends from Newcastle to Jervis Bay, New South Wales (dark rose).

Maugean Molluscan Province

Named for Rene Mauge, an 18th century French naturalist who explored southern Australia and Tasmania, the Maugean Province encompasses the southeastern corner of the Australian continent, extending from Jervis Bay, New South Wales to Cape Jaffa, South Australia, and including the entire coast of Victoria State, the Bass Strait, and the entire island of Tasmania. Because of the differences in water temperatures of the Zeehan Current and Tasman Front, the Maugean Province can be divided into two separate subprovinces: a coastal Victorian Subprovince and an insular Tasmanian Subprovince (both discussed and described in the following sections). A large and highly endemic gastropod fauna has evolved within the Maugean Province, including classic index taxa such as the cowries *Umbilia hesitata* and its white form *howelli*, the volute *Amoria undulata australiae*, and the conid *Parviconus rutilus smithi* (all shown here on Figure 9.7). Of special interest within the Maugean Province is a species radiation of the cowrie genus *Notocypraea*, containing at least ten taxa, with the most widespread being *N. angustata*, *N. angustata molleri*, *N. declivis*, *N. dissecta*, and *N. hartsmithi* (all shown here on Figure 9.7).

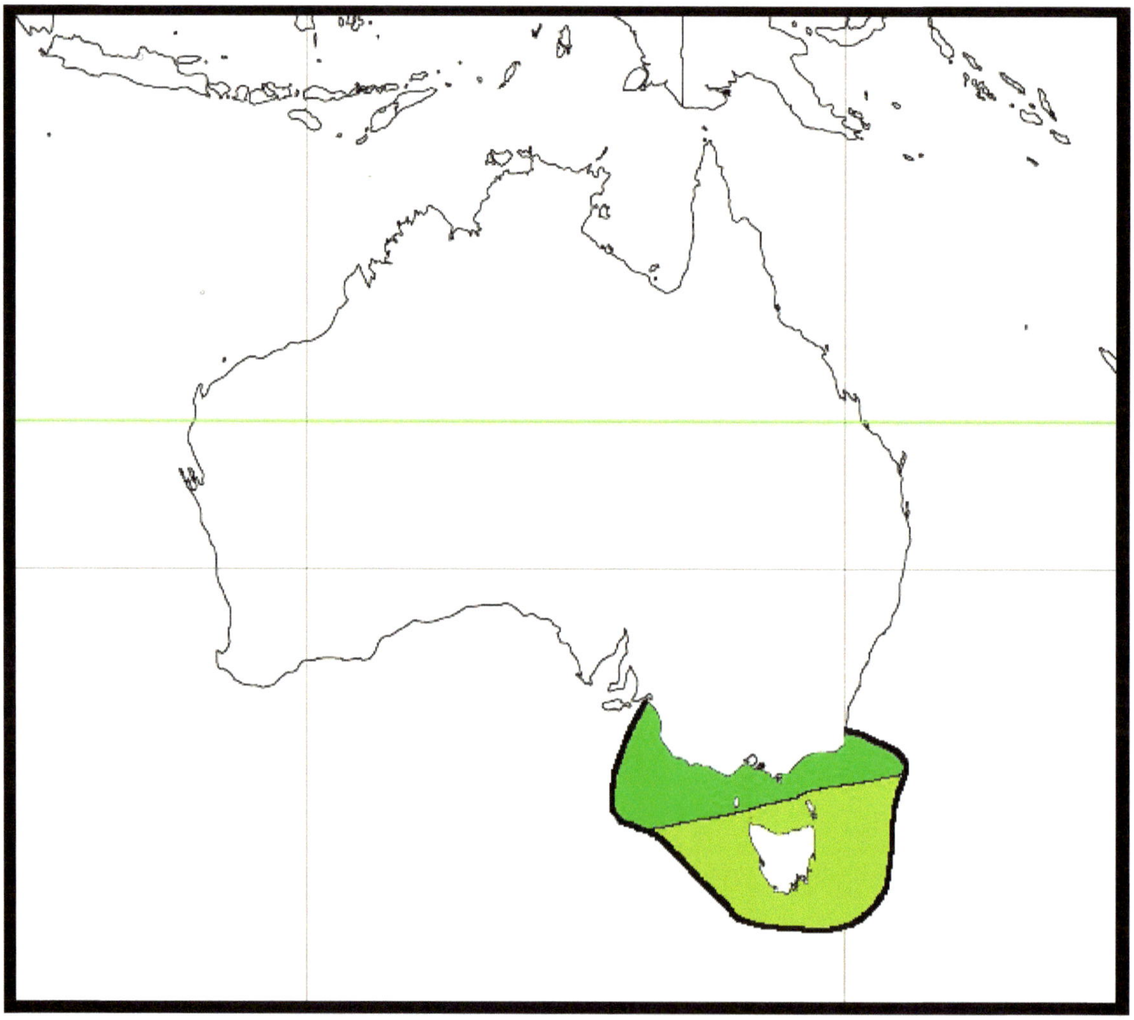

Figure 9.3 Map of the Maugean Molluscan Province, showing the areal extent of its subprovinces: the **Victorian Subprovince** (green) and the **Tasmanian Subprovince** (lime green).

Victorian Subprovince

Named for the Australian State of Victoria (and also for Queen Victoria), the Victorian Subprovince extends from Jervis Bay, New South Wales to Cape Jaffa, Victoria and represents the Tasman Sea and northern Bass Strait component of the subprovincial area. Containing primarily a standard Peronian Province malacofauna, the Victorian area also houses a small number of interesting endemic gastropods, some of which include the cowries *Notocypraea comptoni wilkinsi* and *Notocypraea comptoni mayi*, and the cone shells *Floraconus anemone carmeli* and *Floraconus anemone singletoni* (all shown here on Figure 9.8).

Tasmanian Subprovince

Named for the island State of Tasmania, and also for Abel Tasman, the 17th century Dutch seafarer and explorer who was the first European to explore the island, the Tasmanian Subprovince encompasses not only the main island of Tasmania, but also the central and southern Bass Strait and the Bass Strait groups of islands, such as the Hogan Group, the Kent Group (with Deal Island), the Furneaux Group (with Flinders, Cape Barren, and Clarke Islands), and the isolated King Island. Being surrounded by the warm temperate Zeehan Current, the Tasmanian marine environments remain marginally paratropical year round and these support an unusual gastropod fauna with a high degree of endemism. Some of the more conspicuous Tasmanian endemic taxa include the cypraeids *Notocypraea declivus dennyorum*, *Notocypraea subcarnea*, and *Notocypraea albata*, the volute *Amoria sclateri* (confined to the islands in the Bass Strait), and the conids *Floraconus anemone saundersi* and *Parviconus macleayanus* (all illustrated here on Figure 9.8).

Flindersian Molluscan Province

Named for Matthew Flinders, a British navigator who explored the southern coast of Australia in the late 18th century, the Flindersian Province extends from Shark Bay, Western Australia to Cape Jaffa, Victoria, and encompasses the immense embayment of the Great Australian Bight. Under the influence of the eastward-flowing Leeuwin and South Australian Currents, the oceanic climate of the entire coastline remains warm temperate to temperate throughout the year, creating stable marine environments. These warmer currents often mix with the westward-flowing Flinders Counter-Current, producing a series of eddies and small gyres, particularly off the southern areas of Western Australia. The coastline from Shark Bay south to Cape Leeuwin is also under the influence of the northward-flowing West Australian Current, which mixes cool subantarctic water with the water of the warm Leeuwin Current, producing pockets of cold water off the Western Australia coast. Because of this complex oceanography and the presence of discrete water masses, the Flindersian Province contains three distinct subprovinces: the eastern Adelaidean Subprovince which is under the influence of the South Australian Current, the central Euclean Subprovince which encompasses the cool water environments of the Great Australian Bight, and the western Perthian

Subprovince, which is under the influence of the mixed water masses of the Leeuwin and West Australian Currents.

The stable paratropical Flindersian Province has allowed for the evolution of a large fauna of widespread endemic gastropods. Some of these, which can be used to define the provincial boundaries, include the cypraeids *Austrocypraea reevei*, *Notocypraea comptoni*, *Notocypraea piperita*, and *Umbilia armeniaca*, the volutes *Amoria exoptanda*, *Melo miltonis*, and *Notovoluta verconis*, and the conids *Austroconus clarus*, *Klemaconus klemae*, and *Parviconus rutilus* (all shown here on Figure 9.9). Of special interest within the Flindersian Province is a species radiation of high-spired vase shells of the genus *Altivasum*, with at least four known species. These include *Altivasum flindersi* (Adelaidean Subprovince), *A. hedleyi* (Euclean Subprovince), *A. clarksoni* (Perthian Subprovince), and *A. profundum* (deep water, Perthian Subprovince). These are shown here on Figures 9.10, 9.11, and 9.12. For detailed reviews of the Flindersian endemic genus *Altivasum*, see Dekkers and Maxwell, 2018 and Maxwell and Dekkers, 2019.

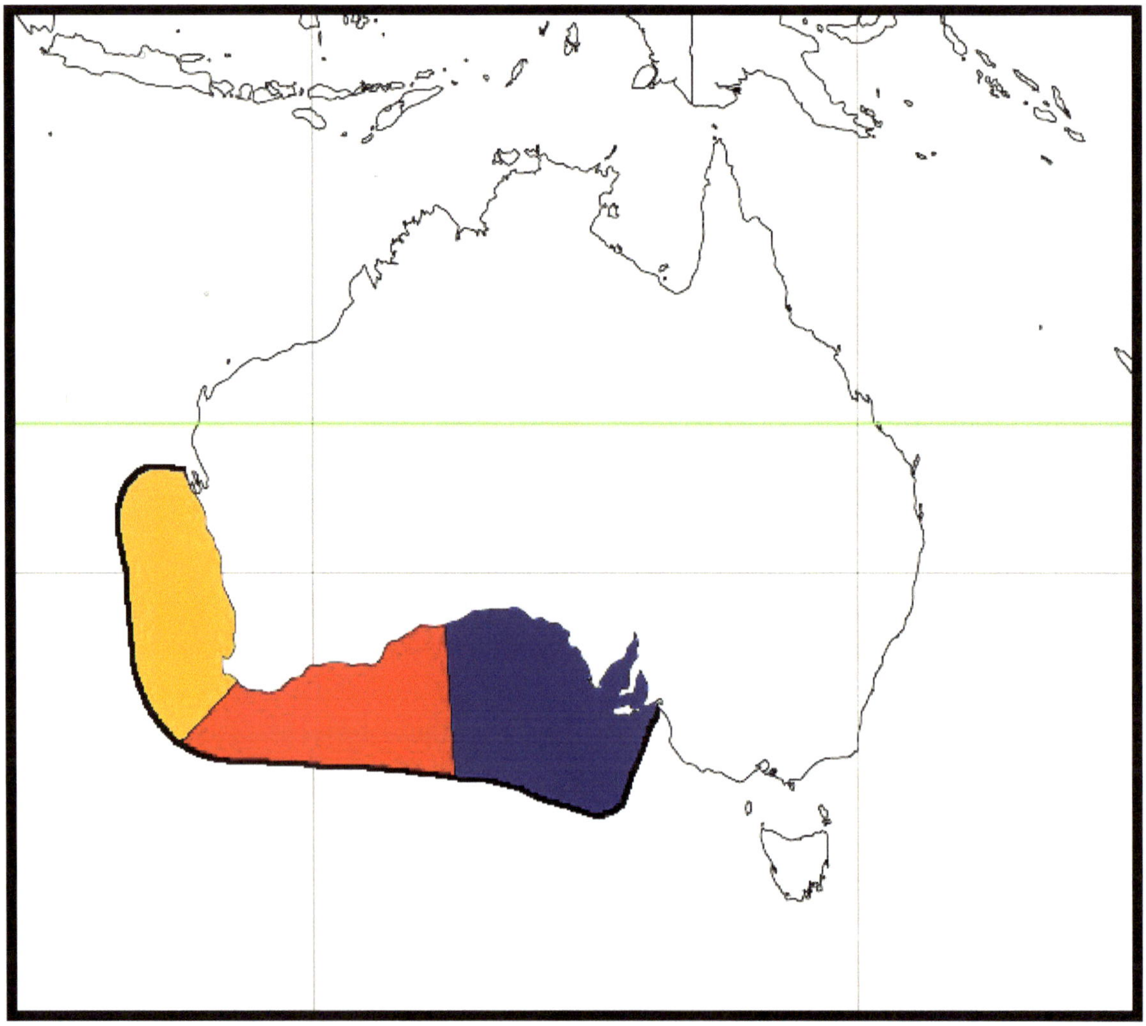

Figure 9.4 Map of the Flindersian Molluscan Province, showing the areal extent of its subprovinces: the **Adelaidean Subprovince** (blue), the **Euclean Subprovince** (red), and the **Perthian Subprovince** (gold).

Adelaidean Subprovince

Named for Adelaide, the capital of South Australia, the Adelaidean Subprovince extends from Cape Jaffa to Fowlers Bay, South Australia, and encompasses the St. Vincent Gulf and Kangaroo Island, along with the Spencer Gulf and its many small islands, such as Thistle, Wedge, Spilsby, Lusby, Partney, Roxby, and Wardang Islands and Ile Castiglione. The rocky coastlines and myriad of small bays found in this area offer a wide variety of habitats for gastropods and a large fauna has evolved here that is rich with endemic species and subspecies. Some of the classic Adelaidean endemic index taxa include the cypraeids *Notocypraea angustata verconis*, *Notocypraea comptoni casta*, and *Notocypraea comptoni griffithi*, the volutes *Amoria undulata angasi* and *Mitraelyria grockeae* (endemic to the Streaky Bay area), and the high-spired cone shell *Floraconus anemone incinctus*, and the elongated vase shell *Altivasum flindersi* (all shown here on Figure 9.10). Of special interest in the deeper offshore areas of the Adelaidean Subprovince is a fauna of large cowrie shells, some of which include rarely-seen species such as *Umbilia armeniaca diprotodon* (endemic to the Thorny Passage area), *Umbilia hesitata suprastrata* (the westernmost subspecies of the *hesitata* species complex), and *Zoila orientalis raybaudii*, along with the more frequently-seen and more common *Zoila thersites* (all illustrated here on Figure 9.10).

Euclean Subprovince

Named for the Eucla area of the Nullarbor desert coast of Western Australia, the Euclean Subprovince extends from Fowlers Bay, South Australia to Cape Esperance, Western Australia and encompasses most of the area of the Great Australian Bight. Within this giant embayment, the broad and shallow Eucla Shelf often extends out over 200 km from the sandy beaches of the Nullabor and Eucla shorelines, forming the widest continental shelf area found anywhere along southern Australia. The Eucla Shelf narrows at Cape Arid and the coastline from there westward to Cape Esperance is characterized by small rocky headlands and isolated bays. The many available habitats found within this subprovince have allowed for the evolution of a rich gastropod fauna with a high degree of endemism. Some of the offshore, deeper water endemic gastropods of the Eucla Shelf include the cowries *Austrocypraea reevei bishopi*, and *Umbilia armeniaca andreyi*, the muricid *Pterochelus webbi*, the fasciolariid *Marmarofusus bishopi*, and the spiny vase shell *Altivasum hedleyi* (all shown here on Figure 9.11). The shallow bays and rocky headlands of the western half of the subprovinces also house a rich endemic cowrie fauna, some of which include *Cribrarula fallax*, *Notocypraea occidentalis*, *Notocypraea piperita bicolor*, *Notocypraea pulicaria*, *Zoila marginata albanyensis*, and *Zoila friendi insulata* (Figure 9.11).

Perthian Subprovince

Named for the city of Perth, capital of Western Australia, the Perthian Subprovince extends from Cape Esperance northward to Shark Bay, Western

Australia. The oceanography of the Perthian Subprovince is complex, primarily due to the confluence of the opposite-flowing Leeuwin and West Australian Currents. The warm Leeuwin Current combines with the Eastern Gyral Current from the tropical northeastern Indian Ocean to create a warmer marine climate than that found within the Great Australian Bight area. Giant eddy systems and gyres constantly form and reform off the southwestern coast of Australia, mixing the warm coastal waters with colder and more nutrient-rich water from the offshore subantarctic West Australian Current. As a result of this complex oceanographic mixing, rich planktonic resources occur all along the Perthian Subprovince and these support immense beds of sponges that literally carpet the sea floor from shallow water to depths of over 1000 m. The area off the Perth coast is often referred to by sponge experts as being the "Hub of Sponge Diversity", as over 1000 species of sponges (including Demospongia, Calcarea, and Hexactinellida) have been collected there, making this the richest sponge fauna anywhere on Earth. As a result of this extremely rich sponge diversity, a large radiation of sponge-eating cowries has evolved within the Perthian Subprovince, supported entirely by the wide variety of food resources. Primary among these is a spectacular radiation of over 20 species and subspecies, all in the spongivorous genus *Zoila*. Some of these large and colorful cowries include *Zoila marginata*, *Z. venusta venusta*, *Z. friendi*, *Z. friendi kostini*, *Z. friendi vercoi*, *Z. jeaniana*, *Z. jeaniana sherylae*, *Z. rosselli*, *Z. raywalkeri regularis*, *Z. venusta sorrentensis*, and *Z. venusta roseopunctata* (all shown here on Figures 9.12 and 9.13). This amazing *Zoila* radiation occurs together with rich assemblages of other endemic gastropods, some of which include the cowries *Austrocypraea reevei lorenzoi*, *Austrocypraea reevei microsphaerica*, *Cribrarula abrolhosensis*, *Cribrarula rottnestensis*, and *Purpuradusta gracilis hilda*, the ficid *Ficus eospila*, the vase shells *Altivasum profundum* and *A. clarksoni*, and the cone shells *Cylinder nodulosus*, *Floraconus anemone compressus*, and *Floraconus cocceus* (all shown here on Figure 9.12).

ICONOGRAPHY OF GASTROPODS OF THE SOUTH AUSTRALIAN PARATROPICAL REGION

(Principal Index Gastropods are shown on Figures 9.5 to 9.13)

Figure 9.5 Widespread Index Gastropods of the South Australian Paratropical Region, occurring in the Peronian, Maugean, and Flindersian Provinces.
A= *Semicassis semigranosum* (Lamarck, 1822), length 50 mm, 30 m depth off Streaky Bay, South Australia, Australia; **B=** *Acutoliva australis* (Duclos, 1835), length 26 mm, low tide, Spencer Gulf, Port Lincoln, South Australia, Australia; **C=** *Austrotriton subdistorta* (Lamarck, 1822), length 51 mm, 30 m depth off King Island, Bass Strait, Tasmania, Australia; **D=** *Pterochelus triformis* (Reeve, 1845), length 57 mm, 50 m depth off Port Lincoln, South Australia, Australia; **E=** *Amoria undulata* (Lamarck, 1804), length 80 mm, 20 m depth off Port Lincoln, South Australia, Australia; **F=** *Livonia roadknightae* (McCoy, 1881), length 74 mm, 100 m depth, Eucla, Western Australia, Great Australian Bight, Australia; **G=** *Mitraelyria mitraeformis* (Lamarck, 1811), length 51 mm, 1 m depth, on sand, Port Fairy, Victoria, Australia; **H=** *Austroharpa exquisita* (Iredale, 1931), length 22 mm, 150 m depth south of Elliston, South Australia, Australia; **I=** *Floraconus anemone* (Lamarck, 1810), length 37 mm, low tide, Whitemark, Flinders Island, Furneaux Group, Bass Strait, Tasmania, Australia; **J=** *Floraconus peronianus* (Iredale, 1931), length 59 mm, Eden, New South Wales, Australia; **K=** *Charonia powelli* Cotton, 1957, length 232 mm, 50 m depth, 1 km due south of Lawrence Rock, Portland, Victoria, Australia.

Figure 9.6 Index Gastropods of the Peronian Province and Macquarian Subprovince.
A= *Umbilia hesitata beddomei* (Schilder, 1930), length 74 mm, 100 m depth off Eden, New South Wales, Australia; **B=** *Tylospira scutulata* (Gmelin, 1791), length 50 mm, 50 m depth off Terrigal, New South Wales, Australia; **C=** *Austroconus sydneyensis* (Sowerby III, 1887), length 32 mm, trawled from 100 m depth off Terrigal, New South Wales, Australia; **D=** *Floraconus papilliferus* (Sowerby II, 1834), length 31 mm, 2 m depth, Little Manly Cove, Manly, Sydney Harbour, New South Wales, Australia; **(Macquarian Subprovince) E=** *Lyreneta laseroni* Iredale, 1937, length 27 mm, found on the beach after a storm, Woolgoolga Head, New South Wales, Australia; **F=** *Amoria benthalis* (McMichael, 1964), length 37 mm, 25 m depth off Tweed Heads, New South Wales, Australia; **G=** *Amoria molleri* (Iredale, 1936), length 86 mm, 50 m depth off Tweed Heads, New South Wales, Australia; **H=** *Cymbiolacca complexa* (Iredale, 1924), length 68 mm, 50 m depth off Cape Byron, New South Wales, Australia; **I=** *Floraconus aplustre* (Reeve, 1843), length 24 mm, low tide, Tweed Heads, New South Wales, Australia; **J=** *Plicaustraconus angasi* (Tryon, 1884), length 37 mm, trawled from 50 m depth off Terrigal, New South Wales, Australia; **K=** *Eremiconus minnamurra* (Garrard, 1961), length 17 mm, 100 m depth off Port Macquarie, New South Wales, Australia; **L=** *Cribrarula gravida* Moretzsohn, 2002, length 27 mm, under rocks in 10 m depth, on Fido's Reef off Fingal Head, Tweed Heads, New South Wales, Australia.

Figure 9.7 Index Gastropods of the Maugean Province.
A= *Notocypraea angustata* (Gmelin, 1791), length 30 mm, washed ashore, West Ulverstone Beach, Tasmania, Australia; **B=** *Notocypraea angustata molleri* (Iredale, 1931), length 33 mm, 70 m depth due south of Cape Otway, Victoria, Australia; **C=** *Notocypraea declivis* (Sowerby II, 1870), length 27 mm, low tide, Port Fairy, Victoria, Australia; **D=** *Notocypraea dissecta* Iredale, 1931, length 22 mm, washed ashore, West Ulverstone, Tasmania, Australia; **E=** *Notocypraea dissecta hartsmithi* (Schilder, 1967), length 18 mm, washed ashore, Lakes Entrance, Victoria, Australia; **F=** *Umbilia hesitata* (Iredale, 1916), length 91 mm, 80 m depth due east of Flinders Island, Bass Strait, Tasmania, Australia; **G=** *Umbilia hesitata* form *howelli* (Iredale, 1931), length 91 mm, 80 m depth south of King Island, Bass Strait, Australia; **H, I=** *Parviconus rutilus smithi* (Angas, 1877), length 7 mm, on the beach, Eden, New South Wales, Australia; **J=** *Amoria undulata australiae* (Cox, 1872), length 80 mm, 50 m depth off Eden, New South Wales, Australia.

Figure 9.8 Index Gastropods of the Victorian and Tasmanian Subprovinces, Maugean Province.
(Victorian Subprovince) A= *Notocypraea comptoni wilkinsi* (Griffiths, 1959), length 26 mm, on the beach, Portland, Victoria, Australia; **B=** *Floraconus anemone carmeli* (Tennison-Woods, 1877), length 45 mm, 10 m depth off Port Lincoln, South Australia, Australia; **C=** *Floraconus anemone singletoni* Cotton, 1945, length 32 mm, 6 m depth off Sorrento, Victoria, Australia; **D=** *Notocypraea comptoni mayi* (Beddome, 1898), length 23 mm, 2 m depth off the mouth of the Black River, Stanley, Tasmania, Australia; **(Tasmanian Subprovince) E=** *Notocypraea declivus dennyorum* Lorenz and Morrison, 2013, length 23 mm, 2 m depth off the mouth of the Black River, Stanley, Tasmania, Australia; **F=** *Notocypraea subcarnea* (Beddome, 1896), length 24 mm, 2 m depth off St. Helens, Tasmania, Australia; **G=** *Notocypraea albata* (Beddome, 1897), length 23 mm, 30 m depth due east of Cape Pillar, Tasmania, Australia; **H=** *Amoria sclateri* (Cox, 1869), length 66 mm, 100 m depth off the west coast of Flinders Island, Bass Strait, Tasmania, Australia; **I=** *Floraconus anemone saundersi* (Cotton, 1945), length 40 mm, on the beach, Goat Island Nature Reserve, West Ulverstone, Tasmania, Australia; **J, K=** *Parviconus macleayanus* (Tennison-Woods, 1875), length 16 mm, on the beach, West Ulverstone, Tasmania, Australia.

Figure 9.9 Index Gastropods of the Flindersian Province.
A= *Austrocypraea reevei* (Verco, 1912), length 39 mm, 6 m depth off Mistaken Island, Albany, Western Australia, Australia; **B=** *Notocypraea comptoni* (Gray, 1847), length 21 mm, low tide, Kangaroo Island, St. Vincent Gulf, South Australia, Australia; **C=** *Notocypraea piperita* (Gray, 1825), length 23 mm, 2 m depth, Spencer Gulf, Port Lincoln, South Australia, Australia; **D=** *Umbilia armeniaca* (Verco, 1912), length 103 mm, 120 m depth due south of Eucla, Western Australia, Australia; **E=** *Amoria exoptanda* (Reeve, 1849), length 104 mm, 30 m depth off Port Lincoln, South Australia, Australia; **F=** *Melo miltonis* (Griffith and Pidgeon, 1834), length 219 mm, 2 m depth off Woodman's Point, Treemantle, Western Australia, Australia; **G=** *Notovoluta verconis* (Tate, 1892), length 35 mm, 40 m depth off Taylor Island, Thorny Passage, Western Australia, Australia; **H=** *Austroconus clarus* (E.A. Smith, 1881), length 29 mm, low tide, northern side of Charley Island, Esperance, Western Australia, Australia; **I=** *Klemaconus klemae* (Cotton, 1953), length 37 mm, low tide in Streaky Bay, South Australia, Australia; **J, K=** *Parviconus rutilus* (Menke, 1843), length 8 mm, low tide in Fowlers Bay, South Australia, Australia.

Figure 9.10 Index Gastropods of the Adelaidean Subprovince, Flindersian Province.
A= *Notocypraea angustata verconis* (Cotton and Godfrey, 1932), length 30 mm, 5 m depth in Venus Bay, South Australia, Australia; **B=** *Notocypraea comptoni casta* Schilder and Schilder, 1963, length 25 mm, low tide, Apollo Bay, Victoria, Australia; **C=** *Notocypraea comptoni griffithi* Lorenz, 2017, length 21 mm, 2 m depth, Fleurieu Peninsula, South Australia, Australia; **D=** *Umbilia armeniaca diprotodon* Lorenz and Beals, 2013, length 107 mm, 35 m depth in Thorny Passage, Port Lincoln, South Australia, Australia; **E=** *Umbilia hesitata suprastrata* Govaert, Deprez, and Vandenberghe, 2015, length 81 mm, 250 m depth due south of Beach Port, South Australia, Australia; **F=** *Zoila orientalis raybaudii* Lorenz, 1992, length 60 mm, on sponges, 25 m depth off Port Lincoln, South Australia, Australia; **G=** *Zoila thersites* (Gaskoin 1849), length 81 mm, on sponges, 3 m depth off Stansbury, South Australia, Australia; **H=** *Amoria undulata angasi* (Sowerby II, 1864), length 70 mm, 10 m depth in Thorny Passage, Port Lincoln, South Australia, Australia; **I=** *Mitraelyria grockeae* Petuch and Berschauer, 2017, length 37 mm, 1 m depth in Streaky Bay, South Australia, Australia; **J=** *Floraconus anemone incinctus* (Feneaux, 1942), length 38 mm, low tide, Victoria Harbour, Melbourne, Victoria, Australia (often incorrectly called "*compressus*", which is a different species from the Perthian Subprovince; see Figure 8.11); **K=** *Altivasum flindersi* (Verco, 1914), length 110 mm, 20 m depth in gravelly sand gutters, off Port Lincoln, South Australia.

Figure 9.11 Index Gastropods of the Euclean Subprovince, Flindersian Province.
A= *Austrocypraea reevei bishopi* Petuch and Berschauer, 2017, length 37 mm, 145 m depth due south of Eucla, Western Australia, Australia; **B=** *Cribrarula fallax* (E.A. Smith, 1881), length 27 mm, 10 m depth off Albany, Western Australia, Australia (pure white color form); **C=** *Altivasum hedleyi* Maxwell and Dekkers, 2019, length 222 mm, 50 m depth due south of Esperance, Western Australia, Australia; **D=** *Notocypraea occidentalis* Iredale, 1935, length 24 mm, 7 m depth, off Augusta, Western Australia, Australia; **E=** *Notocypraea piperita bicolor* (Gaskoin, 1849), length 26 mm, 8 m depth, Esperance Jetty, Esperance, Western Australia, Australia; **F=** *Notocypraea pulicaria* (Reeve, 1846), length 21 mm, low tide, Charley Island, Esperance, Western Australia, Australia; **G=** *Umbilia armeniaca andreyi* Lorenz and Beals, 2013, length 73 mm, 120 m depth due south of Eucla, Western Australia, Australia; **H=** *Zoila friendii insulata* Raybaudi, 1985, length 90 mm, 20 m depth off Charley Island, Esperance, Western Australia, Australia; **I=** *Pterochelus webbi* Petuch and Berschauer, 2018, length 42 mm, 150 m depth southwest of Ceduna, South Australia, Australia; **J=** *Marmarofusus bishopi* Petuch and Berschauer, 2017, length 54 mm, 145 m depth in the Great Australian Bight, due south of Eucla, Western Australia, Australia; **K=** *Mitraelyria grangeri* (Sowerby III, 1900), length 38 mm, low tide, in sand, Two People Bay, Albany, Western Australia, Australia; **L=** *Zoila marginata albanyensis* Raybaudi Massilia, 1985, length 56 mm, 30 m depth off Mondraine Island, Esperance, Western Australia.

Figure 9.12 Index Gastropods of the Perthian Subprovince, Flindersian Province.
A= *Altivasum clarksoni* Maxwell and Dekkers, 2019, length 62 mm, 30 m depth on SSE side of Sandy Hook Island, off Esperance, Western Australia; **B=** *Cribrarula abrolhensis* Lorenz, 2002, length 24 mm, 3 m depth off West Wallabi Island, Houtman Abrolhos Islands, Western Australia, Australia; **C=** *Cribrarula rottnestensis* Lorenz, 2002, length 30 mm, 15 m depth on Roe Reef, Rottnest Island, Western Australia, Australia; **D=** *Purpuradusta macula hilda* (Iredale, 1939), length 18 mm, low tide West Wallabi Island, Houtman Abrolhos Islands, Western Australia, Australia; **E=** *Zoila marginata* (Gaskoin, 1849), length 59 mm, 35 m depth off Stocky Reef, Jurien Bay, Western Australia, Australia; **F=** *Zoila venusta* (Sowerby I, 1847), length 77 mm, 25 m depth in Bremer Bay, west of Albany, Western Australia, Australia; **G=** *Ficus eospila* (Peron and Lesueur, 1807), length 32 mm, trawled from 100 m depth off Geraldton, Western Australia, Australia; **H=** *Floraconus anemone compressus* (Sowerby II, 1866), length 35 mm, on algae-covered rocks, 5 m depth, off Esperance, Western Australia, Australia; **I=** *Floraconus cocceus* (Reeve, 1844), length 33 mm, low tide, Bunbury, Western Australia, Australia; **J=** *Amoria whitworthi* Bail and Limpus, 2001, length 52 mm, low tide in Coral Bay, Western Australia, Australia; **K=** *Cylinder nodulosus* (Sowerby II, 1864), length 42 mm, under coral slabs at low tide, Geraldton, Western Australia; **L=** *Altivasum profundum* Dekkers and Maxwell, 2018, length 77 mm, taken by an ROV from 162 m depth off Augusta, Western Australia.

Figure 9.13 Index Cowries of the Perthian Subprovince, Flindersian Province.
A= *Zoila friendi* (Gray, 1831), length 91 mm, 8 m depth in Geographe Bay, off Busselton, Western Australia, Australia; **B=** *Zoila friendi kostini* Lorenz and Chiapponi, 2007, length 72 mm, 120 m depth off Windy Harbour, Western Australia, Australia; **C=** *Zoila friendi vercoi* Schilder, 1930, length 82 mm, 35 m depth southeast of Bald Island, Western Australia, Australia; **D=** *Zoila jeaniana* (Cate, 1968), length 66 mm, 160 m depth off Geraldton, Western Australia, Australia; **E=** *Austrocypraea reevei lorenzoi* Chiapponi, 2017, length 32 mm, 200 m depth off Albany, Western Australia, Australia; **F=** *Zoila jeaniana sherylae* Raybaudi, 1990, length 67 mm, 33 m depth off Point Quobba, Carnarvon, Western Australia, Australia; **G=** *Zoila rosselli* (Cotton, 1948), length 57 mm, 25 m depth off Horseshoe Reef, Rottnest Island, Western Australia; **H=** *Zoila raywalkeri regularis* Lorenz, 2018, length 53 mm, 110 m depth off Windy Harbour, Western Australia, Australia; **I=** *Zoila venusta sorrentensis* Schilder, 1963, length 52 mm, 20 m depth in Sorrento Bay, Western Australia, Australia; **J=** *Zoila venusta roseopunctata* Raybaudi, 1985, length 76 mm, 50 m depth off Remark Island, Western Australia, Australia; **K=** *Austrocypraea reevei microsphaerica* Raybaudi, 1980, length 36 mm, 100 m depth off Geraldton, Western Australia.

Radiation of *Altivasum* species in southern Australia. *Altivasum flindersi* (Verco, 1914), height 110.3 mm (top left); *A. pauladellaboscae* Cooper and Maxwell, 2020, height 128.2 mm (bottom left); *A. profundum* Dekkers and Maxwell, 2018, height 77.0 mm (top right); *A. hedleyi* Dekkers and Maxwell, 2019, height 222.0 mm (top center); *A. clarksoni* Dekkers and Maxwell, 2019, height 62.0 mm (bottom right).

CHAPTER 10. Southern African Region

The Southern African Region spans the southern tip of Africa, from St. Helena Bay, West Coast Peninsula in the west, around Cape Agulhas, and northward to St. Lucia, northernmost Kwa Zulu-Natal, South Africa in the east (the Ponta do Ouro area of the South Africa-Mozambique border). Although being the smallest of all the biogeographical regions, the Southern African Region is faunistically distinct, containing molluscan faunas unlike any others found in the adjacent Atlantic or Indian Oceans. These malacofaunas appear to be relictual in nature, containing many archaic genera and families, as demonstrated by the presence of the families Strepsiduridae (*i.e. Melapium elatum*) and Procerithiidae (*i.e. Argyropeza divina*), both of which were thought to have become extinct in the late Eocene. Along with radiations of *Athleta* volutes and *Cypraeovula* cowries, the entire Southern African regional fauna has a distinct Eocene appearance and is probably a left-over of the Eocene East Tethyan fauna. Some of the widespread Southern African regional gastropod index taxa include the buccinid *Burnupena pubescens*, the turbinellid *Vasum triangularis*, the volutids *Athleta abyssicola*, *Athleta gilchristi*, and *Callipara bullatiana*, the giant marginellid *Afrivoluta pringlei*, and the conids *Ketyconus tinianus*, *Pictoconus pictus*, *Pictoconus infrenatus*, *Sciteconus mozambicus*, and *Sciteconus simplex*. These are shown here on Figure 10.2.

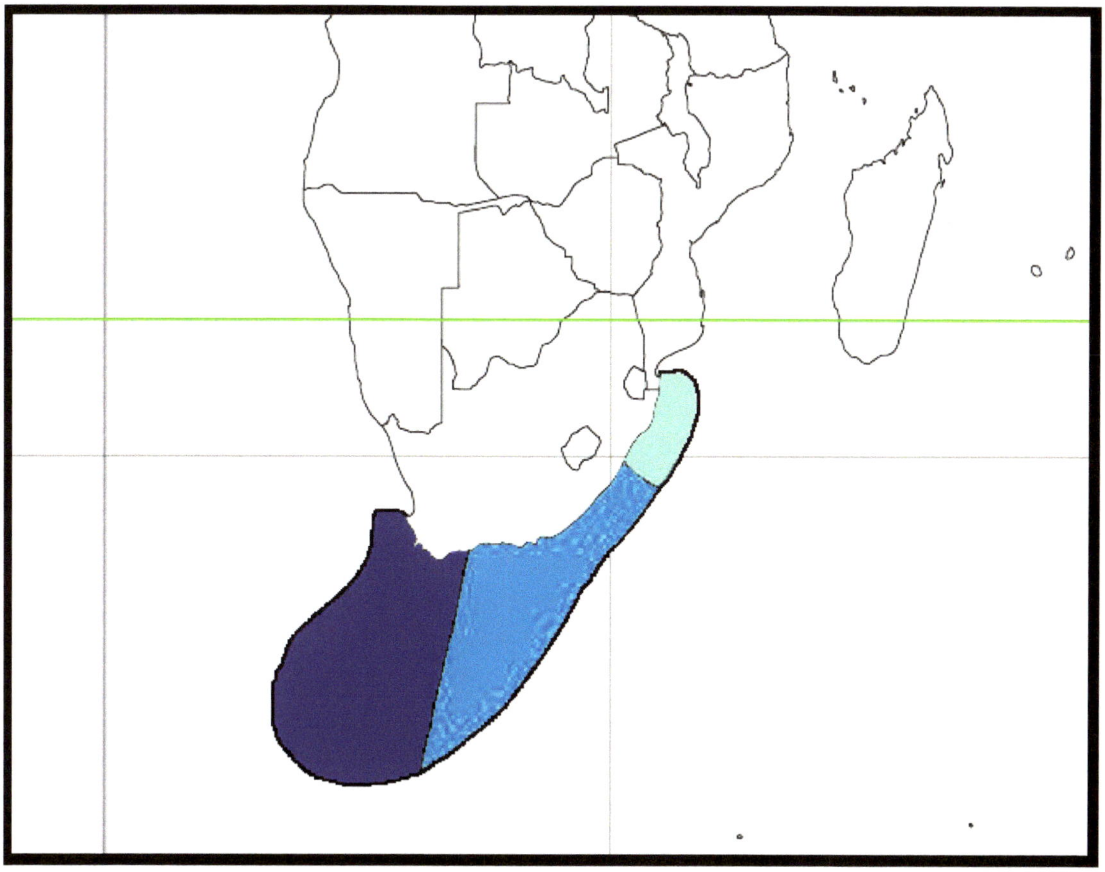

Figure 10.1 Map of the **Southern African Region**, showing the South African Molluscan Province and the areal extent of its subprovinces: the **Capean Subprovince** (dark blue), the **Transkeian Subprovince** (blue), and the **Natalean Subprovince** (pale blue).

The oceanography of the Southern African Region is complex, primarily due to the interactions of two confluent currents: the warm water Agulhas Current, which flows southward through the Mozambique Channel and down the East African coast and curves westward as far as East London and the offshore Agulhas Platform; and the cold water Benguela Current, which flows northward from Cape Province to the Namaqualand Coast (the Good Hope Jet) and along Namibia. The cold Benguela Current derives from an offshoot of the Antarctic Circumpolar Current and creates a temperate marine climate along the western third of the region. The central third of the region has a temperate-to-warm temperate marine climate due to the mixing of the two main currents and through the constant production of warm water eddies and gyres. The eastern third of the region is completely within the influence of the warm Agulhas Current and has a warm temperate marine climate. The offshore Agulhas Platform acts as a barrier to current flow and deflects the Agulhas Current southward (the Agulhas Retroflection), producing a series of large warm water gyres and loops, some of which actually spin into the Southern Atlantic south of the Cape of Good Hope (such as the Agulhas Ring). Based on the overall gastropod fauna, only a single province is recognized within the Southern African Region, the **South African Province**. This large province and its subprovinces are discussed in the following sections.

South African Molluscan Province

Covering the same range as the Southern African Region, the South African Province encompasses the entire southern tip of Africa and also the offshore Agulhas Platform and Agulhas Ridge. As the result of being under the influence of two main current systems and three different marine climates (temperate, temperate-warm temperate mix, and warm temperate), the South African Province is divided into three distinct subprovinces, the Capean, Transkeian, and Natalean Subprovinces, each with its own set of characteristic index species. Due to its oceanographic isolation from the other neighboring provinces, the South African Province has evolved an unusually-rich molluscan fauna, replete with endemic and relictual taxa. The South African Province is also a classic paratropical biotic unit, lacking the high-tropical families Strombidae and Modulidae and the genera *Harpa*, *Turbinella*, and *Lyria*. A number of wide-ranging eurythermal gastropods have established themselves throughout the three marine climatic regimes contained within the South African Province and these can be used as indicators of the provincial boundaries. Some of the more prominent of these widespread index taxa include the abalone *Haliotis parva*, the giant triviids *Triviella aperta* and *Triviella ovulata*, the muricids *Poropteron debruini* and *Poropteron graagae*, the nassariid *Bullia laevissima*, the babyloniid *Zemiropsis papillaris*, the volutid *Neptuneopsis gilchristi*, the marginellid *Roseamarginella rosea*, the marginellonid *Afrivoluta pringlei*, and the conids *Sciteconus algoensis*, *Sciteconus gradatulus* and *Ketyconus tinianus* (all shown here on Figures 10.2 and 10.3).

Capean Subprovince

Named for the Cape of Good Hope area of Western Cape Province, South Africa, the Capean Subprovince extends from St. Helena Bay eastward to Cape Agulhas, and encompasses the West Coast Peninsula, Saldanha Bay, the Cape Peninsula, False Bay, and the Agulhas Peninsula. This subprovincial area is under the influence of the cold Benguela Current and has an only marginally paratropical marine climate. In spite of the cold water conditions, the Capean molluscan fauna is remarkably species-rich with a large component of endemic taxa. Most noteworthy is a large radiation of the archaic cypraeid genus *Cypraeovula*, some of which include *Cypraeovula coronata*, *C. coronata debruini*, *C. mikeharti*, *C. fuscodentata*, *C. fuscorubra*, and *C. (Crossia) atlantica*. These living-fossil species occur along with other classic Capean taxa such as the giant triviid *Triviella magnidentata,* the volutes *Fusivoluta sculpturata* and *Athleta boswellae*, and the cone shell *Sciteconus agulhasi* (all shown here on Figure 10.4).

Transkeian Subprovince

Named for the old Transkei region of the modern Eastern Cape Province, South Africa, the Transkeian Subprovince extends from Cape Agulhas eastward to Port Shepstone, KwaZulu-Natal Province, and encompasses Mossel Bay, Plettenberg Bay, St. Francis Bay, and the East London coast. The area of the subprovince corresponds directly to the warm temperate marine climate caused by the mixing of the Agulhas and Benguela current, and these paratropical water conditions, coupled with high nutrient levels, support a very rich gastropod fauna with a very high level of endemism, approaching 40% in most groups. Of special interest within the Transkeian Subprovince is a very large species radiation of the archaic cowrie genus *Cypraeovula*, with at least 25 species and subspecies having been described to date. Some of these include *Cypraeovula alfredensis*, *C. algoensis*, *C. castanea*, *C. cohenae*, *C. coronata gabrielli*, *C. coronata infantensis*, *C. edentula*, *C. capensis*, *C. capensis profundorum*, *C. alfredensis transkeiana*, *C. amphithales*, *C. immelmani*, *C. (Crossia) volvens*, and *C. (Crossia) connelli* (all illustrated here on Figures 10.5 and 10.6). This remarkable species swarm of relictual cowries occurs along with many other interesting Transkeian endemic taxa, including the muricid *Poropteron graagae*, the turbinellid *Vasum truncatum*, the volutid *Callipara africana*, the cone shells *Nataliconus immelmani*, *Pictoconus transkeianus*, and *Sciteconus bairstowi*, and the clavatulid *Toxiclionella elstoni* (all shown here on Figure 10.6).

Natalean Subprovince

Named for the Natal region of eastern South Africa, the old name for the modern Province of KwaZulu-Natal, the Natalean Subprovince extends from Port Shepstone, KwaZulu-Natal Province to Ponta do Ouro, Mozambique. The entire Natalean area is bathed in the warm waters of the Agulhas Current and the northern end of the subprovince, from Richards Bay to Ponta do Ouro, has water temperatures that are warm enough for tropical Indian Ocean ecosystems to become established along the coast. The molluscan faunas of northern KwaZulu-Natal, however, are ecologically separated from the tropical coastal faunas of Mozambique by the extensive fresh and brackish water lagoon systems of the iSimangaliso Wetlands, Lake St. Lucia, and Kosi Lake and the Mfolozi River Mouth, which line the entire coast from St. Lucia northward to Mahlungulu. These fresh water lagoons and long stretches of open sand beaches act as physiological barriers between the gastropod faunas of the Mozambican Subprovince of the Lemurian Province and those of the Natalean Subprovince of the South African Province. Of special interest along the Umkomaas area of KwaZulu-Natal is the Aliwal Shoal, a 5 km-long coral reef complex built on top of a hardened Pleistocene sand dune. This offshore reef system houses a number of endemic gastropods, including the cowrie *Leporicypraea rosea aliwalensis* and the cone shell *Pionoconus aliwalensis* (shown here on Figures 10.7 and 10.8).

As a faunal transition zone between the Transkeian and Mozambican Subprovinces, the Natalean Subprovince contains a mixed gastropod fauna, with elements of both eutropical Indian Ocean faunas and the paratropical South Africa faunas. This faunal mixture is especially noticeable in the family Cypraeidae, where endemic subspecies of widespread Lemurian taxa occur together with South African endemic relict taxa. Some of the Lemurian-derived cowries include *Bistolida diauges uvangoensis*, *Cribrarula comma fraserorum*, *Ovatipsa chinensis kwazulu*, and *Naria helvola meridionalis*. These occur with South African-derived cowries such as *Cypraeovula (Crossia) connelli peelae* and *Cypraeovula cruickshanki* and the endemic Natalean Subprovince taxa *Talostolida natalensis*, *Purpuradusta durbanensis*, and *Barycypraea fultoni fultoni* (all shown here on Figure 10.7). This rich and unusual cowrie fauna occurs along with an equally rich fauna of other Natalean endemics, such as the muricid *Chicoreus austramosus*, the volutid *Callipara ponsonbyi* (shown here on p. 302), the olivid *Omogymna nitidula leonardi*, and a large cone shell fauna containing species such as *Darioconus natalaurantius*, *Nataliconus natalis*, *Nataliconus royaikeni*, *Sciteconus mpenjatiensis*, *Plicaustraconus visagenus*, *Rhombiconus queketti*, *Lividiconus meyeri*, and *Conasprella lorenzi* (all shown here on Figure 10.8).

ICONOGRAPHY OF GASTROPODS OF THE SOUTH AFRICAN REGION

(Principal Index Gastropods are shown on Figures 10.2 to 10.8)

Figure 10.2 Widespread Index Gastropods of the Southern African Region.
A= *Vasum triangularis* (E.A. Smith, 1902), length 66 mm, 5 m depth off East London, South Africa; **B=** *Athleta abyssicola* (Adams and Reeve, 1848), length 59 mm, 150 m depth off Cape Agulhas, South Africa; **C=** *Athleta gilchristi* (Sowerby III, 1902), length 27 mm, 150 m depth off Cape Town, South Africa; **D=** *Callipara bullatiana* (Weaver and Dupont, 1967), length 60 mm, on the beach, False Bay, Cape Town, South Africa; **E=** *Afrivoluta pringlei* Tomlin, 1947, length 113 mm, 150 m depth off Cape Agulhas, South Africa; **F=** *Ketyconus tinianus* (Hwass, 1792), length 41 mm, 1 m depth in Mossel Bay, South Africa; **G=** *Pictoconus infrenatus* (Reeve, 1848), length 44 mm, on the beach, False Bay, Cape Town, South Africa; **H=** *Pictoconus pictus* (Reeve, 1843), length 40 mm, 70 m depth off Jeffreys Bay, South Africa; **I=** *Sciteconus mozambicus* (Hwass, 1792), length 45 mm, 2 m depth off Kommetjie, Cape Peninsula, South Africa; **J=** *Sciteconus simplex* (Sowerby II, 1858), length 27 mm, 2 m depth off Kommetjie, Cape Peninsula, South Africa; **K=** *Burnupena pubescens* (Kuster, 1858), length 34 mm, 2 m depth in False Bay, Cape Province, South Africa.

Figure 10.3 Widespread Index Gastropods of the South African Province.
A= *Haliotis parva* Linnaeus, 1758, length 26 mm, low tide, False Cape, Cape Province, South Africa; **B=** *Triviella aperta* (Swainson, 1822), length 23 mm, dead on beach, False Bay, Cape Province, South Africa; **C=** *Triviella ovulata* (Lamarck, 1810), length 20 mm, 30 m depth off Betty's Bay, Cape Province, South Africa; **D=** *Poropteron debruini* Lorenz, 1989, length 27 mm, 3 m depth off Betty's Bay, Cape Province, South Africa; **E=** *Poropteron uncinarius* (Lamarck, 1822), length 30 mm, washed ashore after a storm, Jeffreys Bay, Western Cape, South Africa; **F=** *Zemiropsis papillaris* (Sowerby I, 1825), length 37 mm, 75 m depth on the Agulhas Bank, off Cape Agulhas, South Africa; **G=** *Neptuneopsis gilchristi* (Sowerby III, 1898), length 126 mm, 75 m depth on the Agulhas Bank, off Cape Agulhas, South Africa; **H=** *Roseamarginella rosea* (Lamarck, 1822), length 20 mm, washed ashore after a storm, False Bay, Cape Province, South Africa; **I=** *Sciteconus algoensis* (Sowerby I, 1834), length 25 mm, low tide off Simon's Town, Cape Peninsula, Cape Province, South Africa; **J=** *Sciteconus gradatulus* (Weinkauff, 1875), length 71 mm, 140 m depth off Cape Agulhas, Cape Province, South Africa; **K=** *Bullia laevissima* (Gmelin, 1791), length 41 mm, 20 m depth in False Bay, Cape Province, South Africa.

Figure 10.4 Index Gastropods of the Capean Subprovince, South African Province.
A= *Cypraeovula (Crossia) atlantica* (Raybaudi, 1988), length 33 mm, 55 m depth off Kommetjie, Southwest Cape, South Africa; **B**= *Cypraeovula coronata debruini* Lorenz, 2002, length 35 mm, 50 m depth off Hout Bay, Western Cape, South Africa; **C**= *Cypraeovula coronata* (Schilder, 1930), length 28 mm, 30 m depth off Dander Point, Cape Agulhas, Western Cape, South Africa; **D, E**= *Cypraeovula fuscodentata* (Gray, 1825), length 36 mm, 35 m depth off Sesal Point, Jeffreys Bay, Western Cape, South Africa; **F**= *Cypraeovula fuscorubra* (Shaw, 1909), length 34 mm, 35 m depth in Hout Bay, Western Cape, South Africa; **G**= *Cypraeovula mikeharti* Lorenz, 1985, length 24 mm, 10 m depth, Buffelsbay, False Bay, Western Cape, South Africa; **H**= *Triviella magnidentata* (Liltved, 1986), length 20 mm, 3 m depth off Simon's Town, Cape Peninsula, Western Cape, South Africa; **I**= *Fusivoluta sculpturata* Tomlin, 1945, length 36 mm, 250 m depth on the Agulhas Bank, southwest of the Alphard Bank Pinnacles, South Africa; **J**= *Athleta boswellae* (Rehder, 1969), length 64 mm, 200 m depth on the Agulhas Bank, south of Cape Agulhas, South Africa; **K**= *Sciteconus agulhasi* (Coomans, Moolenbeek, and Wils, 1980), length 25 mm, low tide, Cape Agulhas, Western Cape, South Africa.

Figure 10.5 The *Cypraeovula* Species Radiation of the Transkeian Subprovince, South African Province.
A= *Cypraeovula alfredensis* (Schilder and Schilder, 1929), length 35.5 mm, 30 m depth off Phillip's Reef, Algoa Bay, South Africa; **B=** *Cypraeovula algoensis* (Gray, 1825), length 24 mm, 35 m depth in Hout Bay, Western Cape, South Africa; **C=** *Cypraeovula castanea* (Higgins, 1868), length 34.2 mm, 40 m depth off Seal Point, Cape St. Francis, South Africa; **D=** *Cypraeovula cohenae* (Burgess, 1965), length 29 mm, washed ashore after a storm, Aston Bay Tip, Jeffreys Bay, Eastern Cape, South Africa; **E=** *Cypraeovula coronata gabrielli* Lorenz, 1993, length 29 mm, 95 m depth off Seal Point, Jeffreys Bay, Eastern Cape, South Africa; **F=** *Cypraeovula coronata infantensis* Aiken, 2016, length 24 mm, 120 m depth off Port Elizabeth, Eastern Cape, South Africa; **G=** *Cypraeovula (Crossia) iutsui* Shikama, 1974, length 23 mm, 150 m depth off St. Francis Bay, Eastern Cape, South Africa; **H=** *Cypraeovula (Crossia) volvens* Fazzini and Bergonzoni, 2004, length 26 mm, trawled from 110 m depth off the Fish River mouth, Port Alfred, South Africa. **I=** *Cypraeovula edentula* (Gray, 1825), length 22 mm, 15 m depth on Phillip's Reef, Algoa Bay, South Africa; **J=** *Cypraeovula capensis* (Gray, 1828), length 30 mm, 15 m depth on rock reef off Port Elizabeth, Algoa Bay, South Africa; **K=** *Cypraeovula capensis profundorum* Seccombe, 2003, length 27 mm, trawled from 100 m depth off East London, Eastern Cape, South Africa;

Figure 10.6 Index Gastropods of the Transkeian Subprovince, South African Province.
A= *Cypraeovula alfredensis transkeiana* Lorenz, 2002, length 30 mm, washed ashore after a storm, Bonza Bay, Transkei, South Africa; **B=** *Cypraeovula amphithales* (Melvill, 1888), length 30 mm, washed ashore after a storm, Port Elizabeth, Algoa Bay, South Africa; **C=** *Cypraeovula immelmani* Liltved, 2001, length 24 mm, trawled from 110 m depth off Msikaba, Transkei, South Africa; **D=** *Cypraeovula (Crossia) connelli* (Liltved, 1983), length 28 mm, trawled from 110 m depth off Coffee Bay, Transkei, South Africa; **E=** *Callipara africana* (Reeve, 1856), length 54 mm, 10 m depth in Coffee Bay, Transkei, South Africa; **F=** *Vasum truncatum* (Sowerby III, 1892), length 63 mm, 5 m depth on rock platform, Coffee Bay, Transkei, South Africa (Transkei form); **G=** *Nataliconus immelmani* (Korn, 1998), length 43 mm, 15 m depth in Coffee Bay, Transkei, South Africa; **H=** *Pictoconus transkeianus* (Korn, 1998), length 31 mm, trawled from 100 m depth off Coffee Bay, Transkei, South Africa; **I=** *Sciteconus bairstowi* (Sowerby III, 1889), length 32 mm, in shell rubble bed, 5 m depth in Coffee Bay, Transkei, South Africa; **J=** *Toxiclionella elstoni* (Barnard, 1962), length 36 mm, 100 m depth off Coffee Bay, Transkei, South Africa; **K=** *Poropteron graagae* (Coen, 1943), length 19 mm, low tide, East London, Eastern Cape, South Africa;

Figure 10.7 Cowrie Species Radiations of the Natalean Subprovince, South African Province.
A= *Barycypraea fultoni* (Sowerby III, 1903), length 71 mm, trawled in 100 m depth off Umkomaas, Kwa Zulu-Natal, South Africa; **B=** *Bistolida diauges uvongoensis* Massier, 2004, length 28 mm, 1 m depth off Umzumbe, Kwa Zulu-Natal, South Africa; **C=** *Cribrarula comma fraserorum* Lorenz, 2017, length 23 mm, 2 m depth off Richards Bay, Kwa Zulu-Natal, South Africa; **D=** *Cypraeovula (Crossia) connelli peelae* Lorenz and Hubert, 2002, length 28 mm, 110 m depth, Umkomaas, Kwa Zulu-Natal, South Africa; **E=** *Cypraeovula (Crossia) cruickshanki* (Kilburn, 1972), length 30 mm, 500 m depth off The Gate, Kwa Zulu-Natal, South Africa; **F=** *Leporicypraea rosea aliwalensis* Lorenz and Hubert, 2002, length 62 mm, 5 m depth on Aliwal Reef, Natal, South Africa; **G=** *Naria citrina* (Gray, 1825), length 22 mm, 50 m depth in Richards Bay, Kwa Zulu-Natal, South Africa; **H=** *Naria helvola meridionalis* Schilder and Schilder, 1938, length 25 mm, 2 m depth off Park Rynie, Kwa Zulu-Natal, South Africa; **I=** *Ovatipsa chinensis kwazulu* Lorenz, 2017, length 22 mm, 100 m depth, off Durban, Kwa Zulu-Natal, South Africa; **J=** *Purpuradusta durbanensis* (Schilder and Schilder, 1938), length 16 mm, low tide off Umhlanga, Kwa Zulu-Natal, South Africa; **K=** *Talostolida natalensis* (Heiman and Meinis, 2002), length 32 mm, low tide, Park Rynie, Kwa Zulu-Natal, South Africa.

Figure 10.8 Index Gastropods of the Natalean Subprovince, South African Province.
A= *Chicoreus austramosus* E. Vokes, 1978, length 60 mm, 25 m depth in Coffee Bay, Transkei, South Africa; **B=** *Omogymna nitidula leonardi* (Petuch and Sargent, 1986), length 21 mm, 50 m depth in Richards Bay, KwaZulu-Natal, South Africa; **C=** *Conasprella lorenzi* Monnier and Limpalaër, 2012, length 31 mm, 100 m depth off Park Rynie, KwaZulu-Natal, South Africa; **D=** *Darioconus natalaurantius* Veldsman, 2013, length 34 mm, 2 m depth off Port Shepstone, KwaZulu-Natal, South Africa; **E=** *Lividiconus meyeri* (Walls, 1979), length 34 mm, low tide on a reef flat off Park Rynie, KwaZulu-Natal, South Africa; **F=** *Nataliconus natalis* (Sowerby II, 1858), length mm, 2 m depth off Margate, KwaZulu-Natal, South Africa; **G=** *Nataliconus royaikeni* Veldsman, 2010, length 54 mm, 2 m depth off Park Rynie, KwaZulu-Natal, South Africa; **H=** *Plicaustraconus visagenus* (Kilburn, 1974), length 25 mm, 200 m depth off Port Shepstone, KwaZulu-Natal, South Africa; **I=** *Pionoconus aliwalensis* Veldsman, 2018, length 42 mm, 20 m depth on Aliwal Reef, KwaZulu-Natal, South Africa; **J=** *Rhombiconus queketti* (E.A. Smith, 1906), length 32 mm, 3 m depth on Aliwal Reef, KwaZulu-Natal, South Africa; **K=** *Sciteconus mpenjatiensis* Veldsman, 2016, length 21 mm, 25 m depth, off Port Shepstone, KwaZulu-Natal, South Africa.

Living specimen of *Callipara ponsonbyi* (E.A. Smith, 1901) at 42 m on sand near rocky reef, off Port Shepstone in southern KwaZulu Natal, South Africa. This photo was taken by Valda Fraser while diving on April 13, 2016. Photo used with permission.

APPENDIX. Quantitative Methods for Defining Provinces and Subprovinces

The provinces and subprovinces described and proposed throughout this book are largely based upon qualitative data, comprising descriptions of characteristic molluscan assemblages and their most prominent resident index taxa. The conclusions derived from these qualitative analyses, however, are also supported by quantitative numerical analyses, with the methodology shown here in the following appendix. The quantitative support for our biogeographical units is essentially an expansion and refinement of the 50% Rule of Valentine (1973), and was first used by the senior author (Petuch, 2013) to define the provinces and subprovinces of the tropical western Atlantic (discussed in Chapter 1). At the subregional or provincial level, a complete application of the 50% Rule is essentially untenable, as it requires a quantitative overview of every single gastropod family found in the area under study. This would necessitate detailed surveys of many poorly-known microgastropod families, such as those in the superfamilies Rissooidea and Truncatelloidea, and also a large number of families and subfamilies that have never undergone any form of worldwide monographic treatment. The lack of reliable data on these groups is a major impediment to any sort of detailed quantitative biogeographical analysis.

To overcome this quandary, a simple statistical methodology was proposed that incorporates proxy families and subfamilies, primarily those that are wide-ranging in worldwide eutropical and paratropical seas. All of these proxy groups have representative taxa in eutropical areas around the world and the number of endemic species that they contain determines the status and rank of the biogeographical unit. We have modified the original list of proxy taxa proposed by the senior author to comply with recent discoveries and changes in the higher taxonomy of the gastropods. The ten proxy tropical families that we have chosen to define provinces and subprovinces include:

1. **Cypraeidae**
2. **Strombidae**
3. **Modulidae**
4. **Xenophoridae**
5. **Melongenidae**
6. **Pisaniidae**
7. **Turbinellidae**
8. **Harpidae**
9. **Olividae-Olivinae**
10. **Conidae**

We refer to these ten gastropod higher-order taxa as **Provincial Index Taxa**, and a percentage of endemism within these groups that exceeds 50% indicates that the area represents a distinct molluscan faunal province. Each of the Provincial Index Taxa yields a quantity ***T*** (the **Taxon Index**), which is based upon the total number of species within each family found in the subregion and also on how many species are endemic to the area under consideration. These quantities can be derived from the following relationship:

$$T = \frac{n}{N}(100), T > 50$$

where ***N*** is the total number of species and subspecies known from the entire Subregion or Region (with 9 Subregions and 9 Regions being found worldwide) and ***n*** is the number of species and subspecies that are endemic to the area under investigation. An example of the use of Taxon Index proxies is shown here for the definition of the Carolinian Molluscan Province (Northwestern Atlantic Paratropical Subregion; modified from Petuch, 2013: 5). The T values for the Provincial Index Taxa (all listed or illustrated in Chapter 2) are:

T_1 Cypraeidae N= 5; n= 1; T_1= 20
T_2 Strombidae N= 5; n= 1; T_2= 20
T_3 Modulidae N= 6; n= 6; T_3= 100
T_4 Xenophoridae N= 2; n= 1; T_4= 50
T_5 Melongenidae N= 7; n= 5; T_5= 71
T_6 Pisaniidae N= 12; n= 8; T_6= 66
T_7 Turbinellidae N= 4; n= 2; T_7= 50
T_8 Harpidae N= 2; n= 0; T_8= 0
T_9 Olividae-Olivinae N= 12; n= 1; T_9= 83
T_{10} Conidae N= 33; n= 20; T_{10}= 66

These Taxon Indices, each a proxy for the fauna of an entire province, can be added together and averaged to obtain the final determination of provincial status, the quantity ***P*** (the **Provincial Combined Index**), as derived from the relationship:

$$P = \sum_{n=1}^{10} \frac{Tn}{10}, P > 50$$

where ***Tn*** represents the combined individual quantified levels of endemism in the families Cypraeidae, Strombidae, Modulidae, Xenophoridae, Melongenidae, Pisaniidae, Turbinellidae, Harpidae, Olivinae, and Conidae within the Carolinian area (Cape Hatteras to the Yucatan Peninsula). The summation of the quantities from the list of Provincial Index Taxa yields 526, producing a value of P= 52.6%, which is well above the 50% minimum required for full provincial status. The P value would be even higher if the families Cypraeidae and Harpidae did not have so many resident species with wide-ranging, long-lived planktotrophic larvae. With the exception of one endemic Carolinian cowrie (*Macrocypraea (Lorenzicypraea) cervus*), all the other Carolinian resident cypraeids and harpids (Moruminae) have ranges that extend from North Carolina to the Amazon River Mouth or, in some cases, to southern Brazil (Wide-Ranging Western Atlantic taxa; see Chapter 2). The lack of a high n value in these two families lowered the P value for the entire Carolinian Province. High P values are always skewed toward families that predominantly contain genera with direct development and low levels of larval dispersal (low vagility). Some Provinces with abundant low-vagility taxa, such as the Hawaiian, the Marquesan, and the Paulinian, have P values that approach 90%. The presence of an entire endemic family within the Carolinian Province, the Busyconidae (with 6 endemic genera and 17 endemic species and subspecies), outweighs the lack of cypraeids and harpids and underscores its provincial status.

Multiple areas within a province may evolve characteristic localized faunas that are distinctive enough to be separable from the rest of the province. Geographically-separable faunal subdivisions such as these are referred to as **Subprovinces** and they are defined by the 25% Rule (Petuch, 2013: 6; discussed in Chapter 1). To detect subprovincial boundaries within a province, multiple T value analyses should be undertaken on a spatial gradient (latitudinal or longitudinal) and compared, taking into consideration ecological barriers such as river mouths, upwelling systems, and coastal lagoons. Using the same T value calculations as those used for determining P values (T_1-T_{10}), the basic algorithm that is used for determining P is here modified to yield ***S*** (the **Subprovincial Combined Index**), such that:

$$S = \sum_{n=1}^{10} \frac{tn}{10}, S > 25$$

where ***tn*** represents the combined individual quantified levels of endemism in the families Cypraeidae, Strombidae, Modulidae, Xenophoridae, Melongenidae, Pisaniidae, Turbinellidae, Harpidae, Olivinae, and Conidae for smaller areas within a defined province. If the S value for any area along the spatial gradient is greater than 25%, then that area represents a distinct subprovince. A faunal analysis of every subprovince that has been proposed in this book will yield levels of endemicity equal to, or exceeding, 25%.

The red color form of the True Tulip shell, *Fasciolaria tulipa*, crawling on sand in a tide pool on a worm shell reef off Turtle Key, Ten Thousand Islands, Florida.

BIBLIOGRAPHY

Abalde, S., M. Tenorio, C. Afonso, J. Uribe, A. Echeverry, and R. Zadoya. 2017. Phylogenetic relationships of cone snails endemic to Cape Verde based on mitochondrial genomes. *Evolutionary Biology* 17:231 (Chart 1).

Berschauer, D.P. 2019. A review of the *Voluta polypleura-demarcoi* complex. *The Festivus* 51(3):184-197.

Berschauer, D.P. and R.N. Clark. 2018. *Sea Shells of Southern California-Marine Shells of the Californian Province*. San Diego Shell Club, San Diego, California. 135 pp.

Berschauer, D.P., E.J. Petuch, and R.N. Clark. 2018. A new species of *Forreria* (Gastropoda: Muricidae: Ocenebrinae) from the Sea of Cortez, Mexico. *The Festivus* 50(3):139-146.

Briggs, J. 1974. *Marine Zoogeography*. McGraw-Hill Publishers, New York. 452 pp.

Briggs, J. 1995. *Global Biogeography*. Elsevier Press, Amsterdam, Netherlands. 452 pp.

Bright, T.J. and L.H. Pequegnat. 1974. *Biota of the West Flower Garden Bank*. Editors. Gulf Publishing Company, Houston, Texas. 285 pp.

Castelao, R.M. and J.A. Barth. 2006. Upwelling around Cabo Frio, Brazil: The importance of wind stress. *Geophysical Research Letters* 33; LO3602.

Clarke, A.H. 1989. New mollusks from undersea oil seep sites off Louisiana. *Malacology Data Net* 2 (5-6):122-134.

Daughenbaugh, J.D. 2018. The Recent Cypraeidae of Northern New Zealand from the Kermadec Islands to the Poor Knight's Islands, Southwest Pacific Ocean (Mollusca: Gastropoda: Cypraeidae). *The Festivus* 50(1):36-54.

Daughenbaugh, J.D. and M. Beals. 2016. *Talostolida pellucens* (Melvill, 1888) in the Tropical Eastern Pacific: In perspective (Mollusca: Gastropoda: Cypraeidae). *The Festivus* 48(1):4-16.

Dekkers, A.J. and S.J. Maxwell. 2018. *Altivasum* Hedley, 1914 (Gastropoda: Turbinellidae) from South Western Australia. *The Festivus* 50(4):245-255.

Demir, M. 2003. Shells of Mollusca collected from the seas of Turkey. *Turkish Journal of Zoology* 27(2003):101-140.

Doxa, C.K. and P. Divanach. 2013. Feeding of *Charonia seguenzae* (Aradas and Benoit, 1870) on natural prey and alternative foods. *Journal of Molluscan Studies* 79(1):76-78.

Eason, R.L. 2018. *Jaspidiconus acutimarginatus* is part of the Florida cone shell fauna. *The Festivus* 50(1):58-61.

Ebach, M. 2012. A history of biogeographical regionalism in Australia. *Zootaxa* 3392: 1-34.

Ekman, S. 1953. *Zoogeography of the Sea*. Sidgwick and Jackson Publishers, London. 417 pp.

Fraussen, K. and F. Swinnen. 2016. A review of the genus *Euthria* Gray, 1839 (Gastropoda: Buccinidae) from the Cape Verde Archipelago. *Xenophora Taxonomy* (April 2016) 9:9-31.

Gibson-Smith, J. 1976. A new taxon of the genus *Charonia* (Mollusca: Gastropoda) from the Pliocene of Venezuela, and its implications. *Boletin, Asociasion Venezolano de Geologia, Mineria, y Petroleo* (1):1-15, plate 1.

Hernandez, J., E. Rolan, F. Swinnen, R. Gomez, and J. Perez. 2011. *Moluscos y Conchas Marinas de Canarias*. ConchBooks, Hackenheim, Germany. 716 pp.

Houart, R. and M.E. Hendrickx. 2020. Three new species of Muricidae (Ocenebrinae, Pagodulinae) from the Gulf of California, Mexico and update of the living muricids from the area. *Novapex* 21(1):17-33.

Houart, R. and H. Löser. 2020. Description of a remarkable and huge new species of *Zacatrophon* (Muricidae: Ocenebrinae) from the Gulf of California. *Novapex* 21(1):43-48.

Johnston, A. 1856. *The Physical Atlas of Natural Phenomena.* W. Blackwood, Publisher, Edinburgh. 420 pp.

Keen, A.M. 1971. *Sea Shells of Tropical West America.* Second Edition. Stanford University Press. 1064 pp.

Leao, Z.M., R. Kikuchi, and V. Testa. 2003. Corals and Coral Reefs of Brazil. in *Latin American Coral Reefs*. Elsevier Science. Amsterdam, Netherlands. pp. 9-52.

Lecointre, G., K. Tinkler, and H. Richards. 1967. The Marine Quaternary of the Canary Islands. *Proceedings of the Academy of Natural Sciences of Philadelphia* 119 (1967):325-344.

Lorenz, F. 2004. Two new species of Conidae from southern Madagascar. *Visaya* 1(2):19-23.

Lorenz, F. 2017. *Cowries: Guide to the Gastropod Family Cypraeidae. Volume 1-Biology and Systematics.* ConchBooks, Harxheim, Germany. p. 404.

Maxwell, S.J. and A.M. Dekkers. 2019. A new name for *Altivasum typicum* Hedley, 1916 *fide* Dekkers and Maxwell. 2018 and the description of *Altivasum clarksoni* n. sp. *The Festivus* 51(3):171-176.

Meyer, C.P. 1998. *Phylogenetic systematics, biogeography, and diversification patterns in cowries.* University of California, Berkeley. 221 pp.

Meyer, C.P. 2003. Molecular systematics of cowries (Gastropoda: Cypraeidae) and diversification patterns in the tropics. *Biological Journal of the Linnean Society* 79:401-459.

Meyer, C.P. 2004. Toward comprehensiveness: Increased molecular sampling within Cypraeidae and its phylogenetic implications. *Malacologia* 46 (1): 127-156.

Monnier, E., L. Limpalaër, A. Robin, and C. Roux. 2018. *A Taxonomic Iconography of the Living Conidae*. ConchBooks, Harxheim, Germany. pp. 1208 (2 volumes).

Nishihira, M. and J. Veron. 1995. *Hermatypic Corals of Japan.* Kaiyusha Publishers, Tokyo, Japan. 439 pp.

Obura, D. 2015. An Indian Ocean Centre of origin revisited: Palaeogene and Neogene influences defining a biogeographic realm. *Journal of Biogeography* (2015):1-13.

Odum, J. 1971. *Fundamentals of Ecology.* Third Edition. W.B. Saunders, Philadelphia. 574 pp.

Oleinik, A., E.J. Petuch, and W. Aley. 2012. Bathyal Gastropods of Bimini Chain, Bahamas. *Proceedings of the Biological Society of Washington* 125(1):19-53.

Petuch, E.J. 1972. *Strombus costatus* and *Morum dennisoni* collected off the North Carolina coast. *Veliger* 15:32-35.

Petuch, E.J. 1974. A new *Terebra* from the coral reef areas off North Carolina. *Veliger* 17(2):205-208.

Petuch, E.J. 1975a. A review of the small Mauritanian cone shells with the description of a new genus and a species. *Veliger* 17(3):262-263.

Petuch, E.J. 1975b. Two new cone species from Senegal, West Africa. *Veliger* 18(2):180-182.

Petuch, E.J. 1976. An unusual molluscan assemblage from Venezuela. *Veliger* 18(3):322-325.

Petuch, E.J. 1979. New gastropods from the Abrolhos Archipelago and reef complex. *Proceedings of the Biological Society of Washington* 92(3):510-526.

Petuch, E.J. 1980. A new *Falsilyria* and a new *Conus* from Roatan Island, Honduras (Atlantic). *Nautilus* 94(3):115-118.

Petuch, E.J. 1981. A Volutid species radiation from northern Honduras, with notes on the Caloosahatchian Secondary Relict Pocket. *Proceedings of the Biological Society of Washington* 94(4):1110-1130.

Petuch, E.J. 1982a. Geographical heterochrony: Contemporaneous co-existence of Neogene and Recent molluscan faunas in the Americas. *Palaeogeography, Palaeoclimatology, and Palaeoecology* 37:277-312.

Petuch, E.J. 1982b. Paraprovincialism: Remnants of paleoprovincial boundaries in Recent marine molluscan provinces. *Proceedings of the Biological Society of Washington* 95(4):774-780.

Petuch, E.J. 1987. *New Caribbean Molluscan Faunas*. Coastal Education and Research Foundation, Charlottesville Virginia. 158 pp.

Petuch, E.J. 1988. *Neogene History of Tropical American Mollusks*. Coastal Education and Research Foundation, Charlottesville, Virginia. 217 pp.

Petuch, E.J. 1991. A new molluscan faunule from the Caribbean coast of Panama. *Nautilus* 104(2):57-71.

Petuch, E.J. 1992. New mollusks from Los Roques Archipelago, an isolated Caribbean Atoll. *La Conchiglia* 23(262):5-11.

Petuch, E, J. 1994. *Atlas of Florida Fossil Shells (Pliocene and Pleistocene Marine Gastropods)*. Graves Museum of Archaeology and Natural History, Dania, Florida and Chicago Spectrum Press. 394 pp.

Petuch, E.J. 1997. *Coastal Paleoceanography of Eastern North America*. Kendall-Hunt Publishing Company, Dubuque, Iowa. 373 pp.

Petuch, E.J. 1998. The molluscan fauna of the Wawa River region, Miskito Coast, Nicaragua: Ecology, biogeographical implications, and descriptions of new taxa. *Nautilus* 111(1):22-44.

Petuch, E.J. 2000. A review of the conid subgenus *Purpuriconus* Da Motta, 1991, with the description of two new Bahamian species. *Ruthenica* 10(2):81-87.

Petuch, E.J. 2002. New deep water gastropods from the Bimini Shelf, Bimini Chain, Bahamas. *Ruthenica* 12(1):59-72.

Petuch, E.J. 2004. *Cenozoic Seas: The View from Eastern North America*. CRC Press, Boca Raton, London, New York. 308 pp.

Petuch, E.J. 2013. *Biogeography and Biodiversity of Western Atlantic Mollusks*. CRC Press, Boca Raton, London, New York. 234 pp.

Petuch, E.J., R.F. Myers, and D.P. Berschauer. 2015. *The Living and Fossil Busycon Whelks: Iconic Mollusks of Eastern North America*. San Diego Shell Club, San Diego, California. Special Publication. 194 pp.

Petuch, E.J. and D.P. Berschauer. 2019. A review of the Carolinian Province *Americoliva nivosa* Complex (Gastropoda: Olividae) with the description of a new subspecies. *The Festivus* 51(4):284-292.

Petuch, E.J. and D.P. Berschauer. 2020a. A review of the genus *Morum* (Gastropoda: Harpidae) in the Western Atlantic, with the description of two new species from Brazil. *The Festivus* 52(1):60-69.

Petuch, E.J. and D.P. Berschauer. 2020b. New gastropods from Texas, the Western Caribbean, and Southern Brazil. *The Festivus* 52(2):173-183.

Petuch, E.J. and R. Myers. 2014. *Molluscan Communities of the Florida Keys and Adjacent Areas: Their Ecology and Biodiversity*. CRC Press, Boca Raton, London, York. 300 pp.

Petuch, E.J. and C.E. Roberts. 2007. *The Geology of the Everglades and Adjacent* CRC Press, Boca Raton, London, New York. 212 pp.

Powell, A.W.B. 1979. *New Zealand Mollusca (Marine, Land, and Freshwater Shells).* Collins Publishers, Auckland, Sydney, London. 500 pp.

Preece, R. 1995. The composition and relationships of the marine molluscan fauna of the Pitcairn Islands. *In* The Pitcairn Islands: Biogeography, Ecology, and Prehistory. *Biological Journal of the Linnean Society* (1995)56:339-358.

Rabiller, M. and G. Richard. 2019. Conidae Offshore de Guadeloupe: Description du materiel drague lors de l'expedition KARUBENTHOS 2 contenant de nouvelles especies. *Xenophora Taxonomy* 24:3-31.

Rehder, H. 1980. *The Marine Mollusca of Easter Island (Isla de Pascua) and Sala y Gomez*. Smithsonian Contributions to Zoology 289:1-167. Smithsonian Institution.

Rehder, H. and B. Wilson. 1975. New Species of Marine Mollusks from Pitcairn Island and the Marquesas. *Smithsonian Contributions to Zoology* 203:1-16.

Segers, W., F. Swinnen, and R. De Prins. 2009. *Marine Mollusks of Madeira*. Snoeck Publishers, Zwijndrecht, Belgium. 612 pp.

Spencer, T. 1989. *Tectonic and Environmental Histories in the Pitcairn Group, Palaeogene to Present: Reconstructions and Speculations.* Atoll Research Bulletin 1-42. National Museum of Natural History, Smithsonian Institution.

Tenorio, M., S. Abalde, J.R. Pardos-Blas, and R. Zardoya. 2020. Taxonomic revision of West African cone snails (Gastropoda: Conidae) based upon mitogenomic studies: implications for conservation. *European Journal of Taxonomy* 663:1-89.

Tenorio, M., E. Monnier, P. Bouchet, and N. Puillandre. 2018. The cones (Gastropoda) of Madagascar "Deep South": composition, endemism, and new taxa. *Xenophora Taxonomy* 19:25-54.

Thach, N.N. 2018. *New Shells of South Asia*. 48 Hour Books, Akron, Ohio. 173 pp.

Valentine, J. 1973. *Evolutionary Paleoecology of the Marine Biosphere*. Prentice-Hall, Englewood, N.J., 511 pp.

Vermeij, G. 1978. *Biogeography and Adaptation.* Harvard University Press, Cambridge, MA. 332 pp.

Warmke, G. and R.T. Abbott. 1961. *Caribbean Seashells*. Livingston Publishing Company, Narberth, PA. 348 pp.

Waters, J., T. Wernberg, S. Connell, M. Thomsen, G. Zuccarello, G. Kraft, J. Sanderson, J. West, and C. Gurgel. 2010. Australia's marine biogeography revisited: Back to the future? *Austral Ecology* (2010) 35:988-992.

Weisbord, N.E. 1962. Late Cenozoic gastropods of northern Venezuela. *Bulletins of American Paleontology* 42(193):7-672.

Woodward, S.P. 1856. *Manual of Mollusca: or a Rudimentary Treatise of Recent and Fossil Shells*. John Weale, London. 484 pp.

SYSTEMATIC INDEX

All of the species listed in this systematic index are figured in the iconographies at the ends of Chapters 2 through 10. We consider these taxa to be primary index species for the biogeographical units that were described throughout the book and these are arranged here by family. Altogether, 1778 species and subspecies of marine gastropods, in 68 families, were used in the descriptions of the biogeographical units listed throughout this book.

NERITIDAE

OLIVIDAE

A view of the vermetid worm shell reefs at low tide, Turtle Key, Ten Thousand Islands, Florida. These extensive reefs are a unique biohermal feature along the islands of Southwestern Florida. The reef-like bioherms are made up entirely of the black worm shell *Petaloconchus nigricans*, and provide the only hard substrate in the local environment.

BIOGEOGRAPHICAL INDEX

The following is an alphabetical listing of all of the biogeographical units that were used throughout the book. These units, many of which were described here for the first time, include all the Super-Regions, Regions, Subregions, Provinces, Subprovinces, and Infraprovinces found in tropical and warm temperate seas. The individual islands, archipelagoes, and seas found within these units, as well as some of the more important Plio-Pleistocene paleoseas, are also listed here and are discussed under the description of the biogeographical units.

A

B

C

ABOUT THE AUTHORS

Edward J. Petuch was born in Bethesda, Maryland, in 1949. Raised in a Navy family, he spent many of his childhood years collecting living and fossil shells in such varied localities as Chesapeake Bay, California, Puerto Rico, and Wisconsin. His early interests in malacology and oceanography eventually led to BA and MS degrees in zoology from the University of Wisconsin-Milwaukee. During his MS thesis research, Petuch concentrated on the molluscan biogeography of West Africa, traveling extensively in the Canary Islands, Western Sahara, Senegal, Gambia, Sierra Leone, Ivory Coast, and Cameroons. During this time, he also conducted research on the molluscan ecology of both coasts of Mexico and the Great Barrier Reef of Belize.

Continuing his education, Petuch studied marine biogeography and malacology under Gilbert Voss and Donald Moore at the Rosenstiel School of Marine and Atmospheric Sciences at the University of Miami, where he received a full scholarship. During this time, his doctoral dissertation research involved intensive field work in Costa Rica, Colombia, Venezuela, Barbados, the Grenadines, and Brazil, where he often went to sea with the local shrimpers for weeks at a time. After receiving his PhD in oceanography in 1980, Petuch was invited to conduct two years of postdoctoral research, funded by the National Science Foundation, with Geerat Vermeij at the University of Maryland. While there, he also held a research associateship with the Department of Paleobiology at the National Museum of Natural History, Smithsonian Institution, under the sponsorship of Thomas Waller, and conducted field work in the Plio-Pleistocene fossil beds of Florida and North Carolina and the Miocene fossil beds of Maryland and Virginia.

Petuch has also collected and studied living mollusks in Australia, Papua-New Guinea, Fijis, French Polynesia, Japan, the Bahamas, Nicaragua, and Uruguay. This research has led to the publication of over 300 scientific papers and the discovery and description of over 1,400 new species of mollusks and over 90 new genera. His previous 20 books are well-known reference texts in the malacological and paleontological communities, and some of the better known include: *Jewels of the Everglades: The Fossil Cowries of Southern Florida* (2018), *The Living and Fossil Busycon Whelks: Iconic Mollusks of Eastern North America* (2015), *Molluscan Communities of the Florida Keys and Adjacent Areas: Their Ecology and Biodiversity* (2014), *Biogeography and Biodiversity of Western Atlantic Mollusks* (2013), *Molluscan Paleontology of the Chesapeake Miocene* (2010), *The Geology of the Everglades and Adjacent Areas* (2007), *Cenozoic Seas: The View from Eastern North America* (2004), and *Cone Shells of the Okeechobean Sea* (2015). Currently, Petuch is a Professor Emeritus in the Department of Geosciences, Florida Atlantic University in

Boca Raton, Florida where, for thirty years, he taught undergraduate classes in oceanography, paleontology, and physical geology, and graduate classes in paleoecology and paleoceanography. He currently resides in Jupiter, Florida, with his wife Linda, where they both enjoy visits from their three children and their families.

© Morgan Taylor Photography, with permission

David P. Berschauer was born in Rockville Center, New York, in 1964, and spent his youth collecting shells in such varied localities as California, New York, Florida, Washington, and Mexico. His early interests in natural history, malacology, and marine biology eventually led to a BS in biology at the University of California-Irvine, an advanced marine invertebrate zoology course at Washington State University's Friday Harbor Marine Lab, and studies towards the pursuit of a graduate degree in marine biology at Florida State University in Tallahassee, Florida. While still an undergraduate, Berschauer performed field biology research, published a number of research papers and gave scientific presentations at national conferences. He subsequently switched career paths and attended Southwestern University School of Law in Los Angeles, California, earning his Juris Doctorate in 1991. Although having developed a legal career, he has kept malacology as a lifetime avocation and has put together a sizeable research collection and personal museum of molluscan specimens.

Over his entire professional life, Berschauer continued to pursue his passion for marine biology, and collecting and studying marine organisms. In his spare time, he has developed and published a relational database software program to aid in the organization and maintenance of a systematic collection. Although originally designed for malacology, the program is applicable to entomology and other aspects of systematic zoology. Now residing in Laguna Hills, California, Berschauer is an active member of the San Diego Shell Club and is the co-editor of the journal, *The Festivus*. He is also well known for his natural history and shell photography, with a multitude of high-quality examples seen throughout this book. Besides being the author of many important scientific papers on molluscan systematics, Berschauer has described and named over 100 new species of gastropods and is also the co-author of several recent books on mollusks, including *The Living and Fossil Busycon Whelks: Iconic Mollusks of Eastern North America* (2015), *Jewels of the Everglades: The Fossil Cowries of Southern Florida* (2018), and the well-received *Sea Shells of Southern California: Marine Shells of the Californian Province* (2018). Between court cases and an active law practice, Berschauer and his wife Felicia find time to enjoy traveling on cruise ships and exploring.